28

Topics in Heterocyclic Chemistry

Series Editor: Bert U.W. Maes

Topics in Heterocyclic Chemistry

Series Editor: Bert U.W. Maes

Recently Published and Forthcoming Volumes

Click Triazoles
Volume Editor: J. Košmrlj
Volume 28, 2012

Halogenated Heterocycles
Volume Editor: J. Iskra
Volume 27, 2012

Heterocyclic Scaffolds II: Reactions and Applications of Indoles
Volume Editor: G.W. Gribble
Volume 26, 2010

Synthesis of Heterocycles via Multicomponent Reactions II
Volume Editors: R.V.A. Orru, E. Ruijter
Volume 25, 2010

Anion Recognition in Supramolecular Chemistry
Volume Editors: P.A. Gale, W. Dehaen
Volume 24, 2010

Synthesis of Heterocycles via Multicomponent Reactions I
Volume Editors: R.V.A. Orru, E. Ruijter
Volume 23, 2010

Heterocyclic Scaffolds I: β-Lactams
Volume Editor: B. Banik
Volume 22, 2010

Phosphorous Heterocycles II
Volume Editor: R.K. Bansal
Volume 21, 2009

Phosphorous Heterocycles I
Volume Editor: R.K. Bansal
Volume 20, 2009

Aromaticity in Heterocyclic Compounds
Volume Editors: T. Krygowski, M. Cyrański
Volume 19, 2009

Heterocyclic Supramolecules I
Volume Editor: K. Matsumoto
Volume 17, 2008

Bioactive Heterocycles VI
Flavonoids and Anthocyanins in Plants, and Latest Bioactive Heterocycles I
Volume Editor: N. Motohashi
Volume 15, 2008

Heterocyclic Polymethine Dyes
Synthesis, Properties and Applications
Volume Editor: L. Strekowski
Volume 14, 2008

Synthesis of Heterocycles via Cycloadditions II
Volume Editor: A. Hassner
Volume 13, 2008

Synthesis of Heterocycles via Cycloadditions I
Volume Editor: A. Hassner
Volume 12, 2008

Bioactive Heterocycles V
Volume Editor: M.T.H. Khan
Volume 11, 2007

Bioactive Heterocycles IV
Volume Editor: M.T.H. Khan
Volume 10, 2007

Bioactive Heterocycles III
Volume Editor: M.T.H. Khan
Volume 9, 2007

Bioactive Heterocycles II
Volume Editor: S. Eguchi
Volume 8, 2007

Heterocycles from Carbohydrate Precursors
Volume Editor: E.S.H. ElAshry
Volume 7, 2007

Bioactive Heterocycles I
Volume Editor: S. Eguchi
Volume 6, 2006

Click Triazoles

Volume Editor:
J. Košmrlj

With contributions by

B.R. Buckley · Y. Chen · H.-F. Chow · J.D. Crowley ·
A.H. Flood · H. Heaney · A.S. Jalloh · T. Lecourt · S. Lee ·
C.-M. Lo · D.A. McMorran · L. Micouin · S. Mignani ·
S.H. Rouhanifard · M. Watkinson · P. Wu · T. Zheng ·
Y. Zhou

Springer

The series *Topics in Heterocyclic Chemistry* presents critical reviews on "Heterocyclic Compounds" within topic-related volumes dealing with all aspects such as synthesis, reaction mechanisms, structure complexity, properties, reactivity, stability, fundamental and theoretical studies, biology, biomedical studies, pharmacological aspects, applications in material sciences, etc. Metabolism will also be included which will provide information useful in designing pharmacologically active agents. Pathways involving destruction of heterocyclic rings will also be dealt with so that synthesis of specifically functionalized non-heterocyclic molecules can be designed.
The overall scope is to cover topics dealing with most of the areas of current trends in heterocyclic chemistry which will suit to a larger heterocyclic community.
As a rule, contributions are specially commissioned. The editors and publishers will, however, always be pleased to receive suggestions and supplementary information. Papers are accepted for *Topics in Heterocyclic Chemistry* in English.
In references, *Topics in Heterocyclic Chemistry* is abbreviated *Top Heterocycl Chem* and is cited as a journal.

Springer www home page: springer.com
Visit the THC content at springerlink.com

Topics in Heterocyclic Chemistry ISSN 1861-9282
ISBN 978-3-642-29428-0 ISBN 978-3-642-29429-7 (eBook)
DOI 10.1007/978-3-642-29429-7
Springer Heidelberg Dordrecht London New York

Library of Congress Control Number: 2012936667

Printed on acid-free paper

Springer is part of Springer Science+Business Media (www.springer.com)

Topics in Heterocyclic Chemistry Also Available Electronically

Topics in Heterocyclic Chemistry is included in Springer's eBook package *Chemistry and Materials Science*. If a library does not opt for the whole package the book series may be bought on a subscription basis. Also, all back volumes are available electronically.

For all customers who have a standing order to the print version of *Topics in Heterocyclic Chemistry*, we offer free access to the electronic volumes of the Series published in the current year via SpringerLink.

If you do not have access, you can still view the table of contents of each volume and the abstract of each article by going to the SpringerLink homepage, clicking on "Chemistry and Materials Science," under Subject Collection, then "Book Series," under Content Type and finally by selecting *Topics in Heterocyclic Chemistry*.

You will find information about the

- Editorial Board
- Aims and Scope
- Instructions for Authors
- Sample Contribution

at springer.com using the search function by typing in *Topics in Heterocyclic Chemistry*.

Aims and Scope

The series *Topics in Heterocyclic Chemistry* presents critical reviews on present and future trends in the research of heterocyclic compounds. Overall the scope is to cover topics dealing with all areas within heterocyclic chemistry, both experimental and theoretical, of interest to the general heterocyclic chemistry community. The series consists of topic related volumes edited by renowned editors with contributions of experts in the field.

Chapters in volumes are cited like journal articles.

Example:
Top Heterocycl Chem (2012) 27: 1–32
DOI 10.1007/7081_2011_63
© Springer-Verlag Berlin Heidelberg 2011
Published online: 23 September 2011

Preface

Discovered by Sharpless-Fokin and Meldal a decade ago, the copper-catalyzed azide–alkyne cycloaddition that forms 1,4-disubstituted 1,2,3-triazole, also referred to as Click-Triazole, has found a truly widespread application in diverse areas of chemistry. Some of these are illustrated in this volume of *Topics in Heterocyclic Chemistry*, in seven chapters, written by well-established scientists.

Chapter 1, "Mechanistic Investigations of Copper(I)-Catalysed Alkyne–Azide Cycloaddition Reactions", by Buckley and Heaney is concerned with various mechanistic interpretations of thermal and metal-catalyzed reactions of organic azides and alkynes, which result in the formation of 1,2,3-triazoles.

In Chap. 2, "'Click-Triazole' Coordination Chemistry: Exploiting 1,4-Disubstituted-1,2,3-Triazoles as Ligands", Crowley and McMorran review the enormous literature of 1,2,3-triazole containing nitrogen and carbene ligands and their metal complexes, focusing only on examples in which the coordination mode of the triazole ligand has been structurally characterized.

The ability of 1,2,3-triazole to bind anions with a C–H hydrogen bond is covered in Chap. 3, "Binding Anions in Rigid and Reconfigurable Triazole Receptors" by Lee and Flood. Applications including sensors, ion-selective electrodes, catalysis, anion transport and anion regulation, as well as their use in interlocked molecules, are discussed.

In Chap. 4, a companion to the preceding two chapters, Watkinson documents "Click Triazoles as Chemosensors" for the detection of cations, anions, small molecules and biomolecules. The chapter is focused on sensors in which the triazole plays an intimate role in the sensing event.

The presence of lone pair electrons on the 1,2,3-triazole nitrogen atoms and the acidic C–H can promote intra- and interpolymer chain interactions and enhance gel formation in triazole-based polymers. In Chap. 5, Chow, Lo and Chen survey the chemistry and properties of "Triazole-Based Polymer Gels".

In Chap. 6, "Click Triazoles for Bioconjugation", Zheng, Rouhanifard, Jalloh and Wu describe the application of copper catalysts for conjugating biomolecules, including proteins, nucleic acids, lipids and glycans, with biophysical probes for both in vitro and in vivo studies. Recent development of the strain-promoted azide–alkyne cycloaddition that eliminates the use of copper is also discussed.

Keeping with the formation of 1,2,3-triazole, in Chap. 7, Mignani, Zhou, Lecourt and Micouin comprehensively review "Recent Developments in the

Synthesis 1,4,5-Trisubstituted Triazoles". The focus of this chapter is the synthesis of 1,4,5-trisubstituted triazoles by a variety of methods including condensation reactions and rearrangements, selective preparation of trisubstituted triazoles from disubstituted precursors and formal [3+2] cycloaddition reactions involving internal alkynes.

I would like to express my sincere gratitude to the authors of this book for their outstanding contributions. I am grateful to the reviewers for their willingness to participate in this process and to the people at Springer, particularly Anette Lindqvist, Marion Hertel and Elizabeth Hawkins, for coordinating the project. I thank Slovenko Polanc and Bert Maes for giving me the opportunity to be the editor of this volume.

Ljubljana, Slovenia Janez Košmrlj

Contents

Top Heterocycl Chem (2012) 28: 1–30
DOI: 10.1007/7081_2011_71

Published online: 28 February 2012

Mechanistic Investigations of Copper(I)-Catalysed Alkyne–Azide Cycloaddition Reactions

Benjamin R. Buckley and Harry Heaney

Abstract The chapter concentrates on mechanistic aspects of thermal and metal catalysed reactions of organic azides and alkynes, particularly terminal alkynes, that result in the formation of 1,2,3-triazoles.

Keywords Alkyne · Azide · Catalysis · Click chemistry · Copper(I) · Ligands

Contents

B.R. Buckley and H. Heaney (✉)
Department of Chemistry, Loughborough University, Loughborough, Leicestershire
LE11 3 TU, UK
e-mail: H.Heaney@lboro.ac.uk

1 Introduction: Thermal Reactions

The importance of dipolar cycloaddition reactions in the synthesis of five-membered heterocycles is undoubtedly due to the versatility of the reactions that makes the processes comparable to the use of Diels–Alder reactions in the synthesis of six-membered carbocycles. There are a variety of methods available for the preparation of 1,2,3-triazoles, including, for example recent reports of the reactions of aryl azides with β-keto esters, for example using sodium salts [1], also reactions involving enamides [2], as well as reactions using continuous flow reactor systems [3]. Nonetheless, reactions of azides with alkynes provide the most important and most frequently used method for the preparation of 1,2,3-triazoles. In this chapter, we will be mainly concerned with various mechanistic interpretations of the formation of 1,2,3-triazoles by the interaction of azides with alkynes using metal catalysts and the present understanding of the catalytic cycles involved. The formation of 1,2,3-triazoles by means of thermal reactions of azides with alkynes is a reaction type that has been used over a long period of time and has been studied in detail as a special case of 1,3-dipolar cycloaddition reactions [4–8]. The reaction of dimethyl acetylenedicarboxylate with phenyl azide was first reported by Michael in 1883 [9], and reactions of a number of esters of acetylenedicarboxylic acid with a variety of azides have been recorded since that time. For example, the reaction of 1-azidonaphthalene with dimethyl acetylenedicarboxylate in benzene at 80°C for 5 h gave the triazole shown in Scheme 1 in 95% yield [10], and a number of dialkyl acetylenedicarboxylates gave the highly strained bis-triazole derivatives in high yields in reactions with 1,8-diazidonaphthalene [11, 12].

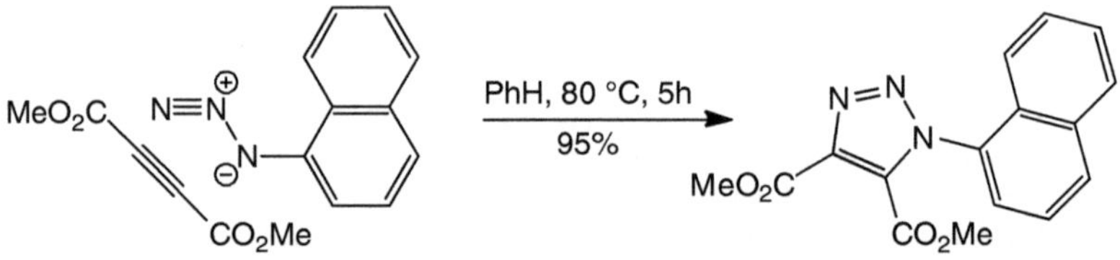

Scheme 1 Reaction of 1-azidonaphthalene with dimethyl acetylene dicarboxylate

In early work on the chemistry of cycloalkynes, cyclononyne was shown to react slowly with phenyl azide to give the cycloadduct in 82% yield; however, the strain energy in cyclooctyne is significantly greater, calculated to be 77.5 kJ mol^{-1}, which accounts for the fact that a reaction of cyclooctyne with phenyl azide was reported to proceed "like an explosion" and resulted in the formation of the 1,2,3-triazole shown in Scheme 2 [13]. The use of cyclooctyne derivatives has been exploited in considerable detail recently, particularly because reactions can be carried out in biological systems where metal-catalysed reactions cannot be used because of toxicity problems (for reviews see [14, 15]; see for example [16–19]).

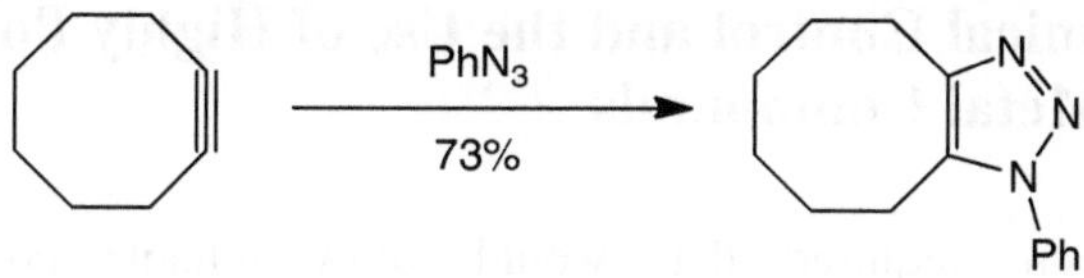

Scheme 2 The reaction of phenyl azide with cyclooctyne

The thermal reactions of azides with unsymmetrically substituted, including terminal alkynes, can result in the formation of regioisomeric 1,4- and 1,5-disubstituted 1,2,3-triazoles. Thus, the reaction of phenyl azide with phenylacetylene results in an almost 1:1 mixture of the two regioisomers [20]; an explanation of that result is provided using a frontier orbital treatment [21]. With more polarized alkynes or azides, there may be a preponderance of one of the regioisomers, as exemplified in the reaction of phenyl azide with 1-(2-chlorophenyl)-3,3,3-trifluoro-1-propyne, shown in Scheme 3 ([22]; see also [23]). The regioselectivity obtained in the reaction shown in Scheme 3 can be explained as a result of the lowering of the energy of both the HOMO and LUMO orbitals of the alkyne and where the larger atomic orbital coefficient in the LUMO in the preferred orientation interacts with the larger HOMO atomic orbital on the terminal nitrogen atom. An alternative simple explanation for the formation of the preferred regioisomer in the reaction of phenyl azide with 1-(2-chlorophenyl)-3,3,3-trifluoro-1-propyne is provided by a consideration of the polarization of the alkyne triple bond by the trifluoromethyl group.

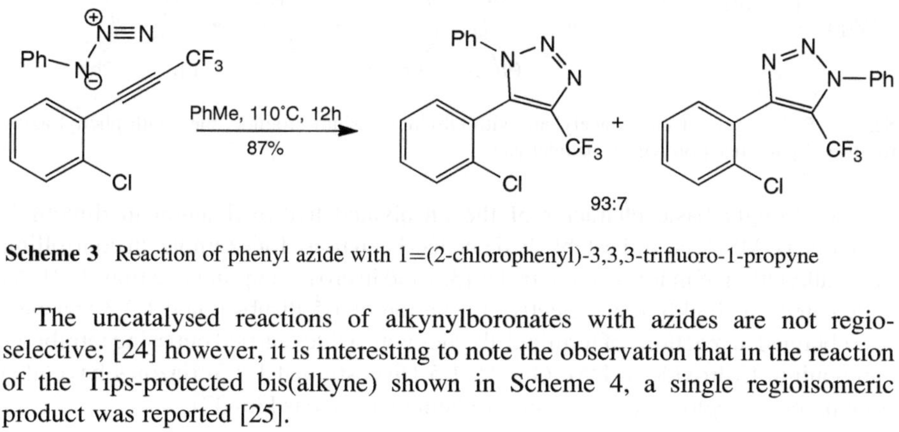

Scheme 3 Reaction of phenyl azide with 1=(2-chlorophenyl)-3,3,3-trifluoro-1-propyne

The uncatalysed reactions of alkynylboronates with azides are not regioselective; [24] however, it is interesting to note the observation that in the reaction of the Tips-protected bis(alkyne) shown in Scheme 4, a single regioisomeric product was reported [25].

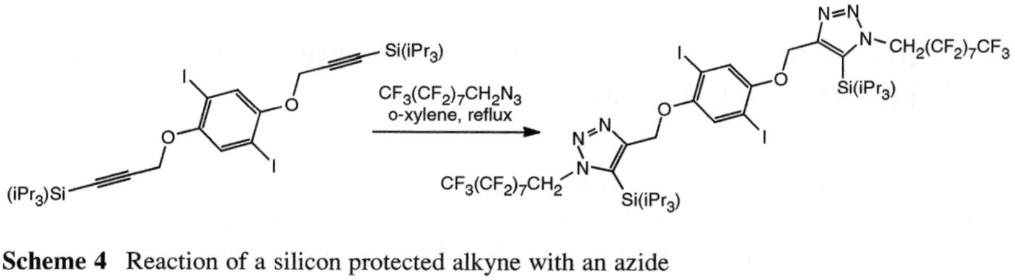

Scheme 4 Reaction of a silicon protected alkyne with an azide

2 Regiochemical Control and the Use of Highly Polarized Alkynyl-Metal Compounds

So methods were required that would allow reliable control over the regioselectivity in the synthesis of substituted 1,2,3-triazoles. It is evident that reactions using nucleophilic derivatives of terminal alkynes, for example involving highly polarized metallic species such as sodium acetylide, should lead to the formation of 1,5-disubstituted 1,2,3-triazoles in reactions with azides [26]. A reinvestigation of reactions of azides with Grignard reagents derived from terminal alkynes showed that 1,5-disubstituted 1,2,3-triazoles can be isolated in excellent yields and that further functionalisation can be achieved by reactions of the first-formed products, the 4-magnesio-derivatives, using a range of electrophiles, an example of which is shown in Scheme 5 [27]; reactions have also been reported using chlorodialkylphosphines [28, 29]. The method has been exploited in the synthesis of *cis*-restricted analogues of combretastatin A-4 [30]. Fischer carbene complexes of arylethynes also react with azides to form fully substituted 1,2,3-triazoles with high regioselectivity [31]. We will return to a discussion of reactions of dimethylphenylethynylaluminium with organic azides when addressing the use of ligands in copper(I)-catalysed reactions.

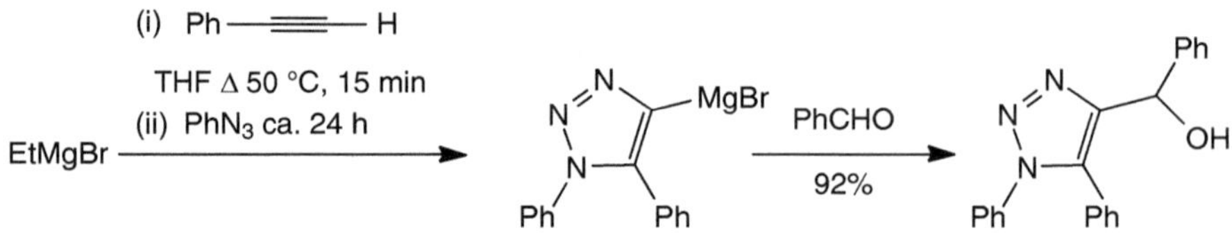

Scheme 5 Reaction of phenylacetylene with ethylmagnesium bromide, then with phenyl azide followed by the reaction with benzaldehyde

The strongly basic character of the unsolvated hydroxyl anion in dimethylsulfoxide (DMSO) is well established and has been used, for example, in controlling the *N*-alkylation of indoles and pyrroles [32] and in cross-coupling reactions [33]. As anticipated, the highly regioselective formation of 1,5-disubstituted 1,2,3-triazoles was obtained in reactions of terminal alkynes with azides carried out using tetrabutylammonium hydroxide in DMSO [34]. 1,5-Disubstituted 1,2,3-triazoles have also been prepared regioselectively using ruthenium catalysis [35–37].

3 Copper Catalysis: The Discovery of Copper(I)-Catalysed Azide–Alkyne Reactions

The rapid development of the use of copper catalysts and reagents in organic synthesis since the middle of the last century is comprehensively documented in a number of books and reviews [38–43]. Probably the most widely studied reaction type involving copper(I) catalysis is the intermolecular alkyne–azide cycloaddition reaction, which was first disclosed as an efficient method for the synthesis of 1,4-disubstituted 1,2,3-triazoles in 2001 [44]. Interestingly, although not strictly a copper(I)-catalysed 1,2,3-triazole-forming reaction, the serendipitous discovery of a low-yielding copper(I)-catalysed reaction that resulted in the formation of 1,2,3-triazoles, was recorded almost 20 years earlier [45]. For example, in an attempt to prepare an allenyl azide, the methansulfonate derived from 3-hydroxy-1,3-diphenylprop-1-yne was allowed to react with lithium azide and copper(I) chloride at low temperature: the reaction resulted in the formation of a 1,2,3-triazole in 24% yield. Subsequently, allenyl azides have been prepared and have been shown to rearrange to triazafulvenes that react with nucleophiles to form 1,2,3-triazoles [46, 47].

Although it is possible to exclude possible mechanistic interpretations as a result of experiment, all that we can do for any organic reaction is suggest the best possible mechanism based on the evidence that is available at any given time. Reactions may display a duality or multiplicity of mechanisms, and it may be possible to bypass certain stages in a catalytic cycle by varying the precise reaction protocol that is used. The view of the present authors is that a number of catalytic cycles may be required to account for all of the information that is currently available in the case of copper-catalysed azide–alkyne cycloaddition (CuAAC) reactions. There are two extreme types of reaction that are identifiable: those reactions that are facilitated by efficient ligands and those that are carried out without the addition of special ligands. Examples that are of intermediate type may exist, but for which no kinetic data are available. A number of reviews on CuAAC reactions have appeared that discuss the question of mechanism (see for example [48–51]).

The discovery of the highly efficient copper(I)-catalysed intermolecular reactions of terminal alkynes with organic azides (CuAAC reactions) was made independently by two groups of workers, and the initial publications in readily accessible sources appeared almost simultaneously. The main features are the high regioselectivity of the reactions and the high tolerance to a wide range of functional groups in both the terminal alkyne and the azide. An example from the synthesis of peptidotriazoles is shown in Scheme 6 [52], in which copper(I) iodide was used by the Meldal group as the source of the copper catalyst, together with Hünig's base: the reactions were shown not to proceed using internal alkynes, and the

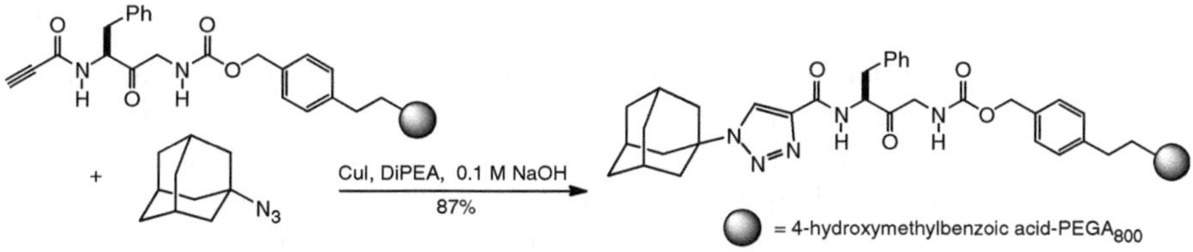

Scheme 6 The solid phase synthesis of a peptidotriazole

intermediacy of a terminal alkyne–Cu(I) complex was suggested. The protocol devised by the Sharpless and Fokin group used copper(II) sulfate together with sodium ascorbate, in which a stepwise mechanism was suggested, also involving copper(I) acetylides; an example is shown in Scheme 7 [53]. Among the very large number of papers that have been published on CuAAC reactions, the copper(II) sulfate–sodium ascorbate protocol is by far the most frequently used experimental method. It is important that we should consider the function of amines in more detail later and also whether sodium ascorbate has a role in addition to that as a reducing agent. Although there is an explicit statement that copper(II) compounds do not catalyse CuAAC reactions [52], there have been a number of reports of the direct use of copper(II) salts ([54–60]; see also [61]). The use of hydrogen peroxide to quench aliquots during a careful kinetic investigation (vide infra) allows us to exclude the possible involvement of copper(II) salts in CuAAC reactions, and alternative explanations are required.

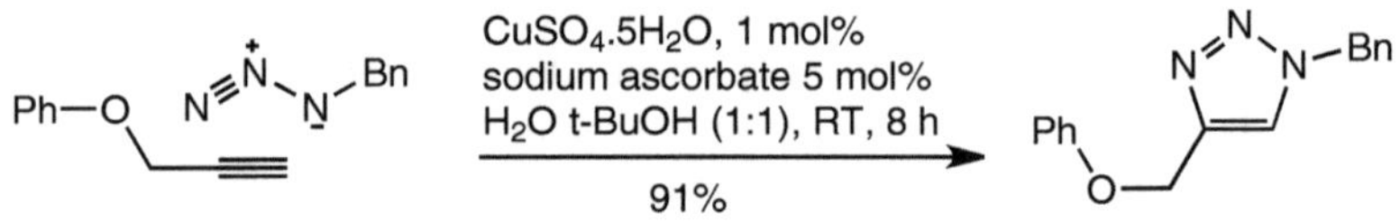

Scheme 7 The $CuSO_4$–Na ascorbate protocol for CuAAC reactions

4 The Variety of Protocols Available for Carrying Out CuAAC Reactions

As well as the many examples where copper(I) salts were used directly ([52]; see for example [62–64]), there are a number of other alternative ways of generating the required copper(I) state. One example involves the comproportionation reaction of metallic copper together with copper sulfate [65]. There are occasions when the presence of sodium ascorbate or its oxidation product may have deleterious effects and the use of metallic copper together with a copper(II) salt provides a valuable alternative protocol [66–68]. The use of a copper nanocluster provides a particularly efficient method with a significantly reduced induction period as compared to the use of other forms of copper(0) and the copper sulfate–sodium ascorbate protocol ([69]; see also [70]). The use of copper(I) oxide "on water," including high-purity material, has been shown to be particularly effective, for example, the reaction of *p*-toluenesulfonyl azide with phenylacetylene gave the anticipated triazole in 91% yield in 3 h at ambient temperature [71]. Other heterogeneous systems have been used effectively, including copper(I) zeolites, [72, 73] a copper–manganese spinel oxide [74], and copper on carbon [75], which is reported to involve a mixture of copper (I) and (II) oxides [76].

5 Early Mechanistic Considerations

Reactions of organic azides with nucleophiles occur at the distal (*N*3) position, as exemplified in Scheme 5, while reactions of azides with soft electrophilic metal derivatives occur by interaction with the proximal (*N*1) nucleophilic nitrogen. Thus, the majority of the discussions concerning the mechanisms involved in CuAAC reactions have been concerned with the nature of the interactions of copper(I) species with the various alkynes. Calculations, using density functional theory (DFT), based on a monocopper(I) acetylide model, gave results that explained the observed regioselectivity and the large rate increase (ca. 10^7) as compared to the results obtained in thermal Huisgen reactions [66]. A number of important features emerged from the calculations. Concerted cycloaddition was excluded. Although terminal alkynes are weak acids and are frequently deprotonated by organometallic species such as Grignard reagents (vide supra) in nonaqueous solvents, π-complex formation was shown to lower the pK_a of terminal alkynes by ca. 10 pK_a units, thereby facilitating deprotonation in aqueous solvent systems. The formation of the alkynylcopper(I) species in an aqueous solvent was also found to be energetically much more favourable than in, for example acetonitrile. An outline early mechanistic interpretation of the catalytic cycle involved in CuAAC reactions, based on those calculations, and involving a copper(III) metallacycle, is shown in Scheme 8.

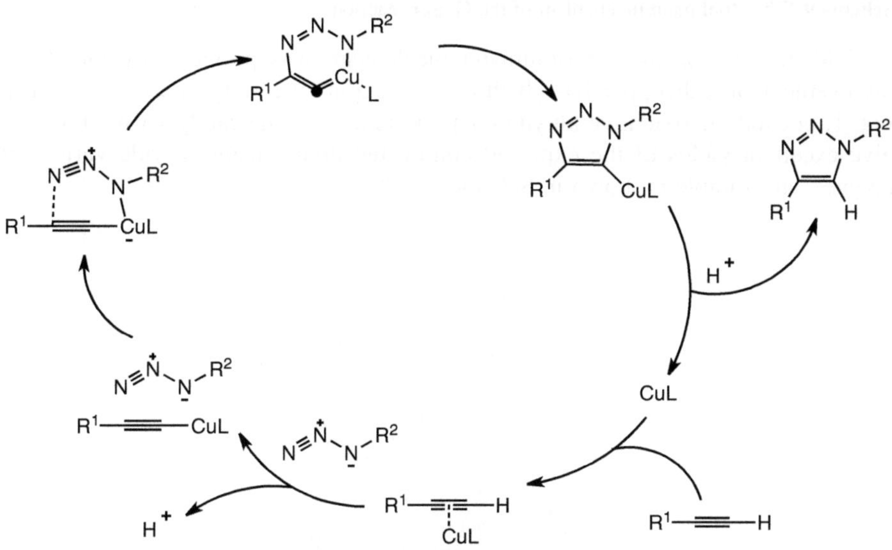

Scheme 8 An early mechanistic interpretation of CuAAC reactions involving copper(I) catalysis

The preparation and characterisation of *N*-heterocyclic carbene (NHC) copper(I) compounds [77, 78], and the use of for example *N*,*N*′-bis(2,4,6-trimethylphenyl) imidazole-2-ylidene and other NHCs in CuAAC reactions [79, 80], led to the successful isolation of a stable copper(I) triazolide with an NHC ligand [81]. Acceleration of CuAAC reactions by ligands such as phenantholine when carried out using NHC copper(I) compounds has also been reported [82].

6 The Glaser Reaction and the Conversion of Copper(II) Pre-catalysts to Copper(I) CuAAC Catalysts

The question of how copper(II) salts can be involved in CuAAC reactions in the absence of an added reducing agent has been addressed by a number of workers and has been shown to involve, in some cases, Glaser reactions. Acetylenic coupling [83], including the copper(II)-catalysed oxidative coupling of terminal alkynes that was discovered by Glaser [84], has been widely studied. The Eglinton variation of the Glaser reaction uses copper(II) acetate and is stoichiometric in copper(II) in the absence of dioxygen [85, 86]. In the widely accepted Bohlmann mechanism of the Glaser reaction, shown in Scheme 9, a dimeric copper(II) complex collapses to the dialkyne, together with copper(I) ([87]; see also [88]), thereby providing a ready explanation for the role of copper(II) salts as pre-catalysts in CuAAC reactions.

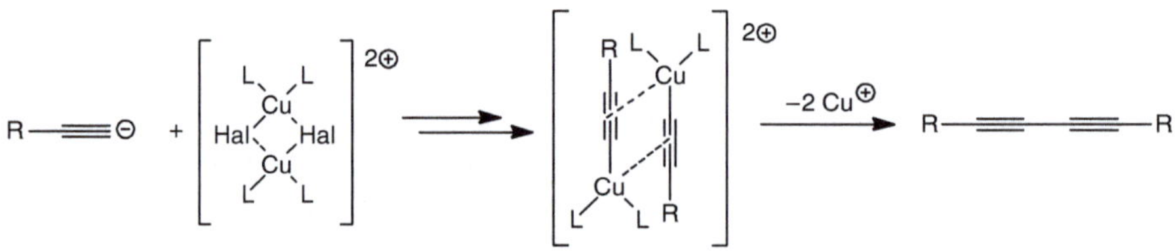

Scheme 9 The Bohlmann mechanism of the Glaser reaction

Evidence supporting the Bohlmann mechanism was provided by the direct involvement of a dicopper(II)-substituted γ-Keggin silicotungstate (**1**), shown in Fig. 1, in catalytic oxidative alkyne-coupling reactions: the catalyst was shown to give excellent yields of the expected conjugated diynes using a wide variety of alkynes, an example is shown in Scheme 10 [89].

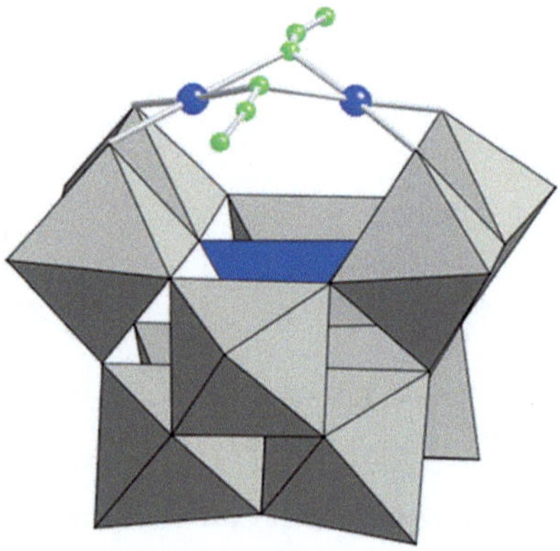

(1)

Fig. 1 A representation of the anion in a dicopper γ-Keggin silicotungstate using a polyhedral and ball and stick model: *blue* and *green balls* representing copper and nitrogen atoms, respectively. Taken from Kamata K, Nkaagawa Y, Kotani M, Yamaguchi K, Mizuno N (2008) Angew Chem Int Ed 47:2407. "Efficient Oxidative Alkyne Homocoupling Catalysed by a Monomeric Dicopper-Substituted Silicotungstate." Copyright Wiley-VCH Verlag GmbH and Co. KGaA. Reproduced with permission

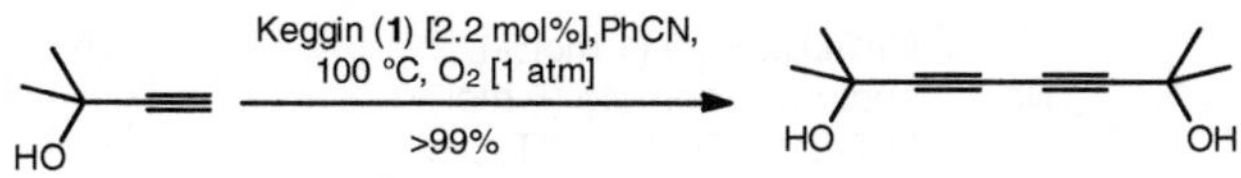

Scheme 10 A Glaser reaction using the Keggin (**1**)

Except when reactions are carried out using specific ligands, the fact that it has been shown that copper(II) salts do not function as catalysts in CuAAC reactions requires that the concomitant formation of Glaser products is expected in reactions that use copper(II) salts in the absence of an added reducing agents: the copper(II) salts act effectively as pre-catalysts. The dicopper(II)-substituted γ-Keggin silicotungstate (**1**) (mentioned above), which had two azide anions as ligands, was used as a pre-catalyst in a reaction of benzyl azide with phenylacetylene, carried out in acetonitrile, and gave the expected product 1-benzyl-4-phenyl-1,2,3-triazole in 98% yield. The reaction is also significant because little of the 1,2,3-triazole was produced during an induction period of ca. 50 min, during which time the absorption band in the visible spectrum at 700 nm, which is assigned to the d-d transition in the Cu(II) species, diminished; simultaneously, an amount of 1,4-diphenylbuta-1,3-diyne was formed in an amount almost equivalent to the amount of pre-catalyst used [90]. The recovered catalytic material, in which the absorption band in the visible spectrum at 700 nm was not present, was reusable and did not require the former induction period: the copper in the recovered material was evidently in the copper(I) state. An example of the method is shown in Scheme 11. The disappearance of the absorption band at ca. 700 nm after ca. 1 min was also noted in reactions of, for example, 2-picolyl azide with propargyl alcohol, using copper(II) acetate in easily oxidized alcohols such as methanol; however, the result using *t*-butanol as a solvent requires and alternative explanation which is discussed (vide infra) in the section concerning ligands [91].

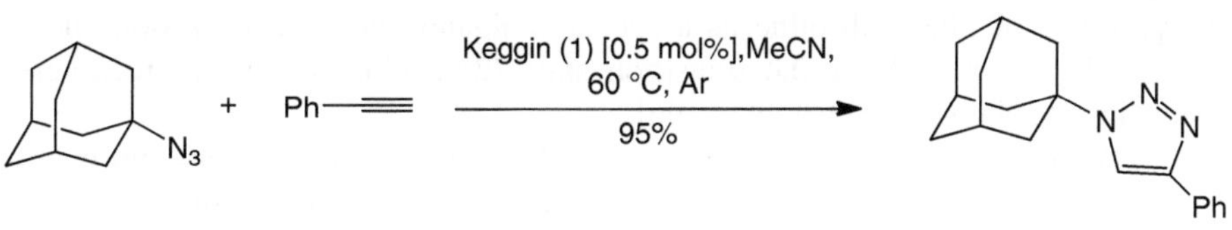

Scheme 11 The reaction of adamantyl azide with phenylacetylene using the Keggin (**1**)

7 Kinetic Studies and the Involvement of Two Copper(I) Centres

The important study of the kinetics of the reaction of benzyl azide with phenylacetylene, mentioned above, was carried out using the copper(II) sulfate–sodium ascorbate protocol in the absence of added nitrogen-based ligands [92]. The data obtained for the reaction shown in Scheme 12 gave rate orders at low

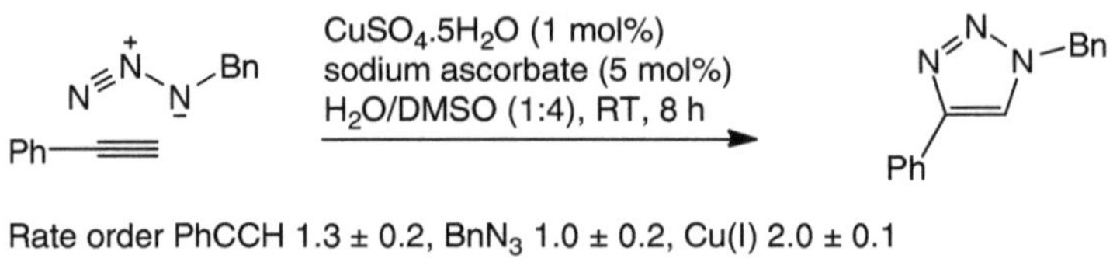

Scheme 12 The kinetics using the $CuSO_4$–Na ascorbate protocol for a CuAAC reaction

concentrations of phenylacetylene and benzyl azide of 1.3 ± 0.2 and 1.0 ± 0.2 respectively, while using a low concentration of the copper(II) pre-catalyst, the rate order for copper was shown to be 2.0 ± 0.1. A second series of kinetic measurements, carried out in the presence of a phenanthroline ligand, also gave a kinetic rate order for copper of 2.0 ± 0.1, with two spectator ligands associated with each copper atom [93]. Kinetics that were second order in copper(I) were also found for early and late stages of fast reactions using water-soluble alkynes and azides, with activation energies of 83.2 and 81.1 kJ mol^{-1}, respectively [94].

The results of the kinetic studies provided an impetus to the search for evidence of an intermediate that involved two copper(I) atoms. Despite being entropically disfavoured, those results clearly suggest that two copper centres are involved in the catalytic cycle. Dinuclear and tetranuclear copper(I) complexes are known [95] and could be involved in CuAAC reactions. Computational studies suggested ways of alleviating the ring strain that is evident in the copper(III) metallacycle shown in Scheme 8.

It is interesting to note that a calculation of the transition state energy involving a mononuclear copper intermediate places the transition state at a higher Gibbs free energy than the transition state involving the uncatalysed Huisgen [3 + 2] cycloaddition reaction. This is due to the high strain energy in the six-membered transition state which would have a bent cumulene-like structure. Calculated interactions using methyl azide together with a simplified tetrameric alkyne–Cu(I) complex, in the presence of phenanthroline as a spectator ligand, gave an activation energy (G) of 78.4 kJ mol^{-1} for the transition state (**2**) leading to the six-membered metallacycle, shown in Scheme 13 [96].

In another DFT study, it was found that the incorporation of a second copper atom into the model resulted in the calculated reactivity of such dinuclear alkynylcopper(I) complexes being dramatically increased as compared to the calculated reactivity of mononuclear alkynylcopper(I) complexes [97]. Calculations were also carried out to locate the positions of copper atoms in dinuclear copper(I) acetylides, transition states, and a metallacycle for a copper(I)-catalysed reaction of propyne with methyl azide: two copper atoms were located, one of which carried as a spectator ligand, for example a prop-1-ynyl anion. A calculated activation energy of 54.0 kJ mol^{-1} was found for the formation of a metallacycle in a reaction of propyne with methyl azide. The diagrams in Scheme 14 show the calculated geometries and distances separating the two copper atoms in a dinuclear copper (I) acetylide in **structure A**, in a calculated transition state in **structure B,** and in a cuprated metallacycle in **structure C**.

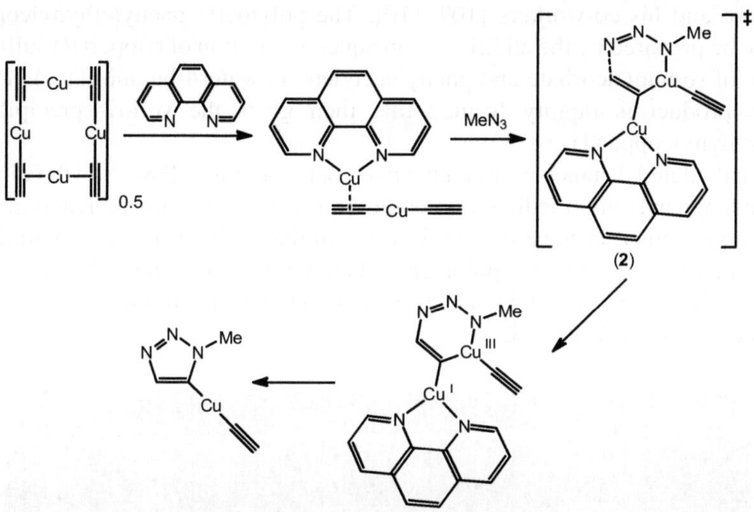

Scheme 13 A computed mechanistic pathway for the formation of a copper triazolide, starting from a tetrameric alkyne–copper complex using phenanthroline as a disaggregating ligand

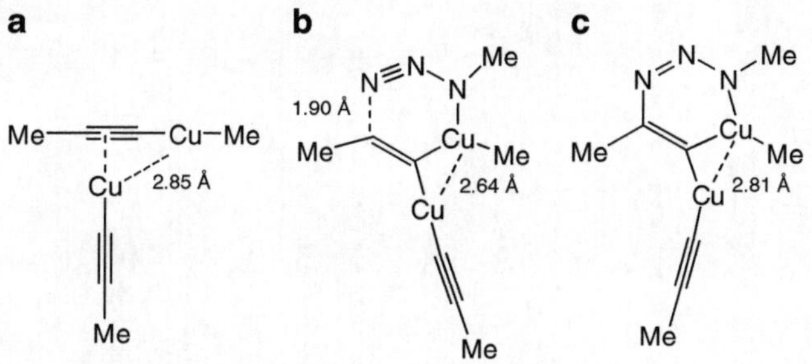

Scheme 14 Calculated structures for a CuAAC reaction of propyne with methyl azide with Cu–Cu distances for (**a**) dinuclear intermediate, (**b**) a transition state, and (**c**) a copper metallacycle

8 The Intervention of Alkynylcopper(I) Polymers in the Absence of Strong Ligands

A number of yellow or orange organocopper(I) compounds are known, some of which are stabilized by ligands; they are characterised by having short inter-Cu (I)–Cu(I) distances [98–101]. Alkynylcopper(I) compounds, including the stable yellow phenylethynylcopper(I), have been known for many years, and preparations have been reported from copper(I) salts using a number of protocols [102–106]. Stoichiometric reactions of phenylethynylcopper(I) have been exploited, especially by Castro and his co-workers [107–110]. The polymeric phenylethynylcopper(I) can also be prepared by the addition of an aqueous solution of copper(II) sulfate to a mixture of sodium acorbate and phenylacetylene in water/t-butanol (1:1): a brown viscous product is rapidly formed that then gives the yellow precipitate of phenylethynylcopper(I).

The calculated distance between the two copper atoms of 2.88 Å in the dinuclear propynylcopper(I) intermediate that could be involved in CuAAC reactions, as in the example shown in Scheme 14a, is very similar to the distances separating the copper atoms in the known polymeric phenylethynylcopper(I), the structure of which is shown in Fig. 2 [111]. The novel ladder-like topology gave refined Cu–Cu distances of 2.49–2.83 Å.

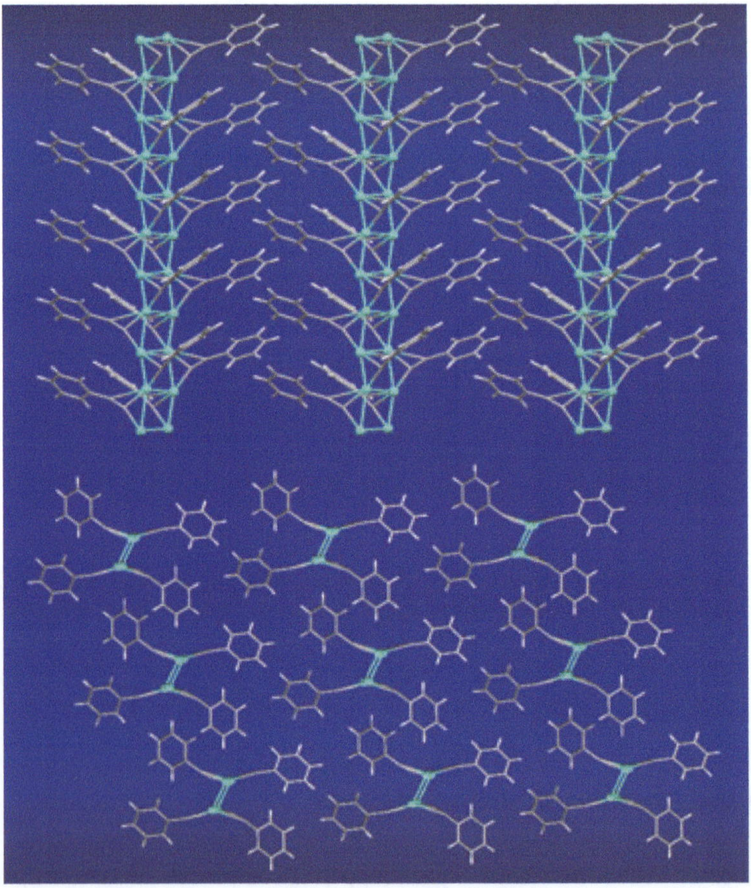

Fig. 2 The *top structure* of the polymeric chains of the *yellow phenylethynylcopper(I)* showing μ,η-1,2-bridging phenylethynyl ligands connected to Cu(I) ions: the *bottom structure* shows the solid state packing of the chains in the [010] direction. Taken from Chui SSY, Ng MFY, Che C-M (2005) Chem Eur J 11:1739. "Structure Determination of Homoleptic AuI, AgI, and CuI Aryl/Alkylethynyl Coordination Polymers by X-ray Powder Diffraction." Copyright Wiley-VCH Verlag GmbH and Co. KGaA. Reproduced with permission

The possibility of using basic copper(II) materials as pre-catalysts in CuAAC reactions has been investigated. Two reasons were adduced: first that the compounds would function as heterogeneous pre-catalysts and second that the hydroxyl ions would facilitate deprotonation of the alkynes. The deprotonation of the terminal proton has been studied using 1-[^{2} H]-phenylacetylene in exchange reactions [112]: efficient exchange was found using copper hydroxide on alumina or titanium dioxide, but ^{1}H/^{2}H exchange hardly proceeded using copper(II) chloride on silicon dioxide. In the event, a number of other protocols that are used in CuAAC reactions reveal that deprotonation of the alkyne is not a major problem. Copper(II) hydroxide, supported on alumina, was used as a heterogeneous pre-catalyst in toluene in CuAAC reactions in which an induction period was again noted as Cu (II) was reduced to Cu(I) by the alkyne [113]. In another investigation of the use of copper(II) salts as pre-catalysts in CuAAC reactions, copper(II) hydroxyacetate was initially chosen because of its potential use as a heterogeneous catalyst in a wide variety of solvents [114]. Reactions in which an excess of either phenylacetylene or its *p*-tolyl analogue and copper(II) hydroxyacetate were allowed to interact in acetonitrile gave the anticipated Glaser products in high yield; interestingly, the polymeric yellow arylethynylcopper(I) derivatives, shown in Scheme 15, were also produced in almost quantitative yields. The arylethynylcopper(I) polymers were identified from their powder X-ray diffraction patterns. Reactions that were carried out using an excess of terminal alkynes and azides, using copper(II) hydroxyacetate (10 mol%) as a pre-catalyst, gave good yields of the CuAAC products, in the absence of added ligands, together with the expected Glaser products and polymeric yellow copper(I) products. Examples of the latter reactions are shown in Scheme 16. A good yield of the CuAAC reaction product was also obtained when using 1 mol% of copper(II) hydroxyacetate in a reaction of phenylacetylene with benzyl azide. An example in which the phenylethynylcopper(I) polymer (10 mol%) was used together with phenylacetylene and an azide is shown in Scheme 17.

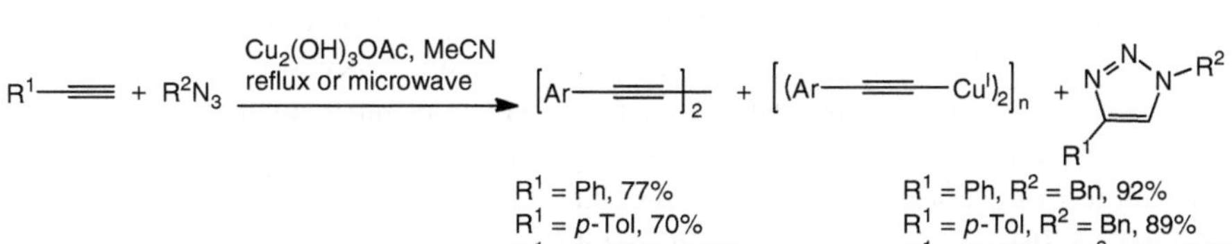

Scheme 15 Glaser reactions using arylethynes and copper(II) hydroxy acetate

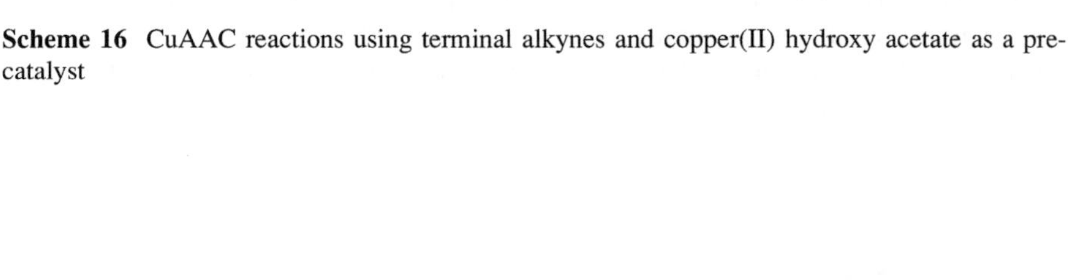

Scheme 16 CuAAC reactions using terminal alkynes and copper(II) hydroxy acetate as a pre-catalyst

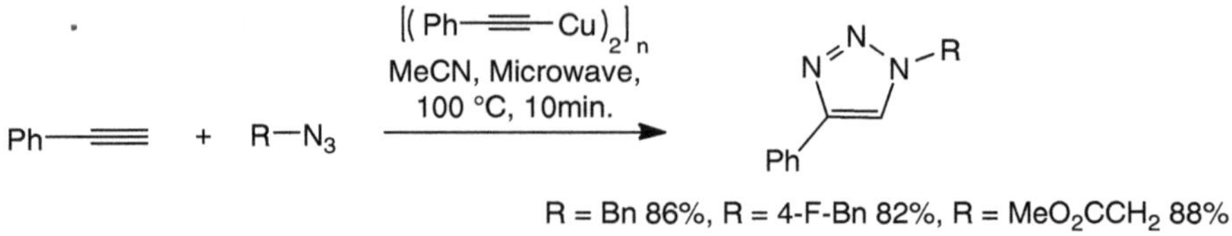

Scheme 17 CuAAC reactions of azides and phenylacetylene using phenylethynylcopper(I) as a pre-catalyst

It should be noted that the recovered alkynylcopper(I) polymers were able to be reused as catalysts, and importantly, in a reaction in which the phenylethynylcopper (I) polymer (10 mol%) was used together with *p*-tolylethyne and benzyl azide, the recovered yellow polymer was shown to be the *p*-tolylethynylcopper(I) compound. The origin of the hydrogen at position-5 in the triazole was investigated by carrying out a reaction in anhydrous acetonitrile using 1-[^{2}H]-2-phenylethyne and benzyl azide together with phenylethynylcopper(I). 1-Benzyl-4-phenyl-5-[^{2}H]-triazole was isolated from the latter reaction with a quantitative incorporation of deuterium, confirming that under aprotic reaction conditions the proton source in the triazole product was derived from the alkyne, for example phenylacetylene. The results described above, and the calculated requirement that two copper(I) atoms are required in the transition state leading to the metallacycle, suggest that acetonitrile is able to disaggregate the polymer under the reaction conditions by functioning as a soft ligand for copper(I). Further the organic azides, although regarded as weak ligands for copper(I), are able to displace acetonitrile and that at a late stage in the catalytic cycle the alkyne involved is reincorporated into the heterogeneous pre-catalyst. Tetrakis(acetonitrile)copper(I) hexafluorophosphate has been used as a source of copper(I) in CuAAC reactions in the presence of a number of stabilizing ligands that can displace acetonitrile, including tris(benzyltriazolylmethyl)amine [115, 116], and in reactions where the 5-cuprio derivative reacts in an oxidation reaction, probably involving a copper(II) intermediate, that incorporates an additional alkynyl residue [117].

The favoured protocol, in which the copper(I) catalyst is generated using sodium ascorbate and copper(II) sulfate in aqueous *t*-butanol (vide supra) in the presence of the terminal alkyne and an organic azide, is of mechanistic interest for several reasons. In the related reactions, where copper(II) acetate and sodium ascorbate in a water/*t*-butanol (10:1) solvent system was used, terminal alkynes were shown to undergo copper(I)-catalysed conjugate addition using alkylidene derivatives of Meldrum's acid: it was suggested that ascorbate may have an important role in addition to that involved in reducing copper(II) to copper(I) [118]. The first dissociation of ascorbic acid gives a pK_a ca. 4.1, which is similar to that of acetic acid (pK_a 4.76) and suggests that both acetate and ascorbate are able to deprotonate terminal alkynes. Visually, the addition of an aqueous solution of copper(II) sulfate to the mixture of alkyne, azide, and sodium ascorbate results in the transient formation of yellow material that is obscured by a white suspension after about

15 s. Since it is known that the formation of the triazole is not rapid under those reaction conditions, it is possible that the white suspended material is copper(I) ascorbate. Phenylethynylcopper(I) was isolated at the end of a reaction of phenylacetylene with benzyl azide using 1 mol% of copper(II) sulfate and sodium ascorbate. Reactions of, for example, benzyl azide with a number of terminal alkynes in *t*-butanol/water (1:1) using the yellow polymeric alkynylcopper(I) compounds were shown to give the anticipated triazoles in good yields, and examples are shown in Scheme 18; in addition, a reaction of benzyl azide with 4-methoxyphenylpropargyl ether in *t*-butanol/water (1:1) gave an excellent yield of 1-benzyl-4-*p*-methoxyphenoxymethyltriazole, shown in Scheme 19 [119].

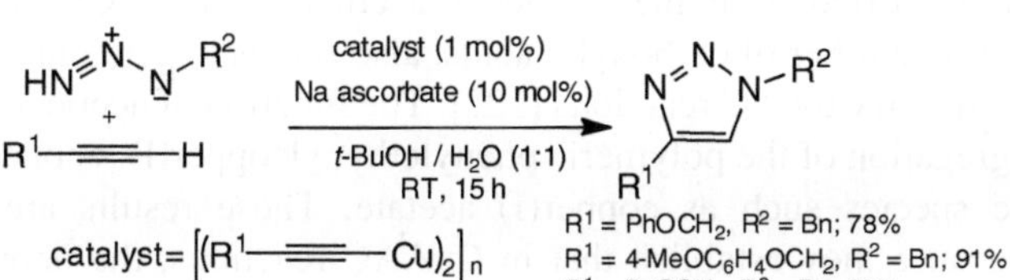

Scheme 18 CuAAC reactions of azides and alkynes using alkynylcopper(I) derivatives as a pre-catalyst and ascorbate as ligand

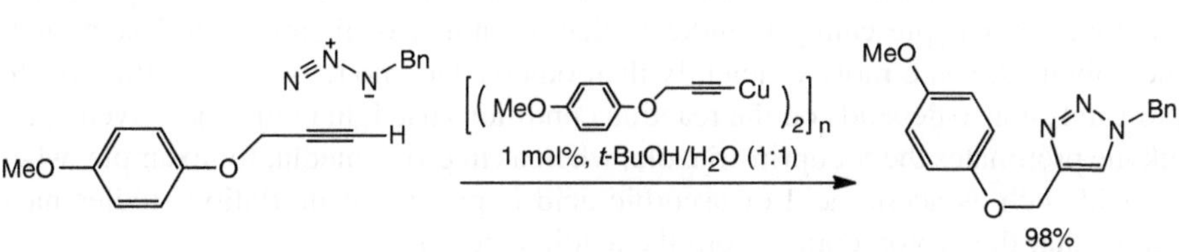

Scheme 19 CuAAC reactions of benzyl azide and alkynes using alkynylcopper(I) derivatives as a pre-catalyst and benzyl azide as ligand

Phenylethynylcopper(I) is also formed rapidly from copper(I) acetate in cyclohexane, and the addition of benzyl azide was then shown to give the expected triazole in an excellent yield, as shown in Scheme 20, while a similar reaction in which copper(II) acetate and sodium ascorbate were substituted for copper(I) acetate proceeded much more slowly. Other copper(I) carboxylates also gave good yields of the CuAAC product formed using phenylacetylene and benzyl azide. The enhanced reactivity of copper(I) acetate was ascribed to the presence of acetic acid that was formed during the reaction forming phenylethynylcopper(I) [120]. It was also found that the addition of an equivalent of acetic acid to an "on water" reaction of phenylacetylene with benzyl azide using phenylethynylcopper(I) as the pre-catalyst resulted in the triazole derivative being produced more rapidly

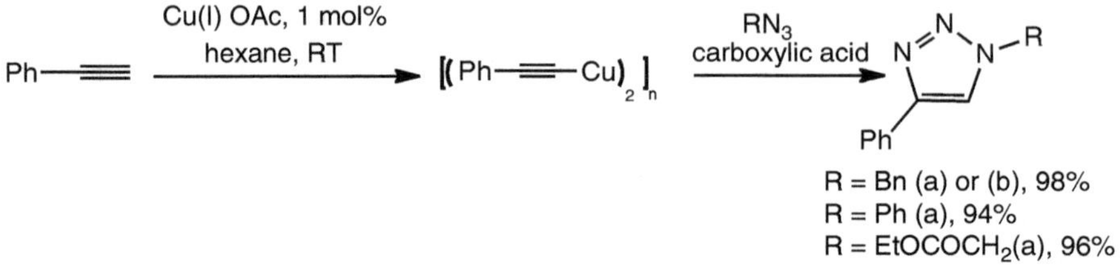

Scheme 20 CuAAC reactions of benzyl azide and phenylacetylene using copper(I) acetate in (**a**) cyclohexane or hexane or (**b**) acetonitrile

than in reactions carried out in the absence of acetic acid [121]. Copper(I) oxide, in the presence of a number of carboxylic acids, also serves to generate active copper (I) species for use in CuAAC reactions [122]. The function of acetic acid may be to aid the disaggregation of the polymeric phenylethynylcopper(I), copper(I) oxide, or even dimeric species such as copper(I) acetate. Those results are particularly interesting because they establish that in CuAAC reactions, the deprotonation of terminal alkynes can be achieved in the presence of carboxylic acids. In contrast to the reactions using phenylethynylcopper(I) in the absence of a proton source, a reaction carried out using benzyl azide in the presence of deuteriated acetic acid gave 1-benzyl-4-phenyl-5-[^{2}H]-triazole with 97% incorporation of ^{2}H [123].

The various studies of CuAAC reactions that have involved polymeric alkynylcopper(I) pre-catalysts indicate that a number of ligands can disaggregate the polymers, some more efficiently than others. The mode of regeneration of the pre-catalyst also depends on the reaction conditions used: in nonprotic solvents, the alkyne protonates the 5-cuprio-triazole, whereas in protic media, for example, when an acid such as acetic acid or ascorbic acid is present, deuteriation studies have shown that the proton comes from the acidic medium.

9 Ligand Acceleration: Introduction

The commonly used copper(II)–ascorbate protocol frequently allows 1,4-disubstituted-1,2,3-triazoles to be formed at rates that are ca. 10^6 times the rate of related reactions carried out without a metal catalyst. A number of examples involving drug discovery [124], and more general biological applications [125–128], were identified at an early stage in the development the use of CuAAC reactions. However, there are a number of applications where there is a requirement for even more rapid CuAAC reactions. Ligand-accelerated catalysis presents such a possibility [129]. The majority of the studies of ligand-accelerated CuAAC reactions have involved soft nitrogen-containing ligands. Early examples gave an indication that certain reactions appeared to be autocatalytic, which suggested that the products of the CuAAC reactions might function as rate-enhancing ligands.

A study using a wide structural range of potential ligands showed that the ligand tris(benzyltriazolylmethyl)amine was the best in a particular set [115, 116]. Tris(benzyltriazolylmethyl)amine was also found to be a useful ligand in bioconjugation reactions [130], while modification of the structure to afford increasing water solubility and stability towards oxidation of copper(I) complexes has led to new methodology [131] and ligand designs [132–135]. At the same time, a study using intramolecular fluorescence quenching of the substituted naphthalene component (the fluorophore) by the diaryldiazo component in the 1,4-disubstituted triazole shown in Scheme 21, allowed kinetic measurements to be made as well as establishing the accelerating effects of a number of ligands; it is interesting to note that the rate increases using different ligands were at different maxima at different pHs [93]. Rate increases of two orders of magnitude were observed, relative to the standard copper(II) sulfate–sodium ascorbate protocol. The results obtained using the ligand shown in Scheme 21 suggested that two copper atoms and two ligands for each copper atom were probably required in order to achieve catalytic turnover. It is worth noting that 1,4-substituted 1,2,3-triazoles function as efficient ligands for a wide range of metal ions [136].

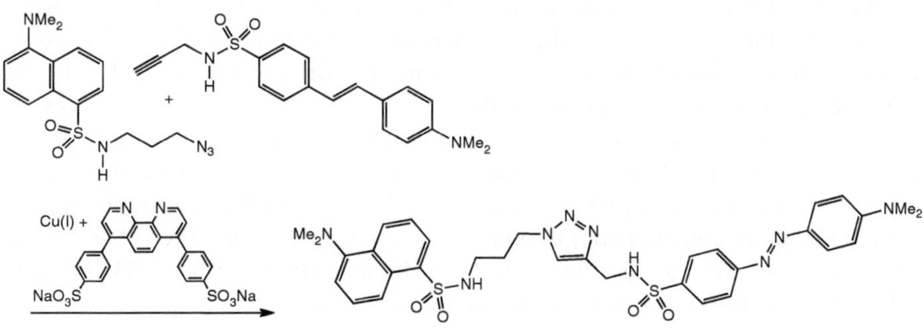

Scheme 21 Fluorescence quenching in a ligand-accelerated CuAAC reaction

10 The Use of Soft Ligands

The position of copper(I) salts as quite soft Lewis acids has resulted in a significant number of publications where soft phosphorus- and sulfur-containing ligands have been used and shown to enhance the catalytic process in CuAAC reactions. A variety of phosphorus-containing ligands has been used to produce significant rate enhancements or improved efficiency in CuAAC reactions. Examples include the use of triphenylphosphine-copper(I) bromide [137] (e.g., in the synthesis of dendrimers [138]), phosphoramidites [139], and bis(triphenylphosphine)-copper(I) carboxylates [140]. In addition, triethylphosphite-copper(I) iodide complexes have been used in sugar chemistry [141].

10.1 The Instability of N-Sulfonyl-5-Cuprio-1,2,3-Triazoles

The copper(I)-catalysed reactions of terminal alkynes with sulfonylazides [142], or phosphoryl azides [143], in the presence of primary and secondary amines that result in the formation of amidines, are mechanistically interesting; they involve the formation of somewhat unstable triazole derivatives. A similar outcome results from the substitution of a terminal ynamide for a simple terminal alkyne [144]. The copper(I)-catalysed cycloaddition reactions of sulfonyl azides has been shown to be complicated by the relatively low-energy barrier to ring opening of the triazole ring system in the products which gives rise to an equilibrium involving breaking the *N*1–*N*2 bond: that process results in the formation of a sulfonyl amide after hydrolysis of the intermediate amidine [145]. 2,6-Lutidine in chloroform was shown to mitigate the problem associated with the ring opening reaction [146], as also have tris(pyrazolyl) ligands [147]. The reactions have been studied in more detail, and evidence for the intermediacy of 5-cuprio-triazoles has been obtained using iodine monochloride as an electrophilic trapping agent, as shown in Scheme 22 [148]. Additional reactions of the ring-opened copper intermediates further confirm the instability of the 5-cuprio-triazoles [149]. A copper(I) oxide–catalysed reaction of phenylacetylene with *p*-toluenesulfonyl azide carried out in deuteriated water resulted in the isolation of the anticipated 5-[^{2}H]-triazole [71]; that result lends support to the calculations that indicate that the 5-cuprio-triazoles are the unstable species in the reactions and that they can be intercepted by an excess of an electrophile. An advantage of using a sulfur-containing ligand, for example, thioanisole, is thought to involve stabilization of the 5-cuprio-1,2,3-triazole involving ligation by sulfur at the copper atom [150]. The use of sulfur-containing ligands [151, 152], including copper(I) thiophene-2-carboxylate, as an efficient and stable source of a copper(I) catalyst, for example, in Ullmann-like reactions [153], has been followed by its use in reactions of sulfonyl azides with alkynes [154].

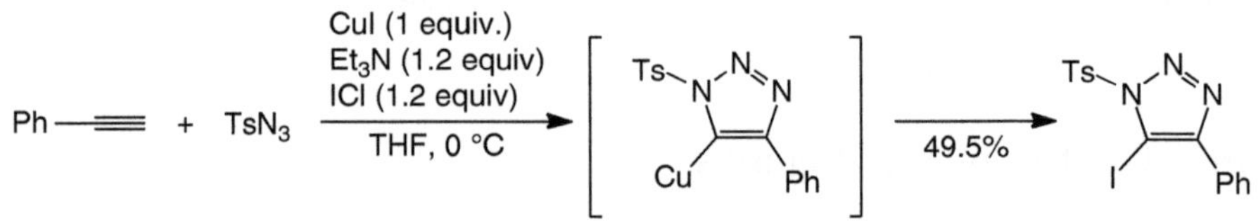

Scheme 22 The trapping of the 5-cuprio-intermediate, formed in the copper(I)-catalysed reaction of phenylacetylene with *p*-toluenesulfonyl azide, using iodine monochloride

11 The Wide Range of Ligands Involved in CuAAC Reactions

It is now be generally agreed that the ligation of copper(I) species is required in the catalytic cycles involved in CuAAC reactions, and yet there are many publications where no ligands are purposefully added; for example in the sodium ascorbate–copper(II) sulfate protocol, illustrated in Scheme 7 [53]. One can speculate that ascorbate, in addition to reducing copper(II), functions as a soft ligand for copper(I). This proposal provides an explanation for the rapid disappearance of the yellow material on the addition of copper(II) sulfate to the mixture of an excess of sodium ascorbate, the terminal alkyne, and the organic azide and on the isolation of, for example, phenylethynylcopper(I) at the end of some reactions. In various mechanistic proposals, an organic azide is shown to displace another ligand at an early stage of a catalytic cycle. It is therefore reasonable to conclude that an organic azide should also be able to act as a soft ligand for copper(I) species, as in the reaction shown in Scheme 19 [119]. In reactions where, for example, copper(II) acetate is used as the pre-catalyst, a copper(II)–azide complex will readily form the catalytic copper(I) species, particularly where there is a suitably positioned additional functional group that can ligate the copper [91]; the result will be that the distal nitrogen will have an increased electrophilicity in the stepwise cycloaddition reaction [155]. We may note that the ease of reduction of copper(II) to copper(I) has been studied using cyclic voltammetry and is particularly easy in the presence of the soft ligands that are found to accelerate CuAAC reactions [115, 116, 156].

In the discussion of reactions of alkynylmetallic derivatives (vide supra), we deferred a discussion of reactions of aluminum derivatives of terminal alkynes. An attempted reaction of dimethylphenylethynylaluminium with benzyl azide at ambient temperature in THF failed to give any product, while a reaction carried out at 150°C resulted in the formation of a mixture of regioisomeric triazoles. Although reactions carried out using 10 mol% of copper(I) salts as catalysts gave, after hydrolytic workup, 1,4-disubstituted 1,2,3-triazoles, the reactions proceeded slowly at room temperature. Some improvement was noted when using 20 mol% of dimethylphenylphosphine as an added ligand in the copper(I) iodide–catalysed reaction of dimethylphenylethynylaluminium with benzyl azide at ambient temperature; however, the best conditions were found to involve the use of the ligand pentamethyldiethyltriamine (PMDTA) in reactions carried out in toluene or methyl *t*-butyl ether [157]. The involvement of the 4-aluminated intermediate, shown in Scheme 23, was confirmed by the products obtained in reactions with a number of electrophiles, including *N*-iodosuccinimide and methyl chloroformate.

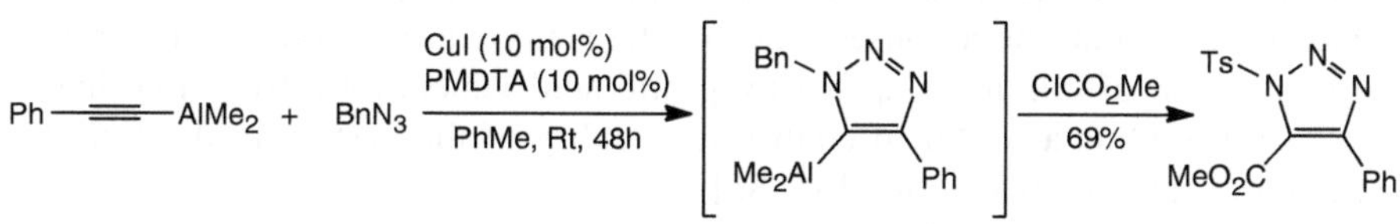

Scheme 23 The copper(I)-catalysed reaction of dimethylphenylethynylaluminium with benzyl azide, followed by reaction with methyl chloroformate

A suggested catalytic cycle is shown in Scheme 24. A number of different ligands have also been used in copper(I)-catalysed reactions of terminal iodoalkynes that lead to the formation of 5-iodo-1,2,3-triazoles: the best ligand was found to be tris(1-*t*-butyltriazolylmethyl)amine, which was found to dramatically reduce the time taken for the reactions to proceed to completion [158].

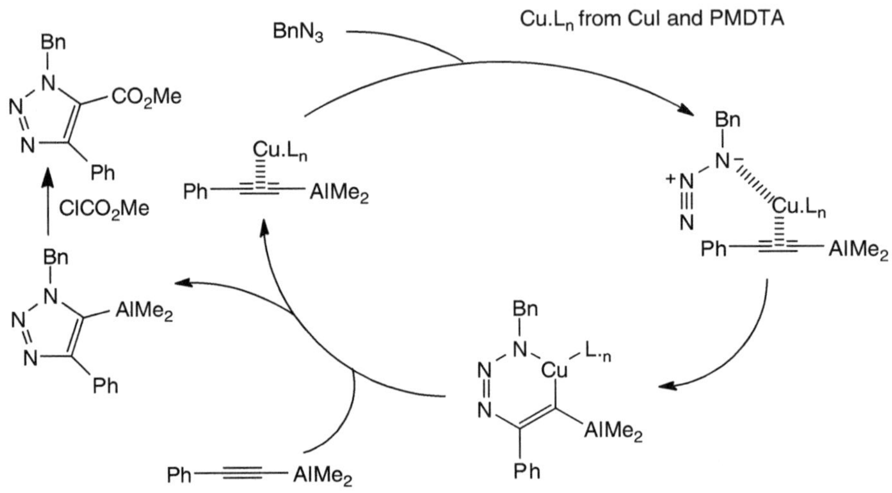

Scheme 24 A possible mechanistic cycle for the copper(I)-catalysed reaction of benzyl azide with dimethylphenylethynylaluminium

11.1 Reactions Involving Chelating Azides

The question of the possible intervention of various copper oxidation states in Ullmann and Goldberg reactions has been discussed from time to time (see the summarising discussion in [159]). Although the reduction of copper(II) to copper (I) involving the formation of conjugated diynes by means of the Glaser reaction and reactions carried out in easily oxidized alcohols (vide supra) provides a ready explanation for a number of investigations carried out using copper(II) pre-catalysts, other studies, particularly when using organic azides that have substituents that are capable of ligating to copper, have merited further investigation. The fact that the Glaser product has not been detected in some CuAAC reactions where copper(II) pre-catalysts have been used, in the absence of a reducing agent, may be a consequence of the analytical methods used or reflect the retardation of alkyne coupling by protic acid generation [160]: pyridine is effectively used as a buffer to control the acidity of the system in the Eglinton variation of the Glaser reaction [85, 86].

A number of *N*-(prop-2-ynyl)-2-diazenecarboxamides were shown to form triazoles very rapidly in reactions with α-amino-ω-azidoalkanes using copper (II) sulfate as a pre-catalyst, as exemplified in Scheme 25 [161]. The addition of particles of metallic copper to reactions did not affect the speed with which the products were formed; however, no product was formed in the presence of dioxygen and a route by which a copper(II) complex pre-catalyst gives rise to the required copper(I) catalyst in those reactions remains to be determined.

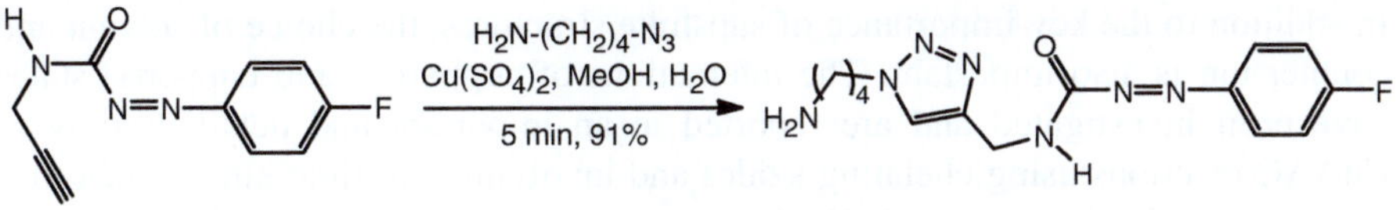

Scheme 25 A reaction using a chelating azide and a copper(II) pre-catalyst

It is interesting to note, for example, the formation of a stable crystalline product from the interaction of copper(II) chloride with *cis*-1-azido-2-(1-pyrazolyl)-cyclohexane [162]. Similarly, blue 2-azidomethylpyridine crystalline copper(II) complexes, λ_{max} ca. 700 nm, are clearly the precursors of the copper(I) species involved in the CuAAC reactions mentioned earlier [91]. Representations of the crystal structures are shown in Fig. 3 [163].

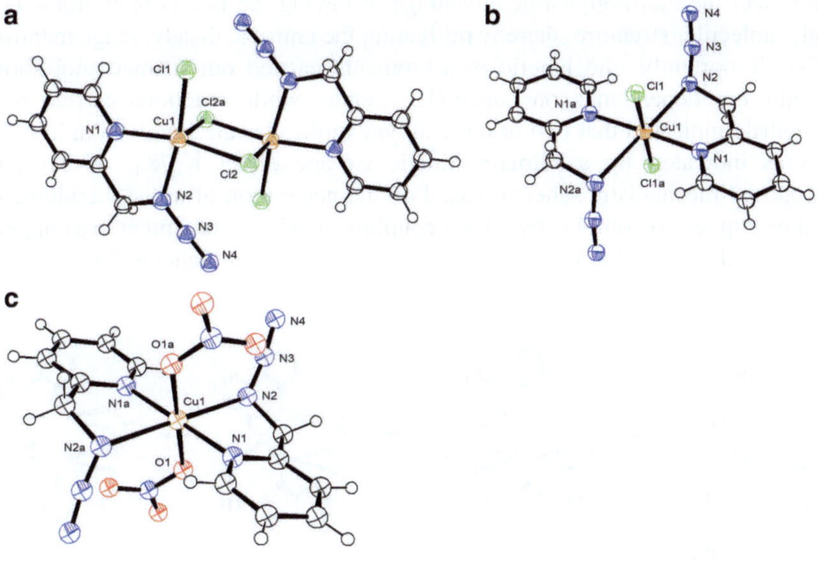

Fig. 3 ORTEP representations (50% elipsoids) of the five-coordinate complex formed by the interaction of copper(II) chloride (**a** and **b**) and copper(II) nitrate (**c**) with 2-azidomethylpyridine: *blue* and *orange balls* representing nitrogen and copper, respectively, and *green* and *red balls* representing chlorine and oxygen, respectively. Taken from Brotherton WS, Guha PM, Phan H, Clark RJ, Shatruk M, Zhu L (2011) Dalton Trans 40:3655–3665. "Tridentate complexes of 2,6-bis (4-substituted-1,2,3-triazol-1-ylmethyl)pyridine and its organic azide precursors: an application of the copper(II) acetate-accelerated azide–alkyne cycloaddition." Reproduced by permission of The Royal Society of Chemistry

The detailed mechanisms involved in CuAAC reactions depend on a number of factors, especially in examples involving chelating azides: routes by which the required copper(I) catalysts are generated differ especially from cases where azides that are not associated with strong chelating functional groups are involved.

In addition to the key importance of substrate structures, the choice of solvent and counter ion is also important. The interactions of copper(II) and copper(I) states have been investigated and are reported in an important and detailed study of CuAAC reactions using chelating azides and involving real-time kinetic measurements, using fluorescence and ^{1}H nmr monitoring techniques, as well as structure, solvent, and counterion variation [164]. Reactions carried out using copper(II) salts as pre-catalysts in oxidisable solvents such as methanol [91, 119], as well as in nonoxidisable solvents such as *t*-butanol and water, were found to proceed most rapidly [164]. In connection with counterion studies, there is a clear parallel in the use of copper(II) acetate in CuAAC reactions with its use in Glaser reactions [85, 86, 165]: both reactions require two copper atoms and the known dimeric structures of copper(II) acetate [166], and copper(I) acetate [167, 168], provide a link. In addition, the fact that copper(II) acetate is now the favoured reagent in alkyne-coupling reactions and is also more frequently reported in CuAAC reactions may well reflect the thermodynamic advantage of having the two copper atoms in the single molecular structure; thereby mitigating the entropic disadvantage mentioned earlier. Importantly, the kinetic measurements carried out in methanol showed second-order dependence on copper(II) acetate, while reactions carried out in acetonitrile indicated that two or more alkyne molecules are involved in important steps as indicated by a primary kinetic isotope effect K_H/K_D of 2.3 [164]. A proposed mechanistic scheme, based on the generation of a bis[2-azidomethylpyridine copper(I)] complex by Glaser coupling, and its involvement in a copper(I)-catalysed alkyne–azide cycloaddition reaction is shown in Scheme 26.

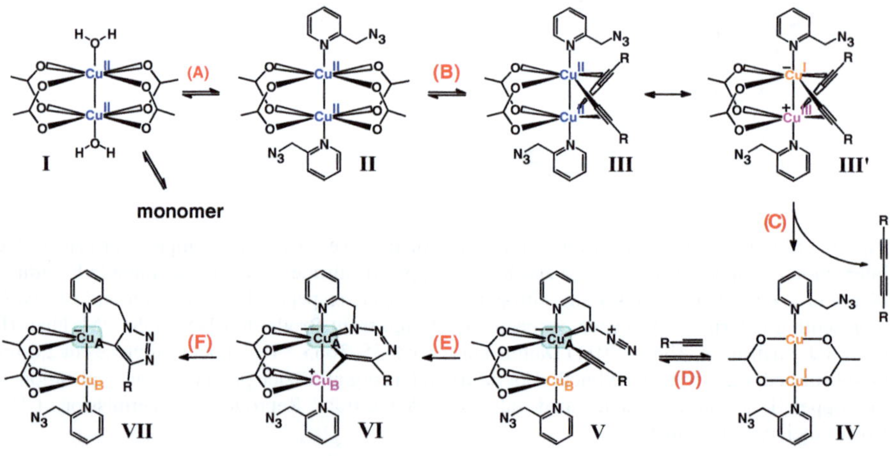

Scheme 26 A proposed mechanistic sequence for the copper-catalysed cycloaddition reaction based on the use of copper(II) acetate and 2-azidomethylpyridine with a terminal alkyne. We thank Dr Lei Zhu for providing Scheme 26, which is an amalgam of Schemes 17 and 18 in the paper "Experimental Investigation on the Mechanism of Chelation-Assisted, Copper(II) Acetate-Accelerated Azide-Alkyne Cycloaddition" by Kuang G-C, Guha PM, Brotherton WS, Simmons JT, Stankee LA, Nguyen BT, Clark RJ, Zhu L (2011) J Am Chem Soc 133:13984

12 Ligand Effects and Mechanism in CuAAC Reactions

It is useful at this point to bring together the most important features of the use of ligand-accelerated reactions, particularly those studies that report kinetic measurements and those that have a specific bearing on mechanistic aspects of CuAAC reactions. The use of soft nitrogen-containing ligands has been discussed at various points earlier in this chapter. Different groups of workers have their own favourite ligand. The toxicity of copper has been mentioned and has led to the exploitation of copper-free triazole synthesis involving cyclooctyne derivatives (vide supra). Whereas a number of a number of known toxic copper complexes kill human hepatoma cells, recent results showed that bis(histidine)copper(II) did not show toxicity effects after 72 h and also that copper(I)-L-histidine is effective in CuAAC reactions involved in labeling live cells [169]. The majority of the compounds used to act as ligands for copper(I) have a basic nitrogen that helps in the deprotonation of the alkyne together with three soft ligating groups that stabilize the copper(I) towards oxidation. No single ligand species has been found that acts as the best pre-catalyst or catalyst under all reaction conditions that are used in CuAAC reactions; however, for preparative scale reactions, a trimethylamine ligand having two pyridinylmethyl residues together with a benzimidizolyl residue was found to be the best among the large group of ligands studied [170]. Tris[(1-*t*-butyl-1-*H*-1,2,3-triazolyl)methyl]amine was the most successful added ligand in controlling the ratio of macrocyclisation to dimerization in reactions carried out in a flow system [171]. In another series of reactions, tris[2-(1-benzyl-1-H-1,2,3-triazol-4-yl)methyl] amine was reported to be the favoured ligand [172]. Rate constants have been evaluated for CuAAC reactions using a variety of ligands that accelerate the reactions, under dilute reaction conditions using aliquot quenching; and also using the more concentrated solutions that are normally used in preparative chemistry using thermochemical measurements: tris(2-benzimidazolylmethyl)amine derivatives were found to be superior to a number of ligands that are frequently used, for example, the tris(triazolylmethyl)amine derivatives that were mentioned earlier in this chapter [173]. The results of a complementary study using a number of different ligands indicated that the CuAAC reaction is mechanistically complex [174]. First-order kinetics were observed for both the alkyne and the azide together with second-order dependence on the chelated copper(I) complex. In an idealized mechanism, the tertiary amino group supplies electron density to the copper(I) while the softer ligands serve to produce a multidentate species that stabilizes the copper in the copper(I) oxidation state. The final proteolysis of the 5-cuprio-triazole was found to occur more rapidly when the ligand used has pendant carboxyl or ester groups. A comprehensive survey of ligands used in CuAAC reactions has been published [175].

Two additional examples serve to illustrate the complexity of copper(I)-catalysed azide–alkyne cycloaddition reactions and the dramatic effect that different ligands can have on the outcome of particular reactions. When the base ammonium carbonate was used in the reaction shown in Scheme 27, the normally

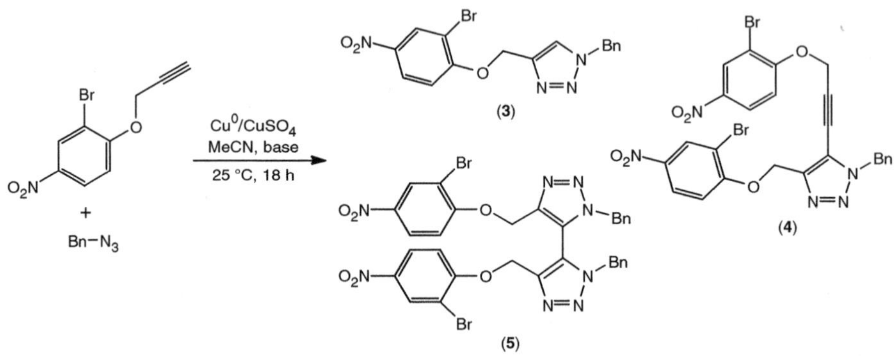

Scheme 27 The copper(I)-catalysed reaction of an arylpropargyl ether with benzyl azide in the presence of sodium carbonate – ratio 1:2:3 = 3:95:2

expected 1,4-disubstituted-1,2,3-triazole (**3**) was the exclusive product. However, changing the base to sodium carbonate resulted in the formation of a mixture of products in which the compound (**3**) comprised only 3% of the mixture and the major product (95% of the mixture) was the interesting axially chiral 5,5′-bi (1,2,3-triazolyl) derivative (**5**) [176].

A divergence in product formation was also observed in reactions of phenylacetylene with the bis(azide), shown in Scheme 28. In the reaction carried out using the original copper sulfate–sodium ascorbate protocol, only trace amounts the product (**6**) were formed; even when using an alkyne to azide ratio of 1:10, the major product isolated was the bis(triazole (**7**) [92]. The reaction was reinvestigated using a wide range of ligands, and variable amounts of the products (**6**) and (**7**) were formed [174]. Reactions using the 2,6-di(benzimidazolyl)-pyridine ligand (**8**) gave increasing amounts of the mono-triazole product (**4**) as the ligand to copper(I) ratio was increased.

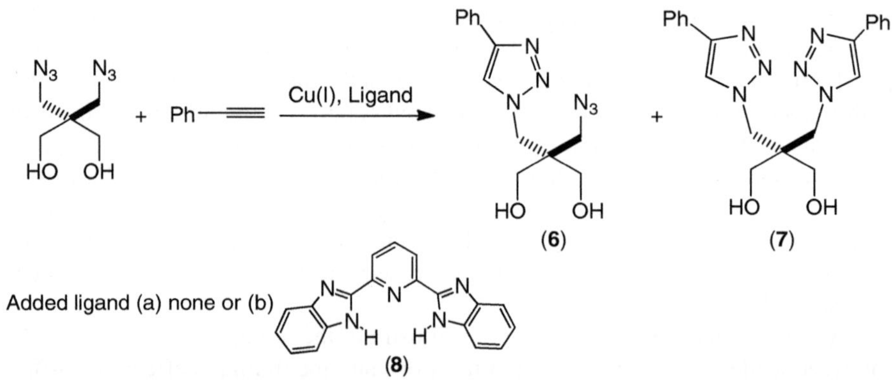

Scheme 28 The copper(I)-catalysed reaction of a bis(azide) with phenylacetylene in the presence or absence of added nitrogen ligand

Although the detailed mechanism involved in any particular CuAAC reaction may vary, it is now clear that copper(I) catalytic species are mandatory and also that the copper(III) metallacyclic intermediate retains exocyclic copper(I) in some form. The ligands that are attached to the two copper atoms depend on the precise protocol that is used. The results of a number of investigations suggest that, in many reactions two copper(I) atoms are present in the catalytic species; while in the presence of multidentate ligands, an equilibrium can exist between ligated copper (I) species that have two copper atoms and species that have only one copper(I) atom.

Acknowledgement We thank Dr Lei Zhu (University of Florida, Tallahassee) for his valuable comments on an earlier version of this chapter.

Note added in proof An interesting short review on the synthesis of 1,2,3-triazoles involving both thermal (Huisgen) and copper(I) catalysed intramolecular alkyne–azide cycloaddition reactions has appeared [178].

References

1. Biagi G, Giorgi I, Livi O, Scartoni V, Betti L, Giannaccini G, Trincavelli ML (2002) Eur J Med Chem 37:565
2. Danence LJT, Gao Y, Li M, Huang H, Wang J (2011) Chem Eur J 17:3584
3. Stazi F, Cancogni D, Turco L, Westerduin P, Bacchi S (2010) Tetrahedron Lett 51:5385
4. Huisgen R (1961) Proc Chem Soc 357
5. Husigen R (1963) Angew Chem Int Ed 2:565
6. Huisgen R (1984) In: Padwa A (ed) 1,3-Dipolar cycloaddition chemistry, vol 1. Wiley, New York, pp 1–176
7. Huisgen R, Szeimes G, Möbius L (1967) Chem Ber 100:2494
8. Padwa A, Pearson WH (eds) (2003) Synthetic applications of 1,3-dipolar cycloaddition chemistry toward heterocycles and natural products. Wiley, Hoboken, New Jersey
9. Michael A (1883) J Prakt Chem 48:94
10. Mitchell G, Rees CW (1987) J Chem Soc Perkin Trans 1 413
11. Honda K, Nakanishi H, Nagawa Y, Yabe A (1984) J Chem Soc Chem Commun 450
12. Nagawa Y, Honda K, Nakanishi H (1987) Synthesis 905
13. Wittig G, Krebs A (1961) Chem Ber 94:3260
14. Jewett JC, Bertozzi CR (2010) Chem Soc Rev 39:1272
15. Debets MF, van der Doelen CWJ, Rutjes FPJT, van Delft FL (2010) ChemBioChem 11:1168
16. Agard NJ, Prescher JA, Bertozzi CR (2004) J Am Chem Soc 126:15046
17. Ning X, Temming RP, Dommerholt J, Guo J, Ania DB, Debets MF, Wolfert MA, Boons G-J, van Delft FL (2010) Angew Chem Int Ed 49:3065
18. Sletten EM, Nakamura H, Jewett JC, Bertozzi CR (2010) J Am Chem Soc 132:11799
19. Plass T, Milles S, Koehler C, Schultz C, Lemke EA (2011) Angew Chem Int Ed 50:3878
20. Kirmse W, Horner L (1958) Liebigs Ann Chem 614:1
21. Fleming I (1976) Frontier orbitals and organic chemical reactions. Wiley, Chichester, p 156
22. Meazza G, Zanardi G (1991) J Fluorine Chem 55:199
23. van Berkel SS, Dirks AJ, Meeuwissen SA, Pingen DLL, Boerman OC, Laverman P, van Delft FL, Cornelissen JJLM, Rutjes FPJT (2008) ChemBioChem 9:1805
24. Huang J, Macdonald SJF, Harrity JPA (2009) Chem Commun 436
25. Englert BC, Bakbak S, Bunz UHF (2005) Macromolecules 38:5868

26. Jiang Y, Kuang C, Yang Q (2011) Tetrahedron 67:289
27. Krasinski A, Fokin VV, Sharpless KB (2004) Org Lett 6:1237
28. Liu D, Gao W, Dai Q, Zhang X (2005) Org Lett 7:4907
29. Dai Q, Gao W, Liu D, Capes LM, Zhang X (2006) J Org Chem 71:3928
30. Odlo K, Hentzen J, Fournier dit Chabert J, Ducki S, Gani OABS, Sylte I, Skrede M, Flørenes VA, Hansen TV (2008) Bioorg Med Chem 16:4829
31. Chakraborty A, Dey S, Sawoo S, Ardash NN, Sakar A (2010) Organometallics 29:6619
32. Heaney H, Ley SV (1973) J Chem Soc Perkin Trans 1 499
33. Yuan Y, Thomé I, Kim SH, Chen D, Beyer A, Bonnamour J, Zuidema E, Chang S, Bolm C (2010) Adv Synth Catal 352:2892
34. Kwok SW, Fotsing JR, Fraser RJ, Rodionov VO, Fokin VV (2010) Org Lett 12:4217
35. Zhang L, Chen X, Xue P, Sun HHY, Williams ID, Sharpless KB, Fokin VV, Jia G (2005) J Am Chem Soc 127:15998
36. Rasmussen LK, Boren BC, Fokin VV (2007) Org Lett 9:5337
37. Boren BC, Narayan S, Rasmussen LK, Zhang L, Zhao H, Lin Z, Jia G, Fokin VV (2008) J Am Chem Soc 130:8923
38. Lipshutz BH, Sengupta S (1992) Org React 41:135
39. Taylor RJK (ed) (1994) Organocopper reagents. Oxford University Press, Oxford
40. Lipshutz BH (2002) In: Schlosser M (ed) Organometallics in synthesis, a manual. Wiley, Chichester
41. Krause N (ed) (2002) Modern organocopper chemistry. Wiley-VCH, Weinheim
42. Ley SV, Thomas AW (2003) Angew Chem Int Ed 42:5400
43. Heaney H, Christie S (2004) Science of synthesis; organometallics, compounds of groups 12 and 11, organometallic complexes of copper. Thieme, Stuttgart, pp 305–662
44. Tornøe CW, Meldal M (2001) Peptidotriazoles: copper(I)-catalysed 1,3-dipolar cycloadditions on solid phase, Peptides 2001, Proc. Am. Pept. Symp. American Peptide Society and Kluwer Academic Publishers, San Diego, pp 263–264
45. L'abbé G, Mahy M, Bollyn M, Germain G, Scheefer G (1983) Bull Soc Chim Belg 92:881
46. Banert K, Hagedorn M (1989) Angew Chem Int Ed 28:1675
47. Fotsing JR, Banert K (2005) Eur J Org Chem 3704
48. Bock VD, Hiemstra H, van Maarseveen JH (2006) Eur J Org Chem 51
49. Meldal M, Tornøe CW (2008) Chem Rev 108:2952
50. Hein JE, Fokin VV (2010) Chem Soc Rev 39:1302
51. Ackermann L, Potukuchi HK (2010) Org Biomol Chem 8:4503
52. Tornøe CW, Christensen C, Meldal M (2002) J Org Chem 67:3057
53. Rostovtsev VV, Green LG, Fokin VV, Sharpless KB (2002) Angew Chem Int Ed 41:2596
54. Reddy KR, Rajgopal K, Kantam ML (2006) Synlett 957
55. Fukuzawa S-i, Shimizu E, Kikuchi S (2007) Synlett 2436
56. Reddy KR, Rajgopal K, Kantam ML (2007) Catal Lett 114:36
57. Bonnamour J, Legros J, Crousse B, Bonnet-Delpon D (2007) Tetrahedron Lett 48:8360
58. Fiandanese V, Bottalico D, Marchese G, Punzi A, Capuzzolo F (2009) Tetrahedron 65:10573
59. Namitharan K, Kumarraja M, Pitchumani K (2009) Chem Eur J 15:2755
60. Fiandanese V, Iannone F, Marchese G, Punzi A (2011) Tetrahedron 67:5254
61. Song YJ, Yoo C, Hong J-T, Kim S-J, Son SU, Jang H-Y (2008) Bull Korean Chem Soc 29:1561
62. Malkoch M, Schleicher K, Drockenmuller E, Hawker CJ, Russell TP, Wu P, Fokin VV (2005) Macromolecules 38:3663
63. Ackermann L, Potukuchi HK, Landsberg D, Vicente R (2008) Org Lett 10:3081
64. Qin A, Lam JWY, Tang L, Jim CJW, Zhao H, Sun J, Tang BZ (2009) Macromolecules 42:1421
65. Appukkuttan P, Dehaen W, Fokin VV, Van der Eycken E (2004) Org Lett 6:4223
66. Himo F, Lovell T, Hilgraf R, Rostovtsev VV, Noodleman L, Sharpless KB, Fokin VV (2005) J Am Chem Soc 127:210

67. Urankar D, Košmrlj J (2008) J Comb Chem 10:981
68. Chandrasekhar S, Seenaiah M, Kumar A, Reddy CR, Mamidyala SK, Kumar CG, Balasubramanian S (2011) Tetrahedron Lett 52:806
69. Pachón LD, van Maarseveen JH, Rothenberg G (2005) Adv Synth Catal 347:811
70. Moore E, McInnes SJ, Vogt A, Voelcker NH (2011) Tetrahedron Lett 52:2327
71. Wang K, Bi X, Xing S, Liao P, Fang Z, Meng X, Zhang Q, Liu Q, Ji Y (2011) Green Chem 13:562
72. Chassaing S, Kumarraja M, Sido ASS, Pale P, Sommer J (2007) Org Lett 9:883
73. Chassaing S, Sido ASS, Alix A, Kumarraja M, Pale P, Sommer J (2008) Chem Eur J 14:6713
74. Yousuf SK, Mukherjee D, Singh B, Maity S, Taneja SC (2010) Green Chem 12:1568
75. Lipshutz BH, Taft BR (2006) Angew Chem Int Ed 45:8235
76. Lee C-T, Huang S, Lipshutz BH (2009) Adv Synth Catal 351:3139
77. Mankad NP, Gray TG, Laitar DS, Sadighi JP (2004) Organometallics 23:1191
78. Goj LA, Blue ED, Munro-Leighton C, Gunnoe TB, Petersen JL (2005) Inorg Chem 44:8647
79. Díez-González S, Correa A, Cavallo L, Nolan SP (2006) Chem Eur J 12:7558
80. Díez-González S, Nolan SP (2008) Angew Chem Int Ed 47:8881
81. Nolte C, Mayer P, Straub BF (2007) Angew Chem Int Ed 46:2101
82. Teyssot M-L, Chevry A, Traïkia M, El-Ghozzi M, Avignant D, Gautier A (2009) Chem Eur J 12:6322
83. Siemsen P, Livingston RC, Diederich F (2000) Angew Chem Int Ed 39:2632
84. Glaser C (1869) Chem Ber 2:242
85. Eglinton G, Galbraith AR (1959) J Chem Soc 889
86. Behr OM, Eglinton G, Galbraith AR, Raphael RA (1960) J Chem Soc 3614
87. Bohlmann F, Schönowsky H, Inhoffen E, Grau G (1964) Chem Ber 97:794
88. Fedenok LG, Shvartsberg MS (2011) Tetrahedron Lett 52:3776
89. Kamata K, Nkaagawa Y, Kotani M, Yamaguchi K, Mizuno N (2008) Angew Chem Int Ed 47:2407
90. Kamata K, Nkaagawa Y, Tamaguchi K, Mizuno N (2008) J Am Chem Soc 130:15304
91. Brotherton WS, Michaels HA, Simmons JT, Clark RJ, Dalal NS, Zhu L (2009) Org Lett 11:4954
92. Rodionov VO, Fokin VV, Finn MG (2005) Angew Chem Int Ed 44:2210
93. Lewis WG, Magallon FG, Fokin VV, Finn MG (2004) J Am Chem Soc 126:9152
94. Kasuga K, Ito M, Onoda W, Nakamura Y, Inokuma S, Matsuda T, Nishimura J (2009) Heterocycles 78:963
95. Olbrich F, Behrens U, Weiss E (1994) J Organomet Chem 472:365
96. Straub BF (2007) Chem Commun 3868
97. Ahlquist M, Fokin VV (2007) Organometallics 26:4389
98. Karlin KD, Hayes JC, Gultneh Y, Cruse RW, McKown JW, Hutchinson JP, Zubieta J (1984) J Am Chem Soc 106:2121
99. Jin K, Huang X, Pang L, Li J, Appel A, Wherland S (2002) Chem Commun 2872
100. Huang X-C, Zhang J-P, Chen X-M (2004) J Am Chem Soc 126:13218
101. Zheng S-L, Messerschmidt M, Coppens P (2005) Angew Chem Int Ed 44:4614
102. Nast R, Pfab W (1956) Chem Ber 89:415
103. Blake D, Calvin G, Coates GE (1959) Proc Chem Soc 396
104. Green MLH (1968) Organometallic compounds, the transition elements, vol 2, 3rd edn. Methuen, London, pp 273–276
105. Sazonova VA, Kronrod NYa (1956) Zh Obshch Khim 26:1876 (Chem Abstr 1957, 51, 4981c)
106. Ito H, Arimoto K, Sensui H-o, Hosomi A (1997) Tetrahedron Lett 38:3977
107. Stephens RD, Castro CE (1963) J Org Chem 28:3313
108. Atkinson RE, Curtis RF, Taylor JA (1967) J Chem Soc C 578
109. Owsley DC, Castro CE (1988) Org Synth Coll 6:916
110. Castro CE, Gaughan EJ, Owsley DC (1966) J Org Chem 31:4071
111. Chui SSY, Ng MFY, Che C-M (2005) Chem Eur J 11:1739

112. Yamaguchi K, Oishi T, Katayama T, Mizuno N (2009) Chem Eur J 15:10464
113. Katayama T, Kamata K, Yamaguchi K, Mizuno N (2009) ChemSusChem 2:59
114. Buckley BR, Dann SE, Harris DP, Heaney H, Stubbs EC (2010) Chem Commun 46:2274
115. Chan TR, Hilgraf R, Sharpless KB, Fokin VV (2004) Org Lett 6:2853
116. Chouhan G, James K (2011) Org Lett 13:2754
117. Gerard B, Ryan J, Beeler AB, Porco JA Jr (2006) Tetrahedron 62:6405
118. Knöpfel TF, Carriera EM (2003) J Am Chem Soc 125:6054
119. Buckley BR, Dann SE, Heaney H (2010) Chem Eur J 16:6278
120. Shao C, Cheng G, Su D, Xu J, Wang X, Hu Y (2010) Adv Synth Catal 352:1587
121. Buckley BR, Dann SE, Heaney H, Stubbs EC (2011) Eur J Org Chem 770
122. Shao C, Zhu R, Luo S, Zhang Q, Wang X, Hu Y (2011) Tetrahedron Lett 52:3782
123. Shao C, Wang X, Xu J, Zhao J, Zhang Q, Hu Y (2010) J Org Chem 75:7002
124. Kolb HC, Sharpless KB (2003) Drug Discov Today 8:1128
125. Fazio F, Bryan MC, Blixt O, Paulson JC, Wong C-H (2002) J Am Chem Soc 124:14387
126. Speers AE, Adam GC, Cravatt BF (2003) J Am Chem Soc 125:4686
127. Link AJ, Tirrell DA (2003) J Am Chem Soc 125:11164
128. Deiters A, Cropp TA, Mukherji M, Chin JW, Anderson JC, Schultz PG (2003) J Am Chem Soc 125:11782
129. Berrisford DJ, Bolm C, Sharpless KB (1995) Angew Chem Int Ed 34:1059
130. Wang Q, Chan TR, Hilgraf R, Fokin VV, Sharpless KB, Finn MG (2003) J Am Chem Soc 125:3192
131. Hong V, Udit AK, Evans RA, Finn MG (2008) ChemBioChem 9:1481
132. Gupta SS, Kuzelka J, Singh P, Lewis WG, Manchester M, Finn MG (2005) Bioconjugate Chem 16:1572
133. Bergbreiter D, Hamilton PN, Koshti NM (2007) J Am Chem Soc 129:10666
134. Candelon N, Lastécouères D, Diallo AK, Aranzaes JR, Astruc D, Vincent J-M (2008) Chem Commun 741
135. Hong V, Presolski SI, Ma C, Finn MG (2009) Angew Chem Int Ed 48:9879
136. Struthers H, Mindt TL, Schibli R (2010) Dalton Trans 39:675
137. Lal S, Diez-González S (2011) J Org Chem 76:2367
138. Wu P, Feldman AK, Nugent AK, Hawker CJ, Scheel A, Voit B, Pyun J, Fréchet JMJ, Sharpless KB, Fokin VV (2004) Angew Chem Int Ed 43:3928
139. Campbell-Verduyn LS, Mirfeizi L, Dierckx RA, Elsinga PH, Feringa BL (2009) Chem Commun 2139
140. Gonda Z, Novák Z (2010) Dalton Trans 39:726
141. Pérez-Balderas F, Ortega-Muñoz M, Morales-Sanfrutos J, Hernández-Mateo F, Calvo-Flores FG, Calvo-Asín JA, Isac-García J, Santoyo-González F (2003) Org Lett 5:1951
142. Bae I, Han H, Chang S (2005) J Am Chem Soc 127:2038
143. Kim SH, Jung DY, Chang S (2007) J Org Chem 72:9769
144. Kim JY, Kim SH, Chang S (2008) Tetrahedron Lett 49:1745
145. Cassidy MP, Raushel J, Fokin VV (2006) Angew Chem Int Ed 45:3154
146. Yoo EJ, Ahlquist M, Kim SH, Bae I, Fokin VV, Sharpless KB, Chang S (2007) Angew Chem Int Ed 46:1730
147. Cano I, Nicasio MC, Pérez PJ (2010) Org Biomol Chem 8:536
148. Yoo EJ, Ahlquist M, Bae I, Sharpless KB, Fokin VV, Chang S (2008) J Org Chem 73:5520
149. Kim SH, Park SH, Choi JH, Chang S (2011) Chem Asian J 6:2618
150. Wang F, Fu H, Jiang Y, Zhao Y (2008) Adv Synth Catal 350:1830
151. Bai S-Q, Koh LL, Hor TSA (2009) Inorg Chem 48:1207
152. Fabbrizzi P, Cicchi S, Brandi A, Sperotto E, van Koten G (2009) Eur J Org Chem 5423
153. Zhang S, Zhang D, Liebeskind LS (1997) J Org Chem 62:2312
154. Raushel J, Fokin VV (2010) Org Lett 12:4952
155. Kuang G-C, Michaels HA, Simmons JT, Clark RJ, Zhu L (2010) J Org Chem 75:6540
156. Donnelly PS, Zanatta SD, Zammit SC, White JM, Williams SJ (2008) Chem Commun 2459

157. Zhou L, Lecourt T, Micouin L (2010) Angew Chem Int Ed 49:2607
158. Hein JE, Tripp JC, Krasnova LB, Sharpless KB, Fokin VV (2009) Angew Chem Int Ed 48:8018
159. Strieter ER, Bhayana B, Buchwald SL (2009) J Am Chem Soc 131:78
160. Thorwirth R, Stolle A, Ondruschka B, Wild A, Schubert US (2011) Chem Commun 47:4370
161. Urankar D, Steinbücher M, Kosjek J, Košmrlj J (2010) Tetrahedron 66:2602
162. Barz M, Herdtweck E, Thiel WR (1998) Angew Chem Int Ed 37:2262
163. Brotherton WS, Guha PM, Pan H, Clark RJ, Shatruk M, Zhu L (2011) Dalton Trans 40:3655
164. Kuang G-C, Guha PM, Brotherton WS, Simmons JT, Stankee LA, Nguyen BT, Clark RJ, Zhu L (2011) J Am Chem Soc 133:13984
165. Berscheid R, Vögtle F (1992) Synthesis 58
166. Hansen AE, Ballhausen CJ (1965) Trans Faraday Soc 61:631
167. Edwards DA, Richards R (1973) J Chem Soc, Dalton Trans 2463
168. Mounts IC, Ogura T, Fernando Q (1974) Inorg Chem 13:802
169. Kennedy DC, McKay CS, Legault MCB, Danielson DC, Blake JA, Pegoraro AF, Stolow A, Mester Z, Pezacki JP (2011) J Am Chem Soc 133:17993
170. Presolski SI, Hong V, Cho S-H, Finn MG (2010) J Am Chem Soc 132:14570
171. Bogdan AR, James K (2010) Chem Eur J 16:14506
172. Michaels HA, Zhu L (2011) Chem Asian J 6:2825
173. Rodionov VO, Presolski SI, Garinier S, Lim Y-H, Finn MG (2007) J Am Chem Soc 129:12696
174. Rodionov VO, Presolski SI, Díaz DD, Fokin VV, Finn MG (2007) J Am Chem Soc 129:12705
175. Díez-González S (2011) Catal Sci Technol 1:166
176. Angell Y, Burgess K (2007) Angew Chem Int Ed 46:3649
177. Brotherton WS, Guha PM, Phan H, Clark RJ, Shatruk M, Zhu L (2011) Dalton Trans 40:3655–3665
178. Majumdar KC, Ray K (2011) Synthesis 3787

Top Heterocycl Chem (2012) 28: 31–84
DOI: 10.1007/7081_2011_67

Published online: 8 February 2012

"Click-Triazole" Coordination Chemistry: Exploiting 1,4-Disubstituted-1,2,3-Triazoles as Ligands

James D. Crowley and David A. McMorran

Abstract Access to readily functionalized ligand architectures is of crucial importance in a range of different areas including catalysis, metallopharmaceuticals, bioimaging, metallosupramolecular chemistry, mechanically interlocked architectures, and molecular machines. The mild and modular Cu(I)-catalyzed 1,3-cycloaddition of terminal alkynes with organic azides (the CuAAC "click" reaction) allows the ready formation of functionalized 1,4-disubstituted-1,2,3-triazole scaffolds, and this has led to an explosion of interest in the coordination chemistry of these heterocycles. The parent 1,4-disubstituted-1,2,3-triazole units can potentially act as monodentate or bridging ligands. Examples of both the monodentate (through either the N3 nitrogen or C5 carbon positions of the 1,2,3-triazole) and bridging (through the N2 and N3 nitrogen atoms) coordination modes have been structurally characterized. A diverse array of bi-, tri-, and polydentate ligands incorporating 1,4-disubstituted-1,2,3-triazole units have also been synthesized and characterized. When the chelate pocket involves coordination through the N3 nitrogen atom of the 1,2,3-triazole, these are called "regular" click ligands. While these are the most common type of "click" chelate, "inverse" ligands in which the 1,2,3-triazole unit coordinates through the less electron-rich N2 nitrogen atom have also been synthesized and characterized. The resulting "click" complexes are beginning to find applications in catalysis, metallosupramolecular chemistry, photophysics, and as metallopharmaceuticals and bioimaging agents.

Keywords Catalysis · Click chemistry · Ligands · Metal complexes · Metallosupramolecular chemistry

J.D. Crowley (✉) and D.A. McMorran
Department of Chemistry, University of Otago, P.O. Box 56, Dunedin, New Zealand
e-mail: jcrowley@chemistry.otago.ac.nz

Contents

Abbreviations

[9]aneS3	1,4,7-Trithiacyclonoane
AAC	Azide-alkyne cycloaddition
Ac	Acetyl
Acac	Acetylacetonate

Ad	Adamantyl
Ar	Aryl
ATH	Asymmetric transfer hydrogenation
ATRP	Atom transfer radical polymerization
Bn	Benzyl
Boc	*Tert*-butoxycarbonyl
Bpy	2,2′-Bipyridyl
Bu	Butyl
Bz	Benzoyl
CAN	Ceric ammonium nitrate
cod	Cyclooctadiene
Cp*	1,2,3,4,5-Pentamethylcyclopentadiene
CuAAC	Cu(I)-catalyzed azide-alkyne cycloaddition
d	Day(s)
DABCO	1,4-Diazabicyclo[2.2.2]octane
DFT	Density functional theory
Dipp	3,5-Diisopropylphenyl
Dmbpy	4,4′-Dimethyl-2,2′-bipyridyl
DME	1,2-Dimethoxyethane
DMEDA	*N,N′*-dimethyl-1,2-ethylenediamine
DMF	Dimethylformamide
DMSO	Dimethyl sulfoxide
Dppb	Bis(diphenylphosphino)butane
DSSC	Dye-sensitized solar cells
ECL	Electrochemiluminescence
EDTA	Ethylenediaminetetraacetic acid
ee	Enantiomer excess
Equiv	Equivalent(s)
ESI-MS	Electrospray ionization mass spectrometry
Et	Ethyl
F_2ppy	2-(2,4-Difluoro)phenylpyridine
Fc	Ferrocene
h	Hour(s)
i-Pr	Isopropyl
LEECs	Light-emitting electrochemical cell
MAO	Methylaluminoxane
Me	Methyl
MeCN	Acetonitrile
Mes	Mesityl (2,4,6-trimethylphenyl)
min	Minute(s)
Nap	Naphthalene
NHC	*N*-heterocyclic carbene
OLEDs	Organic light-emitting diodes
Ph	Phenyl

Pic	Picolinic acid
ppy	Phenylpyridine
Pr	Propyl
PS	Polystyrene
py	Pyridine
RAFT	Reversible addition-fragmentation chain transfer
RT	Room temperature
RuAAC	Ru(II)-catalyzed azide-alkyne cycloaddition
Salen	*N,N′*-ethylenebis(salicylimine)
SCO	Spin crossover
SEC	Size-exclusion chromatography
t-Bu	*Tert*-butyl
TEP	Tolman electronic parameter
Tf	Trifluoromethanesulfonyl (triflyl)
THF	Tetrahydrofuran
TMEDA	*N,N,N′,N′*-tetramethyl-1,2-ethylenediamine
TMS	Trimethylsilyl
TOF	Turnover frequency
Tol	4-Methylphenyl
TON	Turnover number
Ts	Tosyl (4-toluenesulfonyl)

1 Introduction

Many areas of modern coordination chemistry would be enhanced by facile and functional group tolerant synthetic protocols that allow for the rapid generation of functionalized ligand scaffolds. Moreover, reactions that enable the modular tuning of steric and electronic properties would be particularly useful in the development of novel catalysts, materials, and metallopharmaceuticals. The recently discovered Cu(I)-catalyzed 1,3-cycloaddition of terminal alkynes (**A**) with organic azides (**B**) (the CuAAC "click" reaction, Fig. 1) is a synthetic protocol that potentially fulfills these requirements.

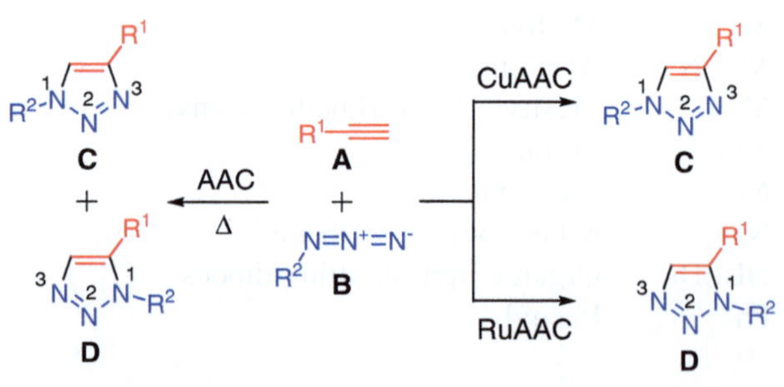

Fig. 1 The generation of 1,4-disubstituted-1,2,3-triazoles using the thermally induced [3+2] AAC, the CuAAC, and the RuAAC

1.1 Cu(I)-Catalyzed 1,3-Cycloaddition of Organic Azides with Terminal Alkynes: "Click" Chemistry

The thermally induced [3+2] cycloaddition between alkynes (**A**) and organic azides (**B**) (the AAC reaction, Fig. 1) was discovered in 1893 [1] and extensively studied by Huisgen and coworkers in the middle of the last century [2–5]. However, the reaction was never widely adopted by synthetic chemists presumably because it usually generates a mixture of the 1,4-disubstituted-1,2,3-triazole (**C**) and 1,5-disubstituted-1,2,3-triazole (**D**) products. That all changed in 2002 when Tornoe and Meldal [6] and Fokin and Sharpless [7] independently discovered the copper(I) catalyzed version of the AAC reaction [8].[1] Compared with the thermal AAC process, the rate of the CuAAC reaction is increased by a factor of 10^7 making it conveniently fast at room temperature [9]. Furthermore, the CuAAC reaction is regioselective providing only the 1,4-disubstituted-1,2,3-triazoles (**C**) in very high yields with few or no side products. The reaction is not significantly affected by steric or electronic factors, and the presence of most organic or inorganic functional groups is well tolerated. Additionally, the reaction proceeds in a wide variety of protic and aprotic solvents, including water. Due to these factors, the CuAAC methodology has become the "flagship reaction" of "click chemistry" [10]. Indeed, the CuAAC reaction is often referred to as "the click reaction," and it has become the synthetic method of choice in many different areas of chemistry [11], including materials [12–16], polymers [17–20], interlocked architectures [21–25], bioconjugation [14, 26–32], and drug discovery [30–33]. Subsequent work by Fokin [34] has shown that ruthenium cyclopentadienyl complexes catalyze the regioselective formation of the 1,5-disubstituted-1,2,3-triazoles (**D**). While the scope and functional group tolerance of the RuAAC reaction are excellent, the reaction is more sensitive to the choice of solvent and the steric demands of the azide reactant than the CuAAC reaction, and as such, it has not yet been applied as widely [35, 36].

Despite their interesting structural and electronic features, initially, the 1,4-disubstituted-1,2,3-triazole groups that are formed during the CuAAC reaction were just viewed as innocent scaffolds or linking units. However, more and more efforts are being made to take advantage of the heterocycle's interesting properties [12]. Early workers recognized that 1,4-disubstituted-1,2,3-triazoles are excellent amide mimics [37, 38], due to the positioning of the three electronegative nitrogen atoms on one side of the heterocycle. This leads to a large 5-Debye dipole and increases the acidity of the C5 carbon's proton. Building on this, a number of groups have recently begun to use 1,4-disubstituted-1,2,3-triazoles [39, 40] and the related 1,4-disubstituted-1,2,3-triazolium salts as novel C–H hydrogen bond donor units in supramolecular systems. The presence of two sp^2-hybridized nitrogen atoms (with accessible lone pairs of electrons) within the 1,2,3-triazole unit means

[1] A referee has suggested to us that L'abbé was the first to report the CuAAC reaction in the literature, see [8].

that they are potentially able to coordinate metal ions. This property has led to an explosion of interest in the complexes of 1,4-disubstituted-1,2,3-triazole ligands.

Herein, we examine the development of 1,4-disubstituted-1,2,3-triazole containing "click" ligands and their metal complexes. This area has been recently reviewed [41]; however, even in the short time since the previous review, there has been considerable additional work. This review will focus on examples in which the coordination mode of the 1,4-disubstituted-1,2,3-triazole ligand has been structurally characterized. As such, the vast amount of work on the use of the CuAAC reaction in the development of metal ion sensors [42] will not be covered here; interested readers are referred to the accompanying chapter by Watkinson for further information. The CuAAC methodology has been exploited as an efficient way of assembling novel ligands [43–54], and it has also been widely applied as a mild method for the postsynthetic modification of metal complexes [50, 55–80]. However, in these cases, the triazole unit acts only as a linking group; it is not involved in coordination to the metal ions; and so, again, these examples will not be discussed further.

2 CuAAC Ligand Synthesis

2.1 General CuAAC Reaction Conditions

The CuAAC reaction is an extremely robust and versatile methodology, and because of this, a vast number of successful experimental variations have been developed. A wide array of different copper sources have been exploited in the CuAAC reaction, and while coligands (with nitrogen, phosphorous, and/or NHC donors) have been used to enhance the rate of reaction, they are by no means required [9, 11, 29, 37, 81]. Cu(I) salts/complexes are often employed as the catalyst; however, because Cu(I) is thermodynamically unstable and can be relatively easily oxidized to catalytically inactive Cu(II), the exclusion of oxygen is usually required. More commonly, the combination of a copper(II) salt, usually Cu $(SO_4)\cdot5H_2O$ or $Cu(OAc)_2$, and a sacrificial reducing agent, such as sodium ascorbate, is used as an alternative to oxygen-free conditions. This "aqueous Cu(II)/ ascorbate" [9] method is a practically simple, economical, and efficient way of generating 1,4-disubstituted-1,2,3-triazoles on a preparative scale, and for these reasons, it is often the method of choice for "click" ligand synthesis. Water, in combination with alcohols (MeOH, EtOH, *t*-BuOH), THF, DMF, and DMSO, can all be employed as cosolvents for the reaction, enabling a wide range of substrates to be accommodated. Additionally, while the CuAAC reaction is usually conveniently fast at room temperature, heating, either conventionally or microwave-assisted, can be used to successfully enhance the rate of the reaction when required.

While the "aqueous Cu(II)/ascorbate" method is ubiquitous, the reducing conditions are not always compatible with sensitive substrates. For this reason, alternative approaches to 1,4-disubstituted-1,2,3-triazole ligands have been developed. Zhu and coworkers [82, 83] found that Cu(II) salts can catalyze the CuAAC reaction in alcoholic solvents without the presence of additional reductants such as sodium ascorbate. Spectroscopic observations suggested that Cu(II) precatalyst undergoes reduction to a catalytic Cu(I) species via either alcohol oxidation or alkyne homocoupling, or both, during an induction period. Additionally, they showed that CuAAC reactions involving 2-picolylazide, or similar motif, are extremely facile because chelation of Cu(II) to this unit further activates the azide for cycloaddition. During the synthesis of a family of azoamide-functionalized "click" ligands, Košmrlj and coworkers have developed a number of mild CuAAC conditions that were compatible with the fragile functionally [64, 84, 85]. They showed that a mixture of $Cu(SO_4){\bullet}5H_2O$ and metallic copper in DMSO or, better, the heterogeneous catalysts copper-on-charcoal, cupric oxide, and cuprous oxide could be exploited to generate the sensitive azoamide-functionalized "click" ligands in excellent yields. The heterogeneous catalysts allowed the authors to generate the desired ligands, in almost quantitative yields, using a simple three-step "stir–filter–evaporate" protocol. The authors found that the resulting 1,4-disubstituted-1,2,3-triazole ligands show no, or negligible, copper contamination.

The synthesis of phosphorous-functionalized "click" ligands also requires special conditions. Phosphine building blocks need be protected as either the corresponding borane complex (**1–2a** and **4–5a**) or oxide (**1–2b** and **4–5b**) in order to prevent an undesirable Staudinger reaction with the azide synthons (Scheme 1). A range of bi- and tridentate P,N "click" ligands have been synthesized, and the free "clickphines" (**3** and **6**) are liberated by a final deprotection of the phosphine group with either DABCO in the case of the borane complexes [86–90] or reduction with trichlorosilane [88–92] in the case of the phosphine oxides.

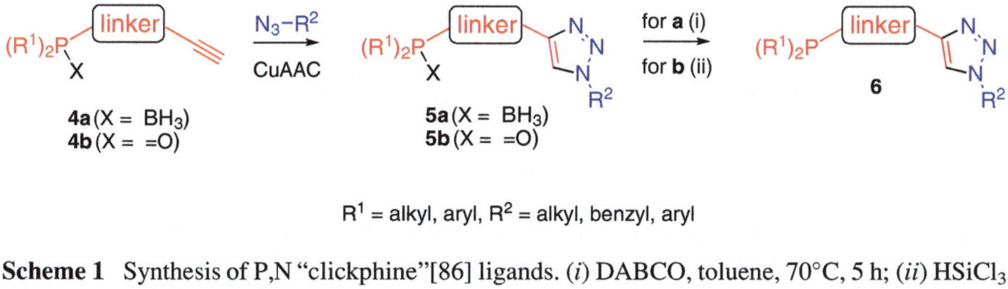

Scheme 1 Synthesis of P,N "clickphine"[86] ligands. (*i*) DABCO, toluene, 70°C, 5 h; (*ii*) $HSiCl_3$, toluene, reflux, 1 h

2.2 *One-Pot CuAAC Reactions for Ligand Synthesis*

In the previously discussed "click" methodologies, the azide and alkyne reactants were synthesized and isolated before their use in the CuAAC reaction. In an effort to further improve the synthetic efficiency and safety of the CuAAC reaction, a number of groups have developed one-pot CuAAC methodologies in which either the alkyne or azide reactant is generated in situ and reacted immediately without isolation and purification.

Fletcher and coworkers [93] have shown that TMS-protected alkynes can be deprotected in situ then "clicked" onto preformed azides to generate bi- (**7**), tri- (**8**), and quadridentate pyridyltriazole ligands in excellent yields (Scheme 2, i). Even more impressively, the commercially available 1,4-bis(trimethylsilyl)butadiyne can be deprotected in situ to furnish 4,4′-bis(1,2,3-triazole) ligands (**9**) under the same reaction conditions, enabling the use of the flammable, explosive parent butadiyne to be avoided. The Crowley [94, 95] and Schubert [96, 97] groups have developed one-pot CuAAC "click" methodologies in which the organic azide is generated in situ then "clicked" onto preformed terminal alkynes. Schubert and coworkers [96, 97] have shown that alkyl- and aryl-substituted 2-(1H-1,2,3-triazol-4-yl)pyridine (**7**) and 2,6-bis(1H-1,2,3-triazol-4-yl)pyridine (**8**) ligands can be generated from the corresponding pyridyl alkynes, sodium azide, and the appropriate alkyl halide or aryl boronic acid, using $Cu(SO_4)\cdot 5H_2O$ and sodium ascorbate in conjunction with microwave irradiation. The Crowley group has also used one-pot CuAAC "click" methodologies for the safe generation of benzyl-, alkyl-, and aryl-substituted 2-(1H-1,2,3-triazol-4-yl)pyridine (**7**) and 2,6-bis(1H-1,2,3-triazol-4-yl)pyridine (**8**) ligands [94, 95]. Exploiting conditions that were developed by Fokin [98], the benzyl- and alkyl-substituted ligands were synthesized from pyridyl alkynes, sodium azide, and the appropriate benzyl or alkyl halide using a $Cu(SO_4)\cdot 5H_2O$/

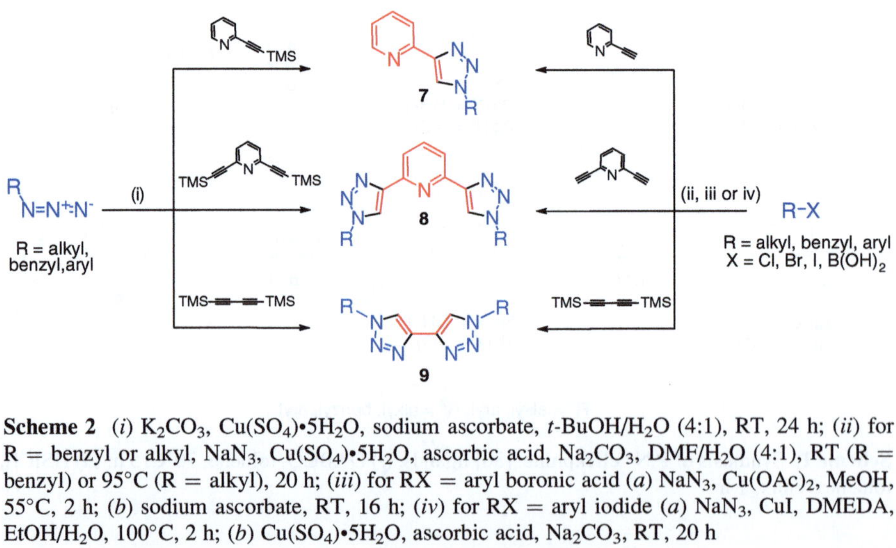

Scheme 2 (*i*) K_2CO_3, $Cu(SO_4)\cdot 5H_2O$, sodium ascorbate, *t*-BuOH/H_2O (4:1), RT, 24 h; (*ii*) for R = benzyl or alkyl, NaN_3, $Cu(SO_4)\cdot 5H_2O$, ascorbic acid, Na_2CO_3, DMF/H_2O (4:1), RT (R = benzyl) or 95°C (R = alkyl), 20 h; (*iii*) for RX = aryl boronic acid (*a*) NaN_3, $Cu(OAc)_2$, MeOH, 55°C, 2 h; (*b*) sodium ascorbate, RT, 16 h; (*iv*) for RX = aryl iodide (*a*) NaN_3, CuI, DMEDA, EtOH/H_2O, 100°C, 2 h; (*b*) $Cu(SO_4)\cdot 5H_2O$, ascorbic acid, Na_2CO_3, RT, 20 h

ascorbate catalyst system in DMF/H_2O (4:1). The benzyl-substituted ligands were obtained in excellent yield after stirring at RT for 20 h; however, the synthesis of the alkyl-substituted ligands required heating at 95°C in order to generate comparable yields within a 20 h period. The aryl-substituted ligands (**7** and **8**) could be synthesized from either aryl iodides or aryl boronic acids, via the in situ generated azide, using methods developed by Guo and Liu [99] (aryl boronic acids) or Liang [100] (aryl iodides). Subsequently, Crowley and coworkers have applied the same methodologies to safely synthesize a family of 2-pyridyl-1,2,3-triazole ligand architectures, containing electrochemically, photochemically, and biologically interesting functional groups [101].

It is clear that the CuAAC "click" reaction can be exploited to rapidly synthesize families of functionalized ligand architectures, and this synthetic versatility could lead to many potential applications. However, before the full potential of these "click" ligands can be realized, an understanding of the coordination properties of 1,4-disubstituted-1,2,3-triazoles is required. In the following sections, we will examine the coordination properties of monodentate, bidentate, tridentate, and polydentate ligands containing 1,4-disubstituted-1,2,3-triazole units.

3 Monodentate and Bridging Coordination Modes in 1,4-Disubstituted-1,2,3-Triazole Ligands

While the coordination chemistry of 1,2,4-triazoles [102–108] and the unsubstituted parent 1H-1,2,3-triazole [102, 108] is rich and extensive, metal complexes of substituted 1,2,3-triazoles remain relatively unexplored. Presumably, this was connected to the fact that before the discovery of the CuAAC reaction, synthetic approaches to substituted 1,2,3-triazoles led to difficult-to-separate isomeric mixtures of products (**C** and **D**). The discovery of the CuAAC reaction has led to an exponential increase in 1,4-disubstituted-1,2,3-triazole "click" ligands. However, despite the ease of their syntheses, there have been very few investigations into the coordination properties of "simple" 1,4-disubstituted-1,2,3-triazole units that are unsupported by other donor groups. This is somewhat surprising as these 1,4-disubstituted-1,2,3-triazoles have the potential to display an extensive array of coordination modes including monodentate binding through the N3 nitrogen (**E**), monodentate coordination through the N2 nitrogen (**F**), bridging of two metals through both the N2 and N3 nitrogens (**G**), and monodentate binding through the C5 carbon (**H**). While there are only a few structurally characterized examples, the coordination modes, **E**, **G**, and **H** have all been observed (Fig. 2).

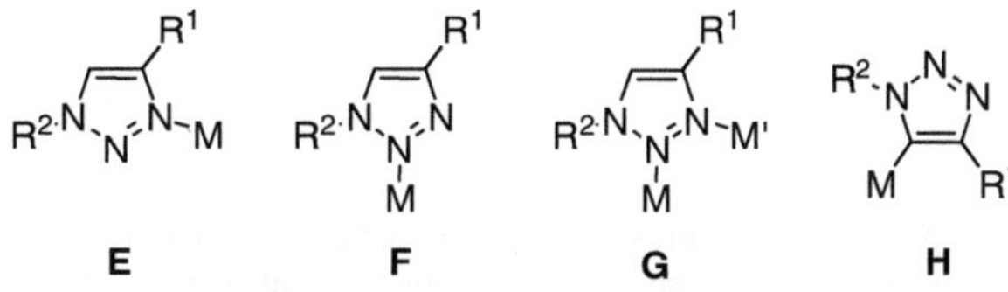

Fig. 2 Potential binding modes for 1,4-disubstituted-1,2,3-triazole ligands

3.1 1,4-Disubstituted-1,2,3-Triazoles as Monodentate N-Donor Ligands

Van Koten and coworkers reported the first structurally characterized metal complexes containing a "simple" 1,4-disubstituted-1,2,3-triazole ligand [109]. The 1,4-disubstituted-1,2,3-triazole complexes **10a–d** were synthesized by the addition of the appropriate triazole **C** to a mixture of the either the Pd or Pt NCN-pincer complex (**11**) and $AgBF_4$ in CH_2Cl_2 (Scheme 3). X-ray crystallography confirmed that the 1,4-disubstituted-1,2,3-triazole unit is coordinated to the Pd or Pt ions through the more electron-rich N3 nitrogen (Fig. 3a). The authors also carried out ^{1}H NMR competition experiments which showed that **C** (R_1 = Ph, R_2 = Bn) is a better ligand than H_2O, DMSO, MeCN, and Et_2S, comparable in strength to aniline, and a weaker donor than pyridine, N-methyl imidazole, and triphenylphosphine. Additionally, they found that the binding strength of the 1,4-disubstituted-1,2,3-triazole ligands could be fine-tuned. Competition experiments suggested that **C** (R_1 = R_2 = *n*-hexyl) is a stronger donor than both **C** (R_1 = Bn, R_2 = Bn) and **C** (R_1 = Ph R_2 = Bn) and has a binding strength comparable to pyridine.

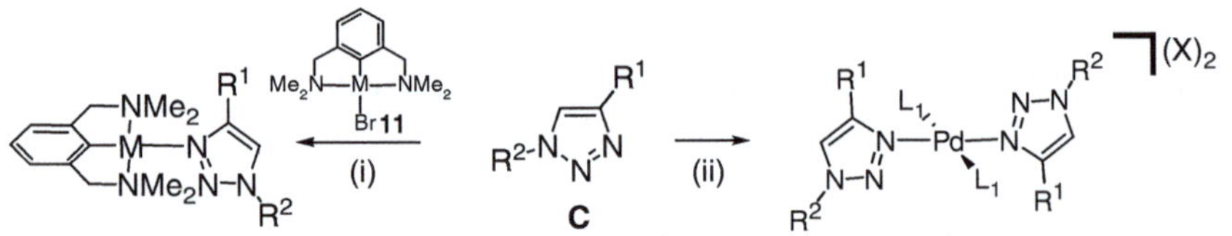

10a (M = Pd, R¹ = Ph, R² = Bn)
10b (M = Pt, R¹ = Ph, R² = Bn)
10c (M = Pt, R¹ = Bn, R² = Bn)
10d (M = Pt, R¹ = C_6H_{13}, R² = C_6H_{13})

12a (R¹ = Fc, R² = 4-MeOBn, L_1 = X = Cl)
12b (R¹ = Ph, R² = Bn, L_1 = X = Cl)
12c $(BF_4)_2$ (R¹ = Ph, R² = Bn, L_1 = 1-benzyl-4-phenyl-1H-1,2,3-triazole, X = BF_4)

Scheme 3 Pd(II) complexes containing monodentate 1,4-disubstituted-1,2,3-triazole ligands; (*i*) $AgBF_4$, CH_2Cl_2, RT, 1 h; (*ii*) for **12a** $Pd(MeCN)_2Cl_2$, toluene, RT, 2.5 h; for **12b** $Pd(MeCN)_2Cl_2$, acetone, RT, 1 h; for **12c** $[Pd(MeCN)_4](BF_4)_2$, acetone, RT, 1 h

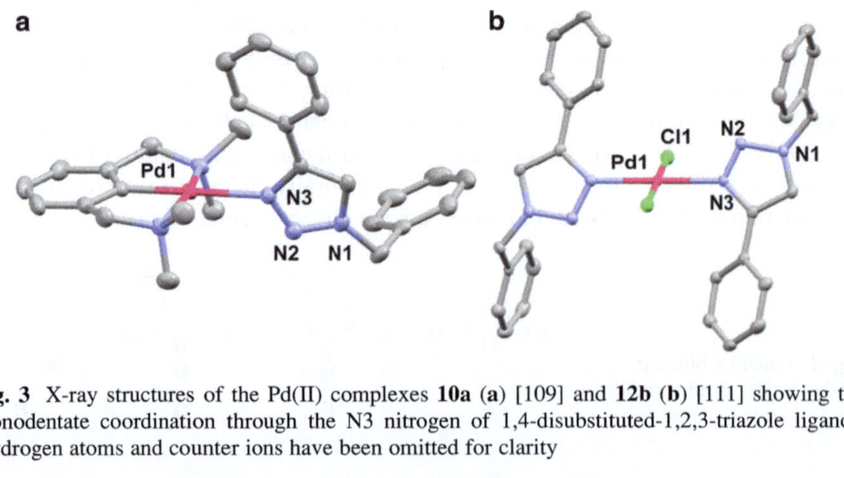

Fig. 3 X-ray structures of the Pd(II) complexes **10a** (**a**) [109] and **12b** (**b**) [111] showing the monodentate coordination through the N3 nitrogen of 1,4-disubstituted-1,2,3-triazole ligands. Hydrogen atoms and counter ions have been omitted for clarity

Following on from van Koten's work, both Astruc [110] and Crowley [111] have found that 1,4-disubstituted-1,2,3-triazole ligands bind to Pd(II) ions in a monodentate fashion through the N3 nitrogen of the triazole. Mixing the 1,4-disubstituted-1,2,3-triazole ligand, **C** and either $Pd(MeCN)_2Cl_2$ or $[Pd(MeCN)_4](BF_4)_2$ leads to the formation of the Pd(II)-1,4-disubstituted-1,2,3-triazole complexes, **12a–c** in excellent yields (Scheme 3). Each of the Pd(II) complexes have been structurally characterized using X-ray crystallography. **12a** and **12b** form the expected *trans*-$[Pd(\mathbf{C})_2Cl_2]$ complexes (Fig. 3), whereas **12c** is the salt $[Pd(\mathbf{C})_4](BF_4)_2$.

Elliott and coworkers [112] have synthesized a series of cationic Re(I) complexes $[Re(bpy)(CO)_3(\mathbf{C})](PF_6)$ (**14a–d**) that contain a 1,4-disubstituted-1,2,3-triazole unit coordinated in a monodentate fashion (Fig. 4). X-ray crystal structures were obtained for **14a** and **14b**, which confirm that the coordination of the Re(I) is through the N3 nitrogen of the 1,2,3-triazole ligand (Fig. 4b). The positioning of the CO bands in the IR spectra suggests that the 1,4-disubstituted-1,2,3-triazole ligands studied are marginally stronger donors than pyridine. The complexes are luminescent in aerated dichloromethane at room temperature, but the emission maxima are blue-shifted relative to the parent chloride complex. Additionally, relatively long luminescent lifetimes are observed for these 1,4-disubstituted-1,2,3-triazole complexes (475 to 513 ns).

In a final example, Osuka, Shinokubo, and coworkers have synthesized the 1,4-disubstituted-1,2,3-triazole appended Zn(II)-porphyrin, **15,** and showed using ^1H NMR studies that it dimerizes in solution. X-ray crystallography confirmed that the dimerization is driven by coordination of the 1,4-disubstituted-1,2,3-triazole donor on one porphyrin to the Zn(II) ion of the adjacent porphyrin (Fig. 5) [113]. As with the previously discussed Pd(II) and Re(I) complexes, the triazole unit interacts with the Zn(II) ions through the N3 nitrogen.

There are a number of other reported 1,4-disubstituted-1,2,3-triazole metal complexes in which monodentate binding is presumed [114–119]. None of these compounds have been structurally characterized, and as such, it is unknown whether the coordination is through the N2 or N3 nitrogen of the 1,2,3-triazole unit. However, on the basis of the previously discussed structurally characterized

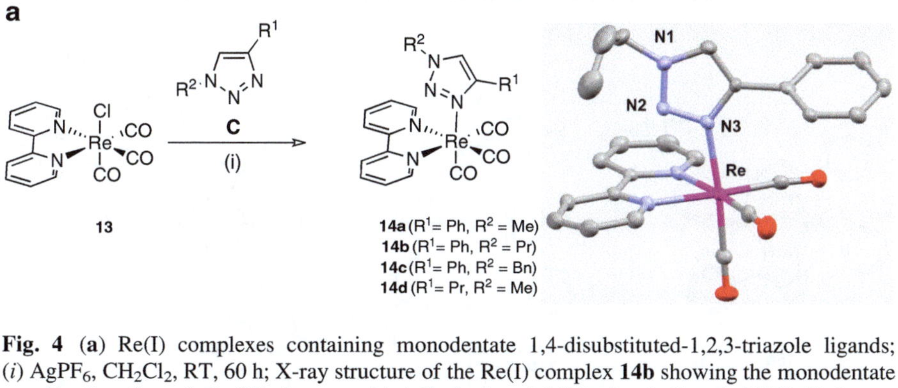

Fig. 4 (**a**) Re(I) complexes containing monodentate 1,4-disubstituted-1,2,3-triazole ligands; (*i*) $AgPF_6$, CH_2Cl_2, RT, 60 h; X-ray structure of the Re(I) complex **14b** showing the monodentate coordination, through the N3 nitrogen of 1,4-disubstituted-1,2,3-triazole ligands [112]. Hydrogen atoms and counter ions have been omitted for clarity

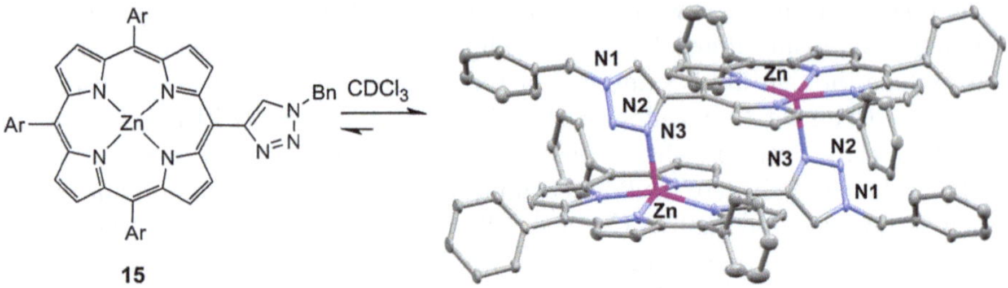

Fig. 5 Osuka's 1,4-disubstituted-1,2,3-triazole-appended Zn(II)-porphyrin, **15,** and the X-ray structure of the dimer $\mathbf{15}_2$ [113]. Hydrogen atoms and the aryl substituents' *t*-Bu groups have been omitted for clarity

examples and a number of theoretical studies (DFT) [120, 121], which have shown the N3 nitrogen of the triazole is the most electron rich, it is expected when a 1,4-disubstituted-1,2,3-triazole unit is coordinated in a monodentate fashion it will be through the N3 not the N2 nitrogen.

3.2 1,4-Disubstituted-1,2,3-Triazoles as Bridging N-Donor Ligands

Somewhat surprisingly, given that related diazole or triazole ligands (e.g., pyrazolates, 1,2,4-triazoles) with side-by-side nitrogen donor atoms are extensively used as bridging ligands, there is only one crystallographically characterized report of 1,4-disubstituted-1,2,3-triazole units acting as bridging ligands. Crowley et al. synthesized the dimeric silver(I) complexes, **16a–b**, by mixing an acetone solution of $AgSbF_6$ with a dichloromethane solution of one of the ligands (**17a** or **17b**) and stirring the resulting solutions at room temperature for 1 h in the absence of light (Scheme 4) [122]. Both the complexes, **16a–b**, were structurally characterized and display the same dinuclear core in which the two silver(I) ions are bridged by

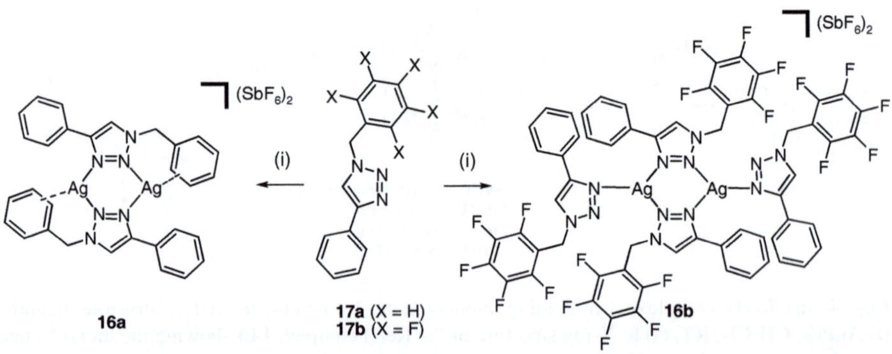

Scheme 4 Ag(I) complexes containing bridging 1,4-disubstituted-1,2,3-triazole ligands; (*i*) $AgSbF_6$, acetone/CH_2Cl_2, RT, 1 h [122]

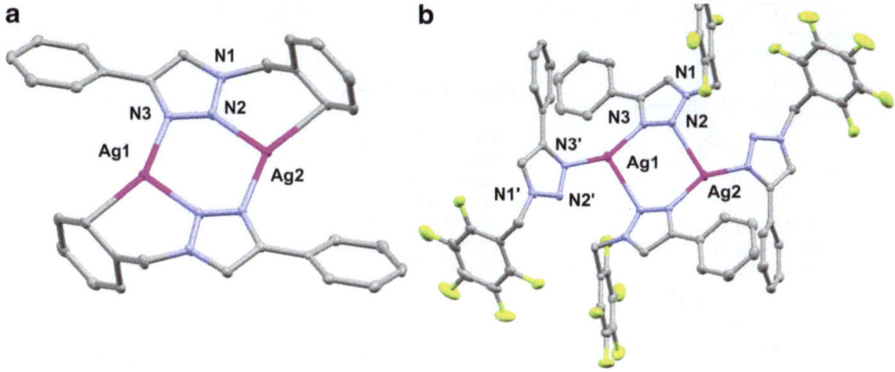

Fig. 6 X-ray structures of the silver(I) dimers **16a–b** containing bridging 1,4-disubstituted-1,2,3-triazole ligands [122]. Hydrogen atoms and counter ions have been omitted for clarity

two 1,4-disubstituted-1,2,3-triazole units. In the case of **16a**, the silver ions are also coordinated to a benzyl substituent through an unusual η^1 benzyl-Ag(I) interaction, giving the Ag(I) ions a trigonal planar coordination sphere (Fig. 6). The molecular structure of **16b** was found to be similar to that of **16a**, the core $[(\mathbf{17b})_2Ag_2]^{2+}$ dication supported by the bridging 1,2,3-triazole ligands, but an additional molecule of **17b** coordinates to the silver ions in a monodentate fashion generating a 2:1 ligand to metal ratio. The Ag(I) ions of **16b** are coordinated to three N donor atoms in a distorted trigonal planar coordination environment. The unusual η^1 benzyl–Ag(I) interaction that was present in **16a** is replaced by the monodentate Ag–N(1,2,3-triazole) interaction presumably because the donor ability of the benzyl ring has been weakened by the presence of the electron-withdrawing fluorine substituents. While these appear to be the only reported examples of a "simple" unsupported 1,4-disubstituted-1,2,3-triazole unit acting as a bridging ligand, this coordination mode has been observed in more complicated "click" ligands (*vide infra*).

3.3 1,4-Disubstituted-1,2,3-Triazoles as Monodentate C-Donor Ligands

While the most obvious coordination sites on the 1,4-disubstituted-1,2,3-triazole unit are the N2 and N3 nitrogen atoms, the acidic nature of the hydrogen on the C5 carbon means that it can be readily deprotonated to provide anionic 1,4-disubstituted-1,2,3-triazolide ligands. Straub and coworkers [123] were able to isolate a stable copper(I) 1,4-disubstituted-1,2,3-triazolide complex (**19a**) by reacting the bulky NHC-stabilized copper(I) acetylide **18a** with azidodi-4-tolylmethane in toluene at RT for 24 h (Fig. 7). **19a** was stable in air and water, and its molecular structure was confirmed by X-ray crystallography (Fig. 7). The copper(I) 1,2,3-triazolide

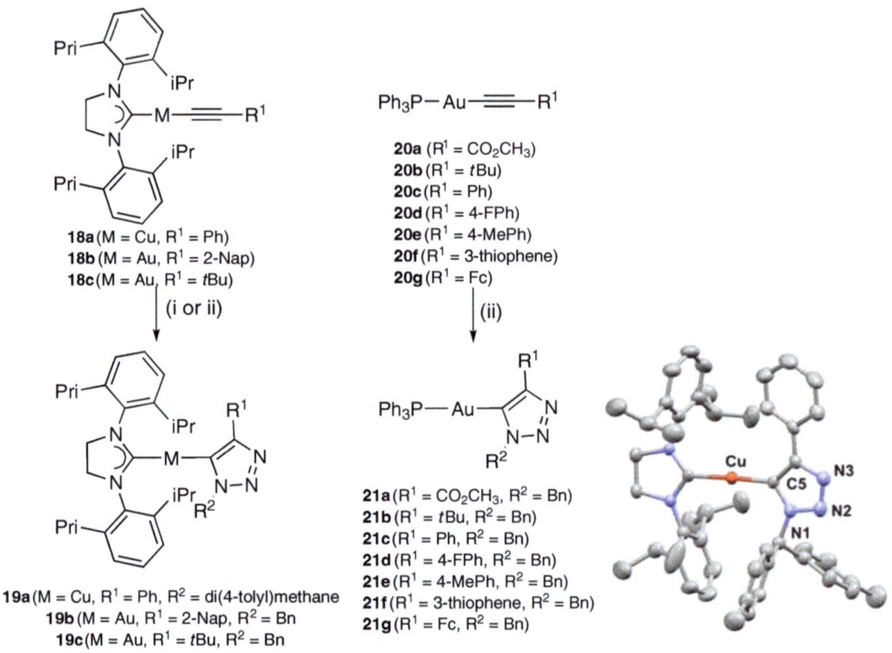

Fig. 7 Synthesis of 1,4-disubstituted-1,2,3-triazolide complexes; (*i*) azidodi-4-tolylmethane, toluene, RT, 24 h; (*ii*) RN_3, $[Cu(MeCN)_4](PF_6)$, MeCN, RT, 12 h; X-ray structure of the copper(I) 1,4-disubstituted-1,2,3-triazolide complex **19a** showing the coordination through the C5 carbon atom of the 1,2,3-triazole unit [123]. Hydrogen atoms and counter ions have been omitted for clarity

species is presumed to be an intermediate in the catalytic cycle of the CuAAC reaction, and Straub showed that treating **19a** with acetic acid in CH_2Cl_2 at room temperature leads to the quantitative formation of the 1,4-disubstituted-1,2,3-triazole and the NHC–Cu–OAc complex. Furthermore, they demonstrated that both **19a** and the NHC–Cu–OAc complex are active CuAAC catalysts, providing strong evidence for the intermediacy of the 1,2,3-triazolide in the reaction's catalytic cycle.

Building on this work, Gray and coworkers [124] have synthesized a family of stable gold(I) 1,4-disubstituted-1,2,3-triazolide complexes (**19b–c** and **21a–f**). They showed that tetrakis(acetonitrile)copper(I) hexafluorophosphate catalyzes the [3+2] cycloaddition of phosphine (**20a–f**) and (NHC)gold(I) alkynyls (**18b–c**) with benzyl azide in yields of up to 96% (Fig. 7). The reaction protocol tolerates a broad range of functionalities on the alkynyl reagent, and the resulting complexes are air and water stable. The gold(I)-1,2,3-triazolide complexes are luminescent, and nine of the family has been characterized crystallographically; however, none of the structures display any gold(I)–gold(I) (aurophilic) interactions presumably because of the steric bulk around the gold center.

The anionic 1,4-disubstituted-1,2,3-triazolides are not the only "click" triazole ligands which display the C5 coordination mode. 1,4-Disubstituted-1,2,3-triazoles (**C**) are readily and regioselectively alkyl/benzylated at the N3 position providing easy

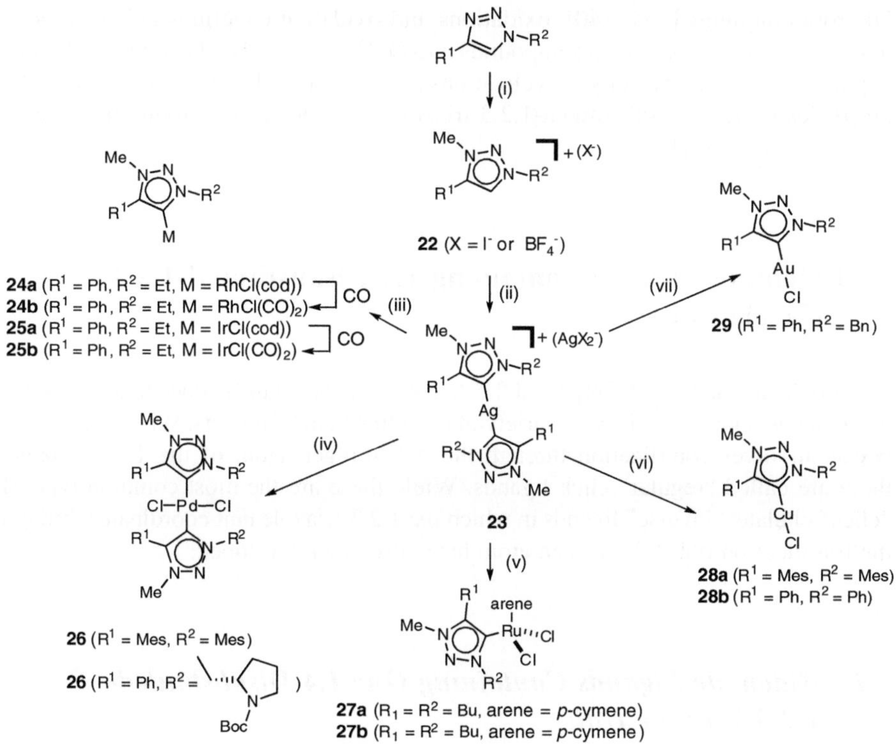

Scheme 5 Synthesis of 1,3,4-trisubstituted-1,2,3-triazolium salts and 1,3,4-trisubstituted-1,2,3-triazol-5-ylidenes complexes; (*i*) either MeI, CH_3CN, 70°C, 12 h or Me_3OBF_4, CH_2Cl_2, RT, 3 d; (*ii*) for X = I, Ag_2O, CH_2Cl_2, RT, 15 h; for X = BF_4, Ag_2O, NMe_4Cl, CH_2Cl_2/MeCN (1:1), RT, 18 h; (*iii*) $[M(cod)Cl]_2$ (M = Rh, Ir), CH_2Cl_2, RT, 24 h; (*iv*) $[Ru(arene)Cl_2]_2$, CH_2Cl_2, RT, 24 h; (*vi*) CuCl, CH_2Cl_2/MeCN (1:1), RT, 6 h; (*vii*) $Au(SMe_2)Cl$, CH_2Cl_2/MeCN (1:1), RT, 2 h

access to the related 1,3,4-trisubstituted-1,2,3-triazolium salts (**22**) (Scheme 5) [125–131]. Albrecht has shown that these triazolium salts can be converted into abnormal or mesoionic N-heterocyclic "click" carbene complexes [132]. Deprotonation of **22** with Ag_2O generates the modestly stable silver(I) 1,2,3-triazolylidenes (**23**) which can be used to synthesize a wide variety (Pd(II) [132–134], Ru(II) [135], Rh(I) [132, 133], Ir(I) [132], Cu(I) [136, 137], and Au(I)[138]) of stable metal 1,2,3-triazolylidene complexes (**24**–**29**) via transmetallation (Scheme 5) [137, 139].[2] The metal complexes of these 1,3,4-trisubstituted-1,2,3-triazol-5-ylidenes have been exploited as catalysts for a variety of reactions including CuAAC cycloadditions [136, 137],

[2] While the mild Ag_2O method is most commonly used for the generation of these 1,3,4-trisubstituted-1,2,3-triazol-5-ylidene complexes, an alternative approach in which the free carbene is generated by deprotonation of the precusor triazolium salt with KO*t*Bu then complexed to a metal salt also leads to the formation of the desired carbene complexes, see [137, 139].

Pd cross couplings [134, 140], oxidations and oxidative couplings [135], carbene transfer reactions from diazo compounds into O–H, N–H, and C–H bonds, hydroalkoxylation of allenes; and enyne cyclizations [138]. A more detailed discussion of the properties of 1,3,4-trisubstituted-1,2,3-triazol-5-ylidenes can be found in a recent review of the field [141].

4 Bidentate Ligands Containing 1,4-Disubstituted-1,2,3-Triazole Units

In 2006, Mindt and Schibli reported the "click-to-chelate"[142] concept, and this has led to the development of a vast variety of chelating "click" ligands. When the chelate pocket involves coordination through the N3 nitrogen atom of the 1,2,3-triazole, these are called "regular" click ligands. While these are the most common type of "click" chelate, "inverse" ligands in which the 1,2,3-triazole unit coordinates through the less electron rich N2 nitrogen atom have also been developed.

4.1 Bidentate Ligands Containing One 1,4-Disubstituted-1,2,3-Triazole Unit

Bidentate ligands containing 1,4-disubstituted-1,2,3-triazole units are by far the most studied class of "click" chelates. Within this subclass of "click" ligands, the pyridyl-1,2,3-triazole ligands 2-(1-R-1H-1,2,3-triazol-4-yl)pyridine **30** and [2-(4-R-1H-1,2,3-triazol-1-yl)methyl]pyridine **31** have received the most attention.

The 2-(1-R-1H-1,2,3-triazol-4-yl)pyridine series of ligands have been extensively examined as readily tuneable bpy analogues. Indeed, for the most part, despite the presence of the additional nitrogen donor atom (N2) within the ligand scaffold, these compounds act as bpy-like N–N five-membered chelates. The ligands coordinate through the pyridyl and N3 nitrogen of the 1,2,3-triazole unit with a wide range of metals ions (**32–40**, Scheme 6) including those with octahedral (Re(I) [143–145], Tc(I) [144], Ru(II) [93, 97, 146–149], Ir(III) [147, 150–152], tetrahedral Ag(I), Cu(I) [95, 153]), and square planar (Pd(II) and Pt(II) [101, 154–157]) coordination geometries (Fig. 8). While **30** forms a range of stable complexes, a binding constant study by Petitjean [153] has shown, unsurprisingly, that they are less stable than the corresponding bpy complexes presumably because the more electron deficient 1,2,3-triazole unit is a poorer σ-donor than py.

The rich photophysical properties displayed by bpy complexes such as $[Ru(bpy)_3]^{2+}$ have encouraged a number of groups to examine the optoelectronic properties of metal complexes of **30**. Obata and Yano [145] and Benoist [143, 144] have synthesized several rhenium(I) complexes of the formulation [Re(**30**)

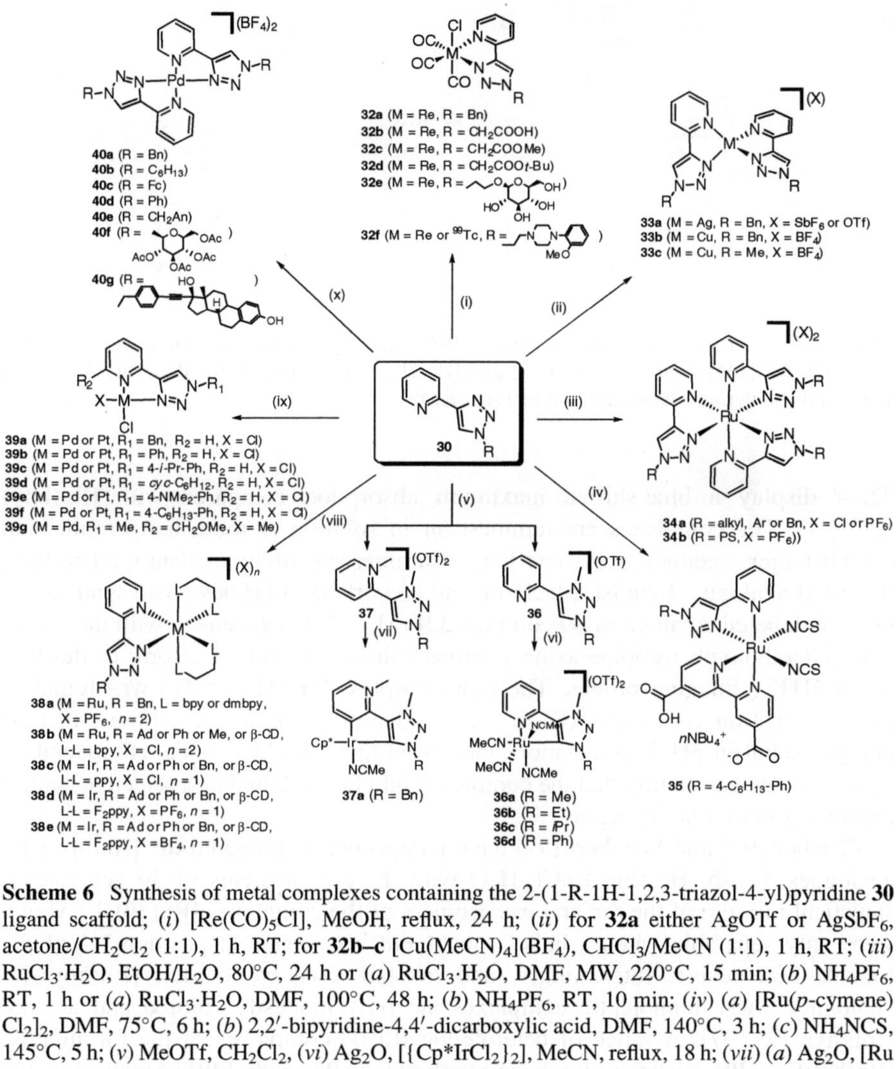

Scheme 6 Synthesis of metal complexes containing the 2-(1-R-1H-1,2,3-triazol-4-yl)pyridine **30** ligand scaffold; (*i*) [$Re(CO)_5Cl$], MeOH, reflux, 24 h; (*ii*) for **32a** either AgOTf or $AgSbF_6$, acetone/CH_2Cl_2 (1:1), 1 h, RT; for **32b–c** [$Cu(MeCN)_4$](BF_4), $CHCl_3$/MeCN (1:1), 1 h, RT; (*iii*) $RuCl_3{\cdot}H_2O$, EtOH/H_2O, 80°C, 24 h or (*a*) $RuCl_3{\cdot}H_2O$, DMF, MW, 220°C, 15 min; (*b*) NH_4PF_6, RT, 1 h or (*a*) $RuCl_3{\cdot}H_2O$, DMF, 100°C, 48 h; (*b*) NH_4PF_6, RT, 10 min; (*iv*) (*a*) [Ru(*p*-cymene) $Cl_2]_2$, DMF, 75°C, 6 h; (*b*) 2,2′-bipyridine-4,4′-dicarboxylic acid, DMF, 140°C, 3 h; (*c*) NH_4NCS, 145°C, 5 h; (*v*) MeOTf, CH_2Cl_2, (*vi*) Ag_2O, [{Cp^*IrCl_2}$_2$], MeCN, reflux, 18 h; (*vii*) (*a*) Ag_2O, [Ru (*p*-cymene)$Cl_2]_2$, CH_2Cl_2, 35 °C, 2 d; (*b*) AgOTf, MeCN, reflux, 18 h; (*viii*) for **38a** (*a*) Ru(bpy)Cl_2 or Ru(dmbpy)Cl_2, EtOH, MW, 125°C, 1 h; (*b*) NH_4PF_6, RT, 2 h; for **38b** (*a*) Ru(bpy)Cl_2 or Ru (dmbpy)Cl_2, 2-methoxyethanol/H_2O, 90°C, 24 h; for **38c–e** (*a*) either $(ppy)_2Ir(\mu\text{-}Cl)_2Ir(ppy)_2$ or $(F_2ppy)_2Ir(\mu\text{-}Cl)_2Ir(F_2ppy)_2$, $CHCl_3$/MeOH (3:1), 40°C, 3 h; (*b*) either NH_4PF_6 or NH_4BF_4, H_2O, 1 h; (*ix*) for **39a–f** (M = Pd) Pd(cod)Cl_2 or $Pd(CH_3CN)_2Cl_2$, CH_2Cl_2 or CH_3CN, RT, 1 h; for **39a–f** (M = Pt) $K_2[PtCl_4]$ H_2O/EtOH (1:1), 60°C, 16 h, or $Pt(DMSO)_2Cl_2$, $MeNO_2$, 5–15 h, reflux; for **39g** [Pd(cod)MeCl], CH_2Cl_2, RT, 1 h; (*x*) [$Pd(CH_3CN)_4$]$(BF_4)_2$, CH_3CN, RT, 1 h

$(CO)_3Cl$] (**32a–f**) by refluxing a stoichiometric mixture of the corresponding pyridyltriazole ligand and [$Re(CO)_5Cl$] in an alcoholic solvent. Spectroscopic methods indicate that the expected *fac* isomers are generated exclusively, and this has been confirmed by X-ray crystallography (Fig. 8c). The electronic spectra of

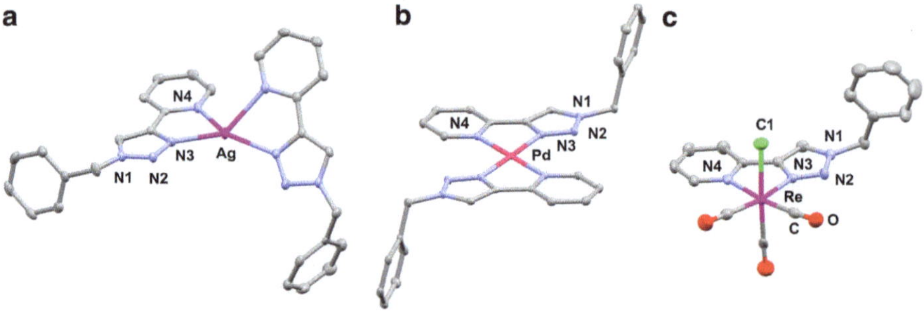

Fig. 8 X-ray structures of tetrahedral **33a** (**a**) [95], square planar **40a** (**b**) [101], and octahedral metal **32a** (**c**) [145] complexes of the 2-pyridyl-1,2,3-triazole ligand **30** (R = Bn). Hydrogen atoms and counter ions have been omitted for clarity

32a–f display a blue-shifted maximum absorption compared with [Re(bpy)$(CO)_3Cl$]. The complexes are luminescent in solution at room temperature and exhibit larger quantum yields and longer luminescent lifetimes than the [Re(bpy)$(CO)_3Cl$] analogue. Benoist, Machura, and coworkers [144] have also synthesized the radiolabeled technetium-99 complex **32f** (M = ^{99}Tc) appended with the bioactive (2-methoxyphenyl)piperazine pharmacophore, as part of efforts to develop novel $5HT_{1A}$ imaging probes. The radiocomplex **32f** (M = ^{99}Tc) was found to have a partition coefficient (P) of 2.34 under physiological conditions (0.1 M phosphate buffer pH 7.4/*n*-octanol), indicating that **32f** (M = ^{99}Tc) is moderately lipophilic and suggesting that the complex might be exploited as a selective central nervous system imaging agent.

Fletcher [93] and Schubert [97] have independently prepared the $[Ru(bpy)_3]^{2+}$ analogues **34a–b**. Heating $RuCl_3{\cdot}H_2O$ with three equivalents of **30** generates a statistical mixture of the *fac* and *mer* isomers of the complexes $[Ru(\mathbf{30})_3]^{2+}$ which were only separable when a bulky substituent was bound at position N1 of the triazole heterocycle. Interestingly, and in stark contrast to their $[Ru(bpy)_3]^{2+}$ analogues, these homoleptic complexes are only modestly colored and are not emissive [93, 97]. A range of heteroleptic Ru(II) complexes of the formulation $[Ru(bpy)_{3-n}(\mathbf{30})_n]^{2+}$ have also been synthesized [97, 148, 149]. Analysis of the photophysical and electrochemical properties of the complexes $[Ru(bpy)_{3\text{-}n}(\mathbf{30})_n]^{2+}$ revealed that as more pyridyl-1,2,3-triazole ligands are incorporated into the complexes, both the UV/vis spectra and the luminescence spectra are blue-shifted, and cyclic voltammograms indicate an increase in the energy band gap. Bauerle, Gratzel, Zakeeruddin, and coworkers have synthesized the heteroleptic Ru(II) complex **35** (Scheme 6) and examined its use as a sensitizer in DSSC [158]. A DSSC consisting of **35** with an acetonitrile-based electrolyte showed a respectable overall power conversion efficiency of 7.8% under full sunlight intensity. Additionally, the authors prepared solid-state devices with 2-mm-thick TiO_2 films which demonstrated power conversion efficiencies of 1.92%.

De Cola et al. [147, 150–152] have synthesized a family of heteroleptic cationic Ir(III) complexes containing ppy or F_2ppy ligands (**38c–e**) and examined their photophysical and electrochemical properties. These complexes displayed excellent emission quantum yields, long excited state lifetimes, and reversible electrochemical behavior. Additionally, the compounds have shown ECL which can be modulated from the green to blue by ligand modification, and the authors suggest that this could be exploited in the development of ECL labels for bioanalytical applications [152]. The same series of compounds have been used to develop LEECs, which are possible alternatives to the more-investigated OLEDs. One of the major drawbacks of LEECs is their long turn-on time (i.e., the time needed to see the emission after the application of the voltage); it often takes hours for emission to occur due to the slow diffusion of charges within the device. The LEECs generated using these Ir(III) "click" complexes display very blue emission along with the fastest response time ever reported for iridium complexes bringing practical LEECs a step closer [151].

A variety of square planar complexes of **30** with Pd and Pt have been synthesized (**39a–g** and **40a–g**) [101, 154–156] and crystallographically characterized. Milani and coworkers [155] have shown that one of these palladium "click" complexes, **39g**, is an active catalyst for styrene carbonylation producing low-molecular-weight oligoketones. Albrecht et al. have also developed catalytic systems based on the **30** [159, 160]. Methylation of **30** with MeOTf produces the mono- and dimethylated pyridyl-1,2,3-triazolium salts, **36** and **37**. Reaction of **36** with Ag_2O and [Ru(*p*-cymene)Cl_2]$_2$ in CH_2Cl_2 at 35°C for 2 days generates the complex [Ru(*p*-cymene)(**36**-H)Cl](OTf) which is readily coverted into the ruthenium(II) complexes, **36a–d**, by halide abstraction with AgOTf and thermal cymene dissociation in refluxing MeCN (Scheme 6). The related iridium(III) complex, **37a**, can be generated by the double C–H bond activation of **37** with Ag_2O followed by transmetallation with [{Cp*IrCl_2}$_2$] (Scheme 6). The unusual C,C-bidentate complex **37a** comprises two different abnormally bound NHC ligands, a 1,2,3-triazoly-5-lidene and a 3-pyridylidene. Albrecht et al. have demonstrated that the ruthenium(II) **36a–d** and iridium(III) **37a** complexes are extremely active water oxidation catalysts in the presence of $(NH_4)_2[Ce(NO_3)_6]$ (CAN) as a sacrificial oxidant. **36a–d** are the most active ruthenium water oxidation catalysts known (TONs > 400, TOFs close to 7,000 h^{-1}), and the iridium(III) complex, **37a**, is even more robust reaching TONs of nearly 10,000 after 5 days. The authors suggest that the high catalytic activity of the complexes may be connected to the mesoionic character of the 1,2,3-triazoly-5-lidene unit which allows the ligand and metal center to act cooperatively, enabling the bond cleavage and oxidation processes to occur simultaneously. This postulate is further supported by the fact that related Ru(II) "normal" NHC complexes, which have less pronounced mesoionic character, are far less active catalysts.

The related, but more flexible, [2-(4-R-1H-1,2,3-triazol-1-yl)methyl]pyridine series of "inverse" ligands, **31**, have also been well studied. Like **30,** this ligand scaffold generally behaves as an N–N bidentate chelate, coordinating through the pyridyl and N2 nitrogen of the 1,2,3-triazole unit forming a six-membered chelate ring. Complexes of octahedral (Cu(II) [82, 83, 95, 121, 161], Co(II) [161, 162],

Ni(II) [162], Ru(II) [121, 163]), tetrahedral (Ag(I)[95, 121] and Zn(II)[162]), and square planar (Ag(I), Pd(II), and Pt(II)[101, 121, 154, 164–166]) coordination geometries have all been characterized crystallographically (Scheme 7 and Fig. 9). The greater flexibility of **31** enables access to a larger number of observed coordination modes for this scaffold. Košmrlj et al. have shown that **31** will bind as a monodentate donor under certain conditions [164]. Reaction of **31** (R = Ph) and the rhodium(II) acetate dimer [$Rh_2(O_2CCH_3)_4$] in hot MeOH/water solution resulted in the formation of the rhodium complex **43** (R = Ph) in which the ligand of **31** (R = Ph) is bound to the rhodium ions in a monodentate fashion through the most electron-rich pyridyl nitrogen atom. The same authors have also shown that the 1,2,3-triazole unit of **31** (R = Ph) can act as a monodentate donor. Košmrlj [121, 165, 166] and Crowley [154] have independently shown that **31** (R = Ph) forms the expected square planar N–N bidentate chelate complexes when reacted with MCl_2

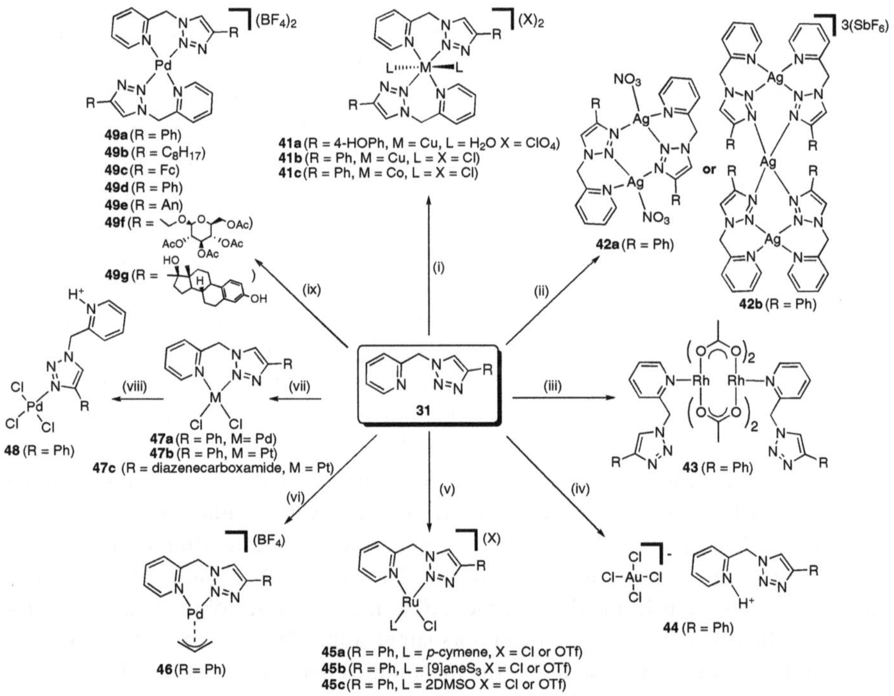

Scheme 7 Synthesis of metal complexes containing the [4-R-1H-1,2,3-triazol-1-yl)methyl]pyridine (**31**) ligand scaffold; (*i*) for **41a–b** either $CuCl_2$, MeOH, RT, 1 h or $CuCl_2$, EtOH/$CHCl_3$, 60°C, 1 h; for **41c** $CoCl_2$, acetone, RT, 24 h; (*ii*) for **42a** $AgNO_3$, MeOH, RT, 24 h; for **42b** $AgSbF_6$, acetone/CH_2Cl_2 (1:1), 1 h, RT; (*iii*) [$Rh_2(O_2CCH_3)_4$], H_2O/MeOH (1:1), 1 h, reflux, 5 min; (*iv*) $AuCl_3$, MeOH, RT, 2 h or $KAuCl_4$, H_2O/MeOH, RT, 2 h; (*v*) for **45a** [Ru(*p*-cymene)$Cl_2]_2$, EtOH, RT, 2 d; for **45b** [Ru([9]aneS3)(DMSO-S)$_2$Cl](OTf), MeOH, reflux, 2 h; for **45c** [*trans*-Ru(DMSO-S)$_4Cl_2$], H_2O, RT, 24 h (*vi*) $AgBF_4$, [Pd(η^3-C_3H_5)Cl]$_2$, MeOH/CH_2Cl_2, RT, 2 h; (*vii*) for **47a** (M = Pd) Pd(cod)Cl_2 or Pd(CH_3CN)$_2Cl_2$, CH_2Cl_2 or CH_3CN, RT, 1 h; for **47b–c** (M = Pt) $K_2[PtCl_4]$ H_2O/EtOH (1:1), 60°C, 16 h, or Pt(DMSO)$_2Cl_2$, CH_2Cl_2 or CH_3CN, RT, 2–13 d; (*viii*) $HCl_{(aq)}$, H_2O, hot, 5 min, (*ix*) [Pd(CH_3CN)$_4$](BF_4)$_2$, CH_3CN, RT, 1 h

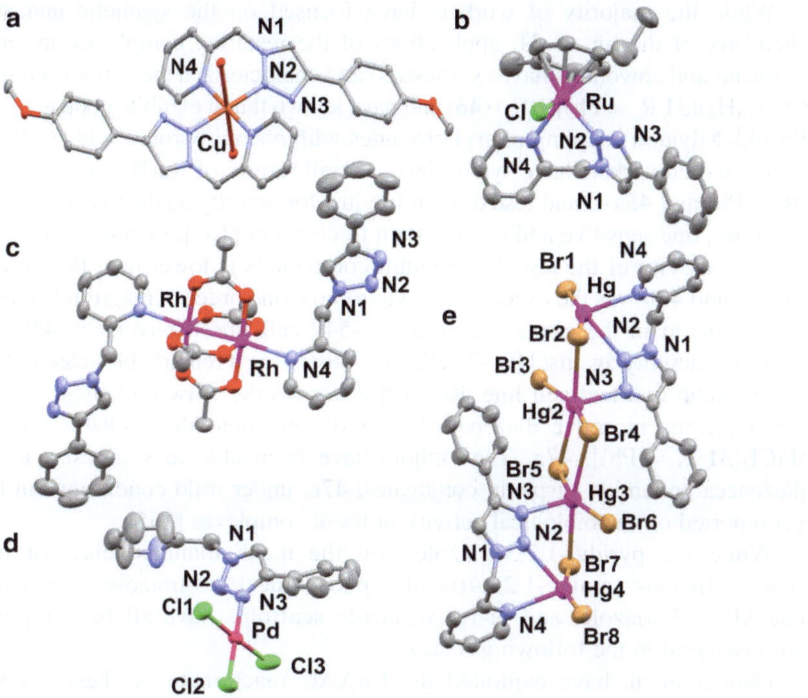

Fig. 9 X-ray structures of **41a** (**a**) [82], **45a** (**b**) [121], **43** (**c**) [164], **48** (**d**) [164], and the $HgBr_2$ adduct of **31** (**e**) [164] showing the range of different coordination modes that the 2-pyridyl-1,2,3-triazole ligand **31** can adopt. Hydrogen atoms and counter ions have been omitted for clarity

(M = Pd or Pt).However, during an attempted recrystallization of **48**, **47a**, from aqueous HCl, Košmrlj and coworkers isolated the complex **48** in which the ligand **31** (R = Ph) is coordinated to the Pd(II) ion through the N3 nitrogen of the 1,2,3-triazole unit. The most basic nitrogen atom (the pyridyl N) within the ligand has been protonated, and the second most electron-rich nitrogen atom (i.e., the triazole N3 nitrogen) coordinates to the palladium [164]. **31** (R = Ph) has also been shown to coordinate through all three of its nitrogen donor atoms simultaneously. Reaction of **31** (R = Ph) with Ag(I) ions leads to the formation of dinuclear ($AgNO_3$), **42a**, or trinuclear ($AgSbF_6$), **42b**, silver(I) complexes, in which **31** displays both chelating and bridging coordination modes. Similar behavior is also observed when **31** (R = Ph) is reacted with $HgBr_2$ or a mixture of $CuCl_2/NaN_3$. These reactions generate oligomeric (in the case of $HgBr_2$) or polymeric ($CuCl_2/NaN_3$) chain complexes in which adjacent metal ions are bridged by the 1,2,3-triazole units of **31** (Fig. 9e). Interestingly, two different research groups have found that the reaction of **31** and Au(III) ions leads to the formation of the salt $[\mathbf{31}H]^+[AuCl_4]^-$ rather than the desired chelated gold(III) complex [154, 164]. These results indicate that the coordination chemistry of **31** with the square planar Au(III) is not as straightforward as that of the related Pd(II) or Pt(II) ions.

While the majority of workers have focused on the synthetic and structural chemistry of the ligand **31**, applications of the resulting complexes are emerging. Scrivanti and coworkers have synthesized and characterized the cationic complex [Pd (η^3-C_3H_5)(**31** R = Ph)](BF_4) (**46**) and have shown that it exhibits good activity in the Suzuki–Miyaura coupling of aryl bromides with phenyl boronic acid [167]. Bratsos, Turel, and coworkers have synthesized a small family of Ru(II) complexes with **31** (R = Ph) and **45a–c** and tested them in vitro for activity against two human cancer cell lines, one sensitive and one resistant to cisplatin [163]. In general, the antiproliferative activity of the tested ruthenium compounds is lower than that of cisplatin. Compound **45b** was the exception; it was almost one order of magnitude more active than cisplatin in the cisplatin-resistant A-549 cell line. Conversely, **45b** does not show any activity against HEp-2 cells, demonstrating a remarkable selectivity toward the cisplatin-resistant cell line. Košmrlj and coworkers used a "click" approach to conjugate the bioactive diazenecarboxamide unit onto the cisplatin-like complex [$PtCl_2$(**31** R = Ph)], **47a**. The authors have been able to synthesize a range of diazenecarboxamide–cisplatin conjugates, **47c**, under mild conditions but have not yet reported on the biological activity of these complexes [165].

While the pyridyl-1,2,3-triazoles are the most abundant class of bidentate "click" ligands, amine-1,2,3-triazole, phosphine-1,2,3-triazole, cyclometallated phenyl-1,2,3-triazole, and bi-1,2,3-triazole scaffolds have all been reported and are discussed in the following sections.

Gautier et al. have exploited the CuAAC reaction to synthesize a series of "regular," **50**, and "inverse," **51**, bidentate amine-1,2,3-triazole ligands and used them to generate cisplatin like platinum complexes [168]. Only the "regular" click ligand, **50**, formed a stable isolable platinum complex (**50a**) when the ligands (**50** and **51**) were reacted with K_2PtCl_4 in water (Scheme 8). A second series of "click" ligands, both regular and inverse, in which the amine donor group was replaced by a carboxylic acid were also synthesized, and they displayed similar coordination properties with only the "regular" ligands providing stable platinum complexes. The authors postulated that the "inverse" click chelates have a lower stability because the coordination motif involves the less electron-rich N2 nitrogen of the 1,2,3-triazole units; therefore, the metal–ligand interaction is weaker; this was also supported by DFT calculations. The stable Pt complexes were tested for their activity toward three cancer cell lines. In agreement with the well-established structure–activity relationships of platinum drugs, only the cis-N_2PtCl_2 coordination complex **50a** showed significant cytotoxicity. In subsequent work, Gautier and coworkers have synthesized the analogous palladium(II) complex, **50b**, then converted these complexes in the diguanosine adducts, **52a–b**, which are models for the interaction of **50a–b** with DNA. The authors also examined the binding of **50a–b** with DNA in vitro. These experiments suggest that the mode of action of **50a–b** is similar to that of cisplatin (i.e., the formation of 1,2-intrastrand adducts with DNA).

As part of efforts to develop novel, more efficient, OLEDs, Eisenberg and coworkers have generated the bidentate amine-1,2,3-triazole ligands, **53a–f**, and used them to synthesize the heteroleptic copper(I) amido-1,2,3-triazole complexes

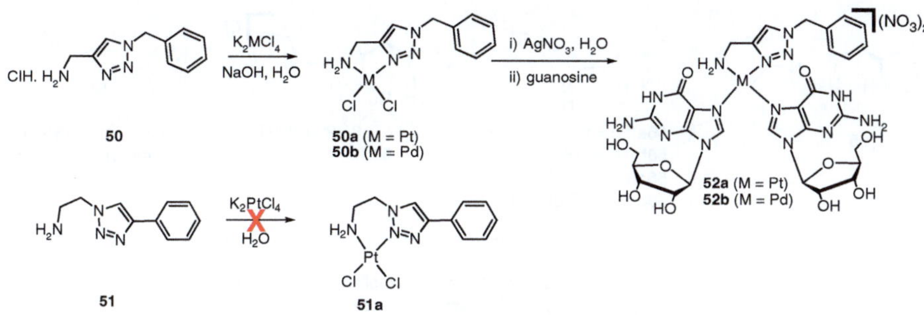

Scheme 8 The synthesis of palladium(II) and platinum(II) amino-1,2,3-triazole complexes (**50a–b** and **52a–b**)

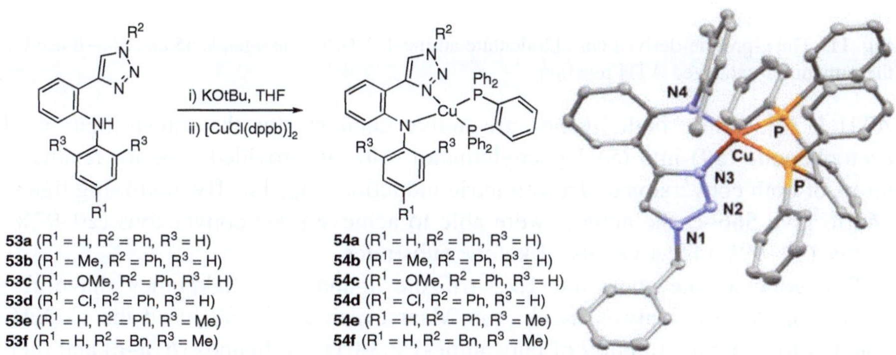

Fig. 10 The synthesis of heteroleptic copper(I) amido-1,2,3-triazole complexes (**54a–f**) and the X-ray structure of the heteroleptic copper(I) amido-1,2,3-triazole complex, **54f** [169]. Hydrogen atoms and counter ions have been omitted for clarity

(**54a–f**) [169]. Complexes **54a–f** were prepared by treating a THF slurry of $Cu_2(\mu^2\text{-}Cl)_2(dppb)_2$ with a mixture of potassium *tert*-butoxide and the "click" ligand (**53a–f**) in THF (Fig. 10). The complexes, while stable, slowly oxidize when exposed to air in the solid state and decompose rapidly when exposed to oxygen in solution. X-ray crystallography confirms that the ligands adopt distorted tetrahedral geometries around the Cu(I) ions of **54a–f** (Fig. 10b). The compounds display properties that could make them useful for the construction of OLED devices, i.e., long-lived photoluminescence with colors ranging from yellow to red-orange in the solid state and quasireversible electrochemistry, but the thermal instability of the complexes means that vapor phase deposition processing via sublimation is currently not feasible. However, the modular CuAAC ligand synthesis should allow for the rapid tuning of the systems, leading to improved thermal stability of the compounds.

Pericas et al. synthesized the L-proline-derived bidentate amine-1,2,3-triazole ligands **55** and **56** and examined their use as chiral ligands in ruthenium-catalyzed

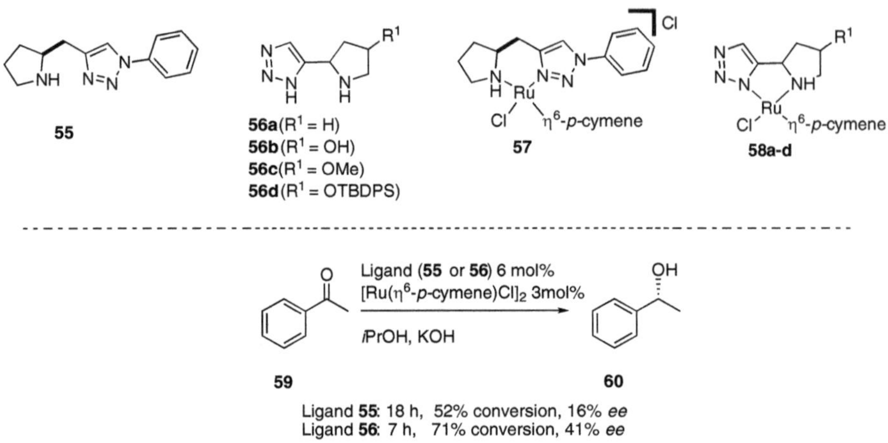

Fig. 11 The L-proline-derived chiral bidentate amine-1,2,3-triazole ligands **55** and **56a–d** used in the ruthenium-catalyzed ATH reaction

ATH [170]. While both ligands are active catalysts for the transformation of acetophenone (**59**) into (S)-1-phenylethanol (**60**), **56** provided superior results in terms of both conversion and asymmetric induction (Fig. 11). By modifying ligand **56a** to give **56b–c,** the authors were able to achieve good conversions (20–97%) and *ee* (34–99%) for a variety of ketone substrates.

Related bidentate phosphine-1,2,3-triazole ligands or "clickphines" have also been exploited in catalysis. Reek, van Maarseveen et al. [86] used the CuAAC reaction to generate a family of phosphine-1,2,3-triazole ligands (**61a–c**) and their corresponding Pd(II)-allyl complexes, **62a–c** and **63a–c** (Fig. 12). Crystal structure analysis of the neutral Pd(II)–allyl chloride complex, **62b** (R^1 = Ph), revealed that the "clickphine" ligand is coordinated in a monodentate fashion through the P-donor. However, ^{1}H and ^{31}P NMR analyses indicate that once **62a–c** has been treated with $AgBF_4$, to abstract the chloride ligand, the "clickphines" exhibit a bidentate coordination mode through the P and N3 nitrogen of the 1,2,3-triazole unit (**63a–c**). These palladium complexes proved to be active catalysts for the allylic alkylation of cinnamyl acetate (**64**) with sodium diethylmethylmalonate. The palladium complexes are highly active and selective for the formation of the linear product **65a** (Fig. 12). Additionally, it was also observed that the cationic palladium complexes generally gave higher initial rates than the neutral analogues, and "clickphines" in which the triazole unit was substituted with electron-donating groups were considerably more active (Fig. 12).

As part of their interest in phospholes, Matano and coworkers have synthesized a family of hybrid bi-, tri-, and polydentate phosphole-1,2,3-triazole ligands [91]. Reaction of the protected 2-ethynylphosphole (**66**) with a range of aryl azides under standard CuAAC conditions followed by deoxygenation with trichlorosilane provided the bidentate phosphole-1,2,3-triazole ligands (**68a–c**) in excellent yields (Scheme 9). These bidentate ligands were then complexed to Pd(II) or Pt(II). When **68a** was reacted with $Pd(cod)Cl_2$, the expected bidentate chelate complex forms,

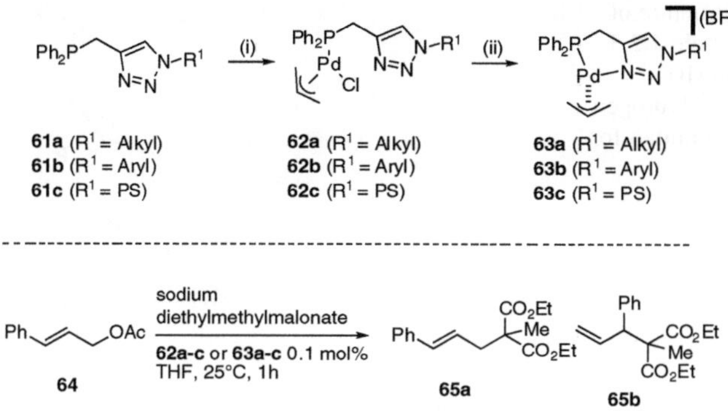

Fig. 12 "Clickphine" Pd(II)-allyl complexes, **62a–c** and **63a–c,** used as catalysts for the allylic alkylation of cinnamyl acetate (**64**) with sodium methyl diethylmalonate

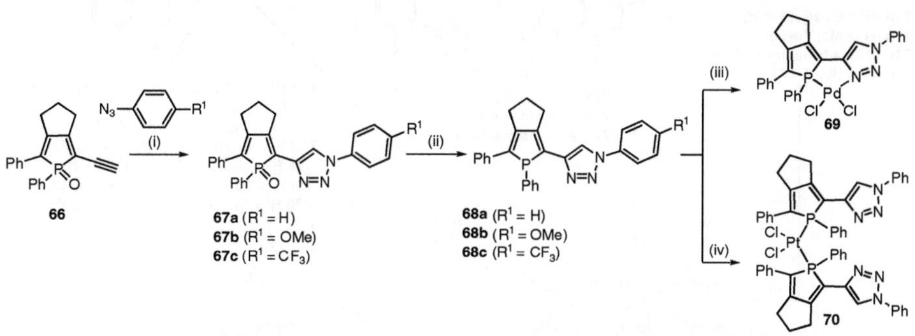

Scheme 9 Synthesis of the hybrid phosphole-1,2,3-triazole ligands **68a–c** and their palladium(II) and platinum(II) complexes, **69** and **70**; (*i*) $Cu(SO_4)\bullet 5H_2O$, Na ascorbate, THF/H_2O, RT, 24 h; (*ii*) $HSiCl_3$, toluene, reflux, 1 h; (*iii*) $Pd(cod)Cl_2$, THF, RT, 24 h; (*iv*) $Pt(cod)Cl_2$, THF, RT, 12 h

and structure of this complex was confirmed by crystallography (Fig. 13). On the other hand, when **68a** was reacted with $Pt(cod)Cl_2$, the product is the bis-ligand complex, **70**, in which only the phosphole P atom binds to the Pt(II) ions.

Spurred on by their potentially interesting photophysical and catalytic properties, a number of groups have examined the use of phenyl-1,2,3-triazoles [150, 171–173] and phenyl-1,2,3-triazol-5-ylidenes [133, 174] in the formation of cyclometallated chelate complexes. Schubert and coworkers [171] have shown that heating $IrCl_3\bullet 3H_2O$ with the "regular" phenyl-1,2,3-triazole ligands (**71a–c**) in 2-ethoxyethanol leads to the formation of the dichloride-bridged diiridium dimers (**72a–c,** Scheme 10). While the ^1H NMR spectra of the compounds were complicated, elemental analysis and mass spectrometry were consistent with the formation of the dimeric structure, as is observed with related ppy ligands. The dimers were treated with series of neutral and anionic bidentate ligands (bpy, **30**, acac, and pic), and this provided the monoiridium complexes **73a–f** good to excellent yields [150, 171]. In stark contrast, efforts to generate analogous complexes with the related "inverse" ligands, under identical conditions, lead to the formation of intractable mixtures

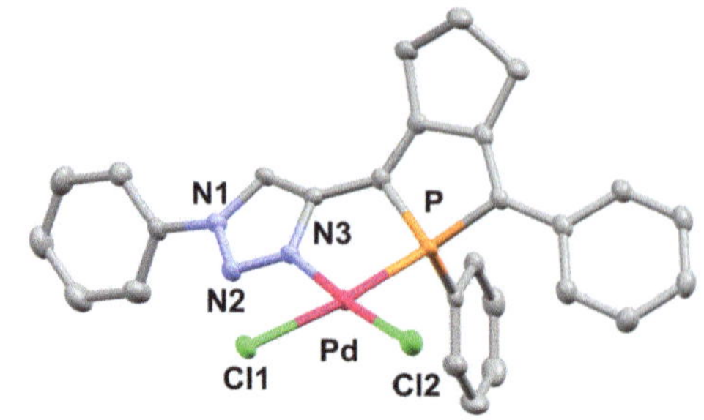

Fig. 13 X-ray structure of the hybrid phosphole-1,2,3-triazole palladium(II) complex, **69** [91]. Hydrogen atoms have been omitted for clarity

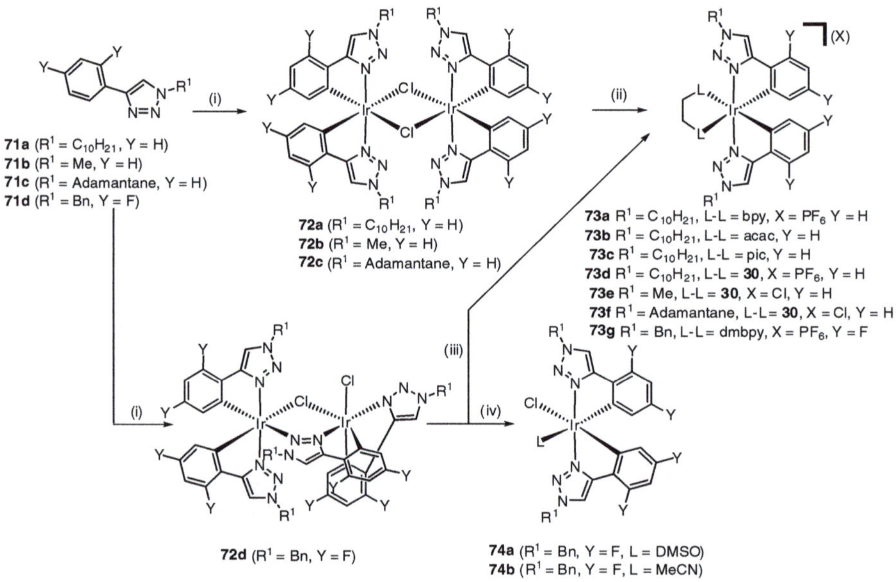

Scheme 10 Synthesis of the cyclometallated iridium(III) complexes (**72a–d**, **73a–g**, and **74a–b**) from phenyl-1,2,3-triazole ligands **71a–d**; (*i*) $IrCl_3{\bullet}3H_2O$, 2-ethoxyethanol or 2-methoxyethanol/H_2O (3:1), reflux, 14–24 h; (*ii*) for **73a** and **73d** (*a*) L–L, CH_2Cl_2/MeOH, reflux, 14 h; (*b*) NH_4PF_6; for **73b** and **73c** L–L, EtOH, Na_2CO_3, reflux, 14 h; for **73e** and **73f** (*a*) L–L, $CHCl_3$/MeOH, reflux, 3 h; (*iii*) (*a*) 2-ethoxyethanol, dmbpy, 120°C; (*b*) NH_4PF_6; (*iv*) d_6-DMSO or d_3-MeCN, RT

which decomposed upon attempted purification. These results suggest that the "inverse" cyclometallated compounds are much less stable than the "regular" complexes presumably because they are coordinated through the less electron-rich N2 nitrogen of the triazole unit. Using very similar conditions to Feiters [150] and Schubert [171], De Cola and coworkers have synthesized and structurally characterized the fluorinated "regular" iridium dimer, **72d** (Scheme 10, Fig. 14). Crystallographic analysis shows that the compound is not an expected dichloride-bridged dimer but is an isomeric dimer in which one of 1,2,3-triazole units acts as a

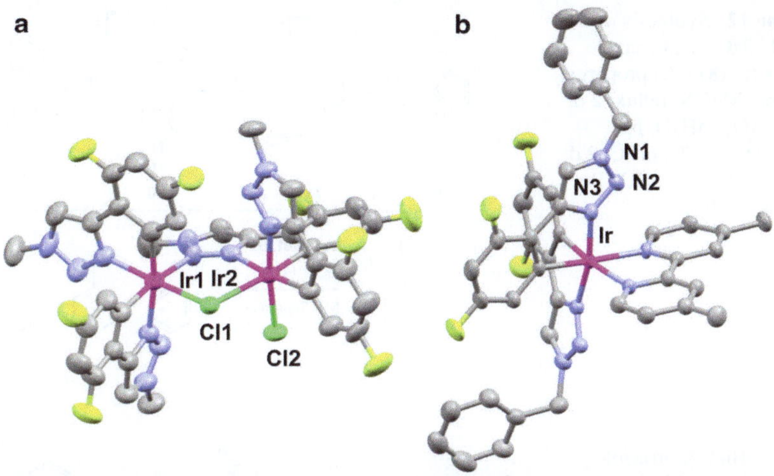

Fig. 14 X-ray structures of the dimeric **72d** (**a**) and monomeric **73g** (**b**) cyclometallated iridium (III) complexes [172]. The benzyl substituents of **72d**, hydrogen atoms, and counter ions have been omitted for clarity

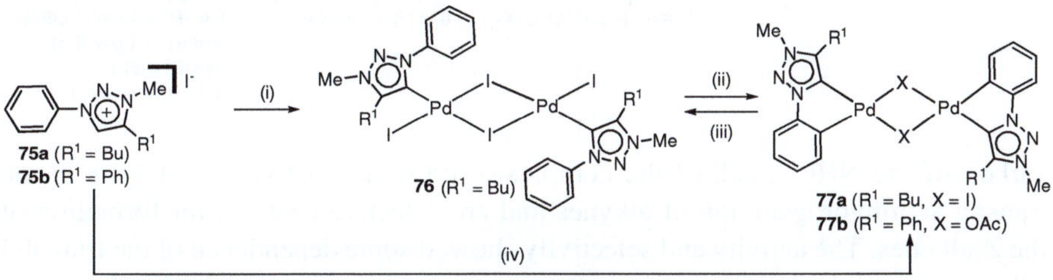

Scheme 11 Synthesis of cyclometallated phenyl-1,2,3-triazol-5-ylidenes complexes (**77a–b**); (*i*) $Pd(OAc)_2$, DMSO, 120°C, 3 h; (*ii*) NaOAc, DMSO, 120°C, 2 h; (*iii*) HI; (*iv*) (*a*) Ag_2O, CH_2Cl_2, RT, 15 h; (*b*) $Pd(OAc)_2$, CH_2Cl_2, RT, 12 h

bridging ligand. The authors were able to cleave the dimer and generate the monoiridium complexes **73g** and **74a–b** which were also structurally characterized.

Albrecht [133] and Sankararaman [174] have independently synthesized the cyclometallated phenyl-1,2,3-triazol-5-ylidenes complexes **77a–b** (Scheme 11). Both groups have found that the cyclometallation is selective for aryl groups in the 1-position of the 1,2,3-triazol-5-ylidenes units presumably because the direct attachment to the nitrogen atom increases the acidity of the 1-substituted aryl group, making the deprotonation more facile. It is noted that **77b** is a catalyst for the stereoselective hydroarylation of alkynes [174]

In related work, Elsevier and coworkers have synthesized a family of hybrid palladium NHC–triazole complexes (**81**, Scheme 12) [175]. Anion exchange to the triflate enabled the X-ray crystal structure of **81** (R^1 = Mes, R^2 = Ad) to be determined, and as expected, the NHC–triazole ligand is coordinated to the Pd(II) ion in a bidentate fashion through the N3 nitrogen of the triazole unit and the C2

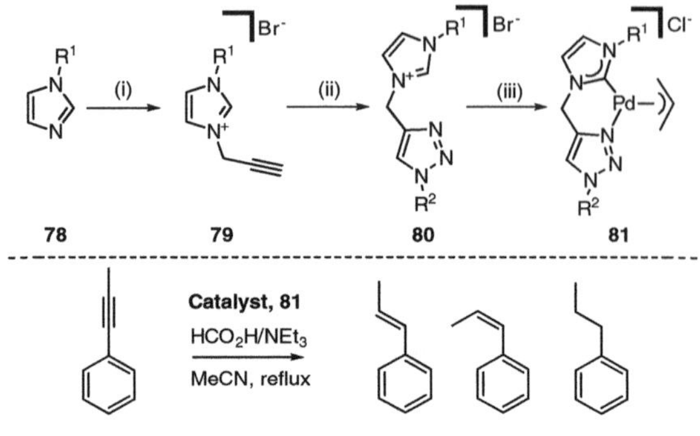

Scheme 12 Synthesis of hybrid NHC-1,2,3-triazole complexes (**81**); (*i*) propargyl bromide, MeCN, reflux, 2 d; (*ii*) $Cu(SO_4)\cdot 5H_2O$, Na ascorbate, MeCN, 60°C, 2 d; (*iii*) $[Pd(allyl)Cl]_2$, KO*t*-Bu, THF, RT, 4 h

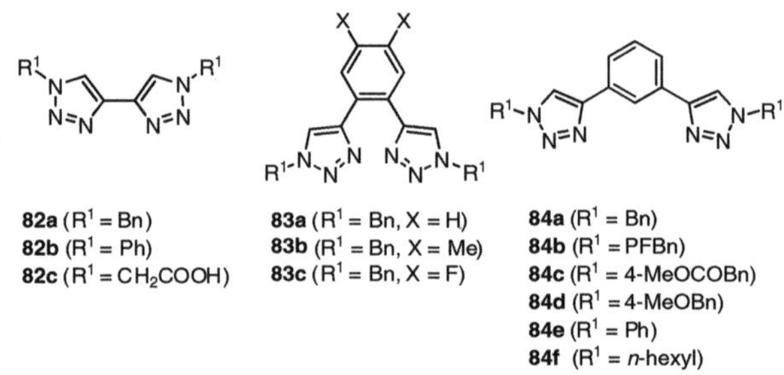

Fig. 15 Bi-1,2,3-triazole "click" ligands **82a–c**, **83a–c**, and **86a–f**

carbon of the NHC. Each of the complexes are highly active precatalysts in the transfer semihydrogenation of alkynes and are selective (>95%) for formation of the Z-alkenes. The activity and selectivity showed some dependence of the triazolyl substituent and the NHC nitrogen substituent, suggesting that the CuAAC reaction could be exploited to further tune these properties in a semi-combinatorial manner.

4.2 Bidentate Ligands Containing Two 1,4-Disubstituted-1,2,3-Triazole Units

The final class of bidentate "click" ligands are the bi-1,2,3-triazoles (Fig. 15). A number of authors have attempted to exploit the symmetrical bis-triazole ligands **82a–c** as readily prepared bpy analogues [93, 176, 177]. König, Monkowius, and coworkers synthesized the bis-triazole ligands **82a–c** and showed that they formed stable complexes with Ru(II), Cu(I), and Re(I) [177]. X-ray crystallography confirmed the bidentate coordination of the ligands **82a–c** through the N3 atoms of the triazoles to form five-membered chelate rings. Interestingly, despite the structural similarities of the bi-1,2,3-triazole complexes with their bpy analogues, the photophysical properties of the complexes were completely different. The bi-1,2,3-triazole complex's electronic absorptions are significantly higher in energy than the corresponding bpy complexes,

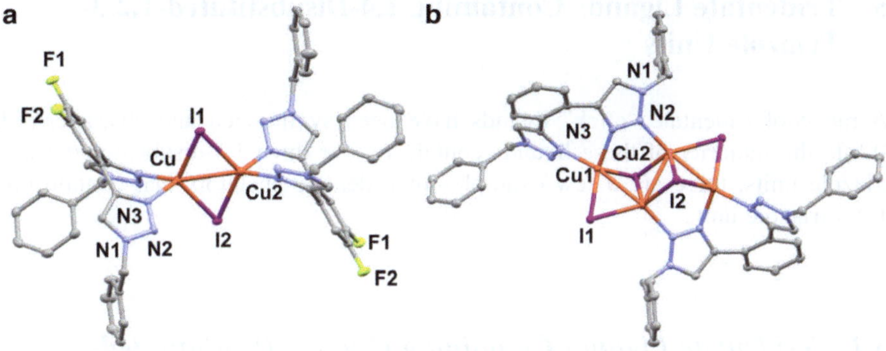

Fig. 16 X-ray structures of the dimeric (**a**) and tetrameric (**b**) CuI complexes of the bi-1,2,3-triazole ligands **83a–c** [178]. Hydrogen atoms have been omitted for clarity

and they are not emissive in solution or the solid state. As such, it has been suggested that these ligands can be viewed as readily modified "innocent" scaffolds.

Eisenberg and coworkers have synthesized the bi-1,2,3-triazole ligands **83a–c** [178]. The authors expected that reaction with CuI would provide dimeric diiodide-bridged complexes in which the ligands act as a bidentate chelate, coordinating to the Cu(I) ions through the N3 nitrogens, and in one case, this motif was observed (Fig. 16a). However, more commonly, the ligands formed tetranuclear copper(I) clusters where both chelating and bridging coordination mode of the 1,2,3-triazole is present (Fig. 16b).

Crowley et al. have synthesized the bi-1,2,3-triazole ligands **84a–f** [111, 122, 179]. While these ligands could potentially form bidentate chelate complexes, they act as bis-monodentate donors forming M_2L_2 metallomacrocycles with Ag(I) ions and M_2L_4 helical cages with Pd(II) which are additionally stabilized by π–π interactions between adjacent ligands (Fig. 17a, b). ^{1}H NMR and ESI–MS experiments indicate that these metallosupramolecular architectures are stable in solution.

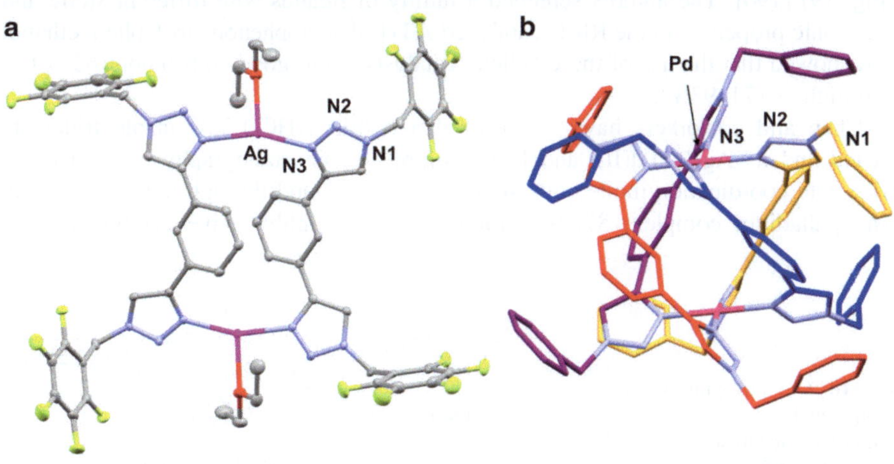

Fig. 17 X-ray structures of the Ag_2L_2 metallomacrocycle [122] (**a**) and Pd_2L_4 helical cage [111, 179] (**b**) complexes formed with the bi-1,2,3-triazole ligands **84a** and **84b**. Hydrogen atoms, counter ions, and solvent molecules have been omitted for clarity

5 Tridentate Ligands Containing 1,4-Disubstituted-1,2,3-Triazole Units

A range of tridentate "click" ligands have been synthesized and characterized. While the majority of these ligands contain two or three 1,4-disubstituted-1,2,3-triazole units, there are a few examples of tridentate chelators that contain one 1,2,3-triazole unit.

5.1 *Tridentate Ligands Containing One 1,4-Disubstituted-1,2,3-Triazole Unit*

Mindt and Schibli, pioneers of the "click-to-chelate"[142] concept, have developed families of amino acid-derived mono-1,4-disubstituted-1,2,3-triazole containing tridentate chelators for Re(I) and ^{99}Tc(I) as part of efforts to develop novel radiopharmaceuticals [142, 180–186]. They have synthesized both "regular" and "inverse" ligands but found that the "regular" complexes are more stable. In a recent report, the authors developed a one-pot procedure that allows the in situ generation of both the ligand and the radiocomplexes, **85**, further improving the synthetic efficiency (Fig. 18) [187]. The mild "click" approach to these ligands has enabled the authors to conjugate a wide range of biotargeting molecules such as carbohydrates, phospholipids, thymidine, bombesin, and folate; a more comprehensive discussion of the area can be found in the author's review [41]. Testament to the success of this methodology, other groups have quickly adopted it for the development of steroid [188] and peptide [189]-conjugated radiopharmaceuticals.

Adolfsson et al. have used related amino acid-derived mono-1,4-disubstituted-1,2,3-triazole containing tridentate chelators, **86**, as ligands in ATH reactions (Fig. 19) [190]. The authors screened a family of ligands with different steric and electronic properties in the Rh(I)-catalyzed ATH of acetophenone to 1-phenyethanol and showed that the use of these "click" catalysts led to good conversion (62–79%) and high *ee* (71–93%).

Chen and coworkers have synthesized a hybrid NHC-1,2,3-triazole tridentate ligand and its Ag(I), Pd(II), and Pt(II) complexes. Crystallography confirmed the tridentate coordination mode with the triazole unit bound through the N3 nitrogen. The palladium complex, **87**, was found to be a highly active catalyst for the

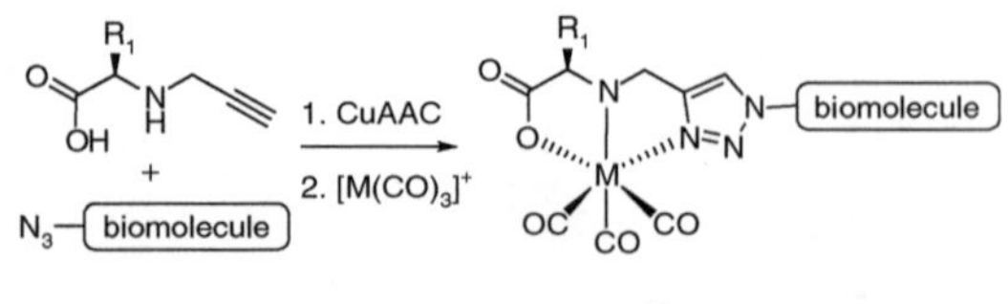

85 (M = Re, ^{99}Tc,)

Fig. 18 One-pot synthesis of biotargeted radiopharmaceuticals complexes, **85**

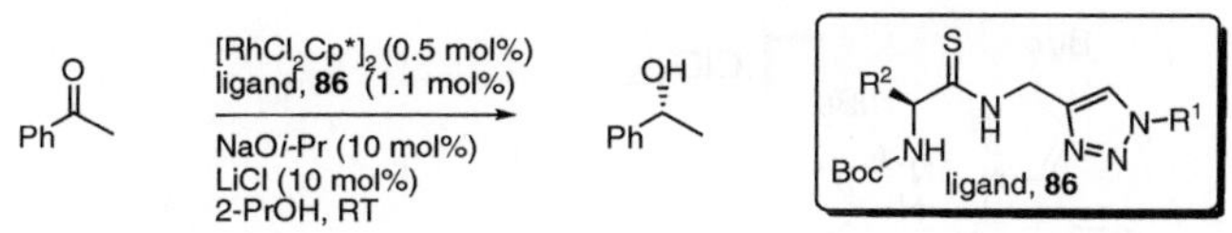

Fig. 19 Amino acid-derived tridentate "click" chelators, **86,** used as ligands in asymmetric transfer hydrogenation (ATH) reactions

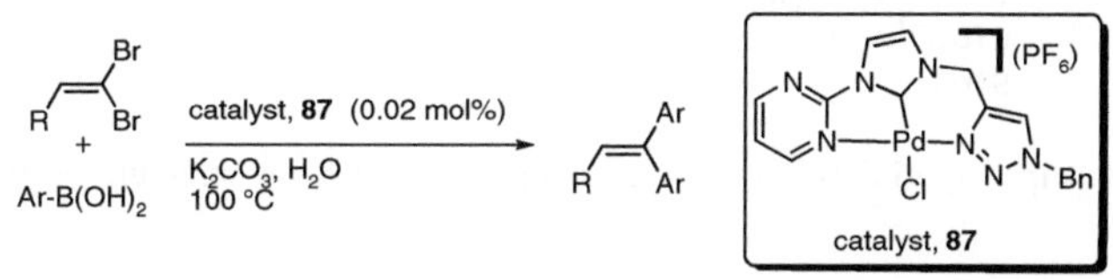

Fig. 20 Suzuki–Miyaura cross-coupling reactions catalyzed by the hybrid NHC-1,2,3-triazole palladium complex, **87**

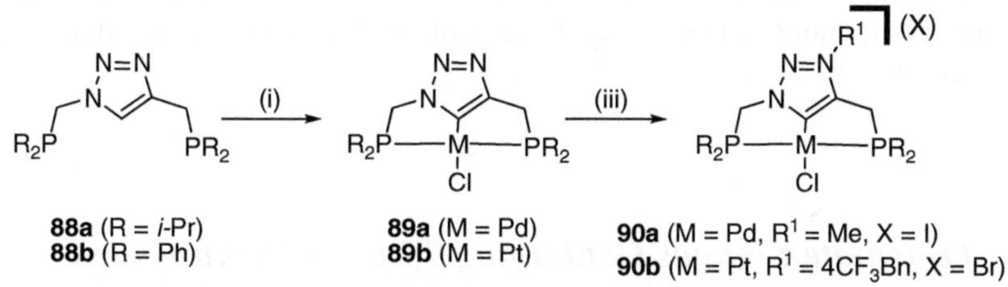

Scheme 13 Synthesis of "click" pincer complexes; (*i*) for **89a** (tmeda)$PdCl_2$, Et_3N, DMF, 70°C; for **89b** (cod)$PtCl_2$, Et_3N, DMF, 70°C; (*ii*) for **90a** MeI, $CHCl_3$, 70°C, 18 h; for **90b** 4-CF_3Bn, MeCN, 70°C, 48 h

Suzuki–Miyaura cross-coupling reactions of aryl bromides and 1,1-dibromo-1-alkenes in neat water under an air atmosphere (Fig. 20) [191].

Gandelman and coworkers have recently reported a series of pincer click ligands, **88a–b**, that incorporate two phosphine donor groups appended to a triazole (Scheme 13) [88–90, 192]. They have shown that the triazole binds to a metal ion as a triazolidene and then, after a postmodification of the coordinated ligand in which the triazole backbone is alkylated, as a carbene. Square planar Pd(II) and Pt(II) complexes of this type have been reported and characterized (**89–90a,b**). X-ray structures confirm the N-alkylation, and the expected changes in metal-donor bond lengths are observed. Palladium(II) complexes of this type have been shown to be excellent catalysts for the Heck reaction.

Chandrasekher and coworkers have prepared Fe(II) complexes, **91a–c**, of so-called "super-hybrid" ligands – tridentate ligands which contain pyridyl,

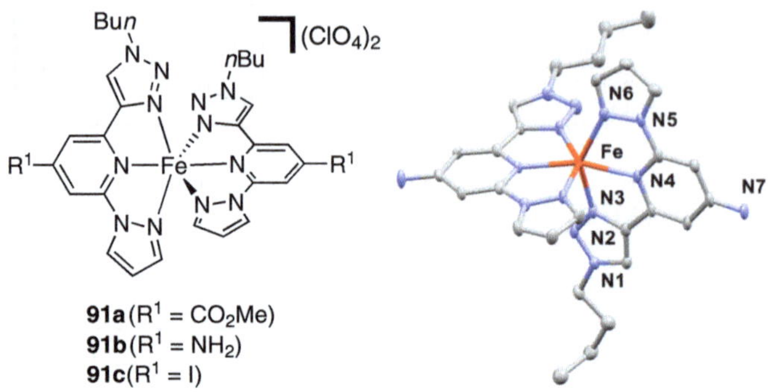

91a (R^1 = CO_2Me)
91b (R^1 = NH_2)
91c (R^1 = I)

Fig. 21 Fe(II) "super-hybrid" spin crossover complexes and X-ray structure of **91b**

pyrazolyl, and triazolyl N donors (Fig. 21a) [193]. For **91a** and **91b**, the X-ray crystal structures (Fig. 21b) showed the octahedral Fe(II) complexes to be low spin at 100 K (the temperature of the X-ray data collection) and deep red, while **91c** was high spin and yellow. **91a** (X = ClO_4^-) showed reversible spin crossover behavior with a wide hysteresis loop, with $T_{1/2}$ at 287 K, while the other two low-spin compounds appeared to have similar behavior but at much higher temperatures.

5.2 Tridentate Ligands Containing Two 1,4-Disubstituted-1,2,3-Triazole Units

In the same way that triazole-appended pyridines have been used as readily tunable analogues of bipyridyl ligands, so too have 2,6-bis(1-R-1,2,3-triazol-4-yl)pyridines been explored as analogues of terpyridine. For the most part, despite the presence of the additional N2 nitrogen donor atoms within the ligand scaffold, these compounds act as terpy-like tridentate chelators, coordinating through the pyridyl and N3 nitrogens of the 1,2,3-triazole units with a wide range of metals ions (**93–100**, Fe (II) [194–196], Ru(II) [196], Cu(II) [95, 197], Zn(II) [196], Eu(III) [194, 195], and Ag(I) [95], Scheme 14, Fig. 22). Interestingly, Crowley and coworkers have shown that when Ag(I) is reacted with the 2,6-bis(1-R-1,2,3-triazol-4-yl)pyridine ligand (R^1 = Bn), the 1,2,3-triazole units bridge two silver(I) ions generating a more complicated tetrameric structure (**100**), which is further stabilized by π–π interactions [95].

Flood [194] and Hecht and Limberg [195] simultaneously reported the first preparation of ligands **92**, along with representative examples of their coordination to Fe(II) and Eu(III). More recently, a more thorough exploration of such ligands with various electron-donating and electron-withdrawing substituents on the central pyridine ring or the triazole rings or both has been reported by Hecht

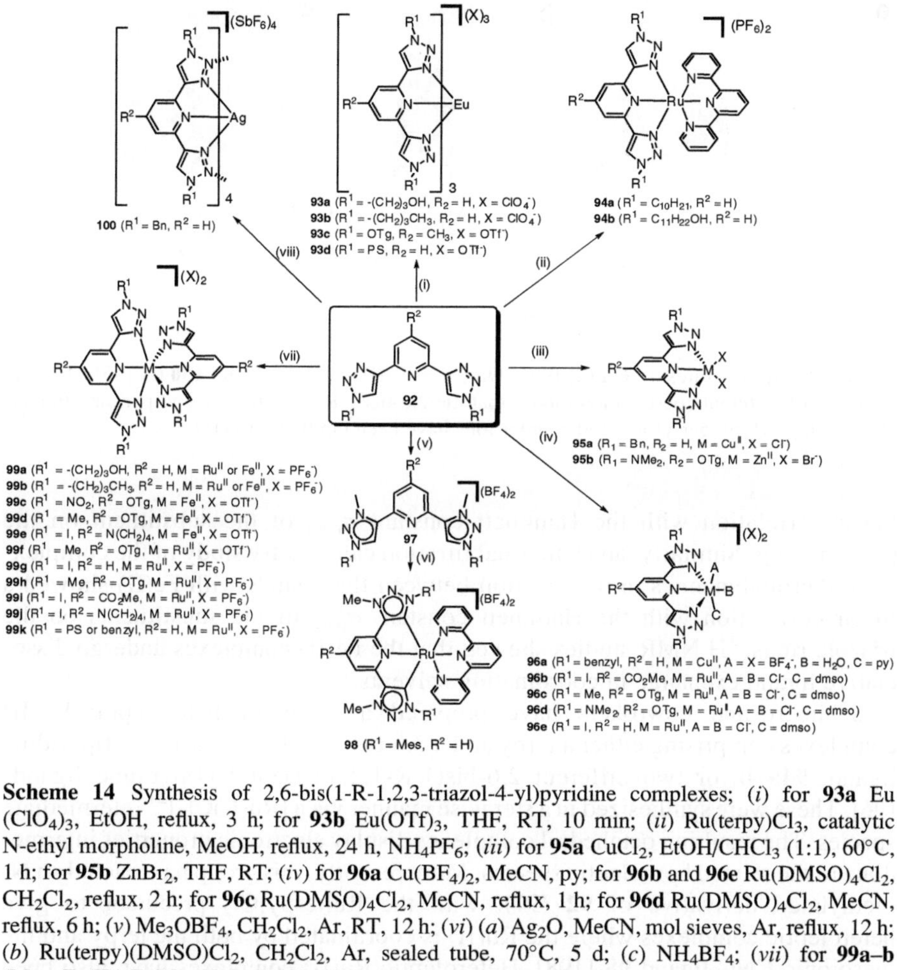

Scheme 14 Synthesis of 2,6-bis(1-R-1,2,3-triazol-4-yl)pyridine complexes; (*i*) for **93a** $Eu(ClO_4)_3$, EtOH, reflux, 3 h; for **93b** $Eu(OTf)_3$, THF, RT, 10 min; (*ii*) $Ru(terpy)Cl_3$, catalytic N-ethyl morpholine, MeOH, reflux, 24 h, NH_4PF_6; (*iii*) for **95a** $CuCl_2$, EtOH/$CHCl_3$ (1:1), 60°C, 1 h; for **95b** $ZnBr_2$, THF, RT; (*iv*) for **96a** $Cu(BF_4)_2$, MeCN, py; for **96b** and **96e** $Ru(DMSO)_4Cl_2$, CH_2Cl_2, reflux, 2 h; for **96c** $Ru(DMSO)_4Cl_2$, MeCN, reflux, 1 h; for **96d** $Ru(DMSO)_4Cl_2$, MeCN, reflux, 6 h; (*v*) Me_3OBF_4, CH_2Cl_2, Ar, RT, 12 h; (*vi*) (*a*) Ag_2O, MeCN, mol sieves, Ar, reflux, 12 h; (*b*) Ru(terpy)(DMSO)Cl_2, CH_2Cl_2, Ar, sealed tube, 70°C, 5 d; (*c*) NH_4BF_4; (*vii*) for **99a–b** (M = Fe(II)) $FeSO_4$, H_2O, EtOH, 55°C, 30 min, KPF_6; for **99a–b** (M = Ru(II)) $Ru(DMSO)_4Cl_2$, ethylene glycol, MeOH, H_2O, 100°C, 2.5 h, NH_4PF_6; for **99c–e** $Fe(OTf)_2(MeCN)_2$, MeCN, RT, 10 min; for **99f** $Ru(DMSO)_4Cl_2$, AgOTf, DMF, reflux, 12 h; for **99g–h** $Ru(DMSO)_4Cl_2$, ethylene glycol, reflux, 2 h, NH_4PF_6; for **99i** $Ru(DMSO)_4Cl_2$, DMF, reflux, 2 h, NH_4PF_6; for **99j** Ru $(DMSO)_4Cl_2$, ethylene glycol, 180°C, 30 min, NH_4PF_6; for **99k** $RuCl_3$, DMF, 145°C, 3 h, $NaPF_6$; (*viii*) $AgSbF_6$, acetone, CH_2Cl_2, RT, 1 h

and Limberg [196]. Reactions (1:1) with Cu(II) and Zn(II) salts gave five coordinate complexes, **95**, while 2:1 reactions with Cu(II), Fe(II) and Ru(II) generated a series of octahedral complexes, **96** and **99**, the natures of which were confirmed by X-ray crystallography [196]. Various spectroscopic and magnetic studies were carried out to assess the extent to which the electron-donating/withdrawing natures of the substituents affected the properties of the complexes. Variations in the oxidation potentials of the octahedral Ru(II) complexes showed a clear

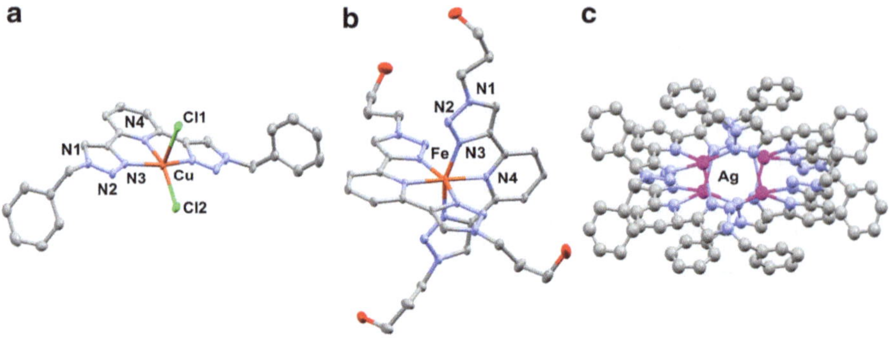

Fig. 22 X-ray structures of **95a** (**a**) [95], **99a** (M = Fe(II)) (**b**) [194], and **100** (**c**) [95], showing the range of different coordination modes that the 2,6-bis(1-R-1,2,3-triazol-4-yl)pyridine ligands **92** can adopt. Hydrogen atoms and counter ions have been omitted for clarity

linear correlation with the Hammett constant σ_{para} of the substituent on the pyridine ring. Similarly, an isothermal titration calorimetry study of the complexation thermodynamics in the reaction between the ligands with Fe(II) showed a linear correlation with the Hammett constant σ_{para} of the substituents on the triazole rings. ^{1}H NMR studies showed that the Fe(II) complexes undergo dissociative ligand exchange in coordinating solvents.

Schubert and coworkers have prepared a series of heteroleptic Ru(II) complexes comprising either a terpy and a 2,6-bis(1-R-1,2,3-triazol-4-yl)pyridine ligand, **94a–b**, or two different 2,6-bis(1-R-1,2,3-triazol-4-yl)pyridine ligands [96]. These were synthesized in a stepwise manner via a $[Ru(L)Cl_3]^{2+}$ intermediate species. The resulting octahedral complexes display electronic properties intermediate to the related homoleptic species. More recently, this group has reacted an N-methylated derivative of **92** (**97**), with the Ru(terpy)Cl_2 precursor to give heteroleptic complexes where the Ru(II) is coordinated by both the terpy and the mesoionic CNC ligand **98** [198]. Heteroleptic Ru(II) complexes have also been reported by Yao, Zhong, and coworkers, who reacted Ru(terpy)Cl_3 with the bis (triazolyl)benzene ligand, **83e,** to give octahedral cycloruthenated complex [199].

2,6-Bis(1-R-1,2,3-triazol-4-yl)pyridines have been used to prepare metallostar compounds. Obata, Kakuchi, and coworkers prepared ligand **92** (R^1 = PS) by first preparing, using ATRP methodologies, a polystyrene polymer with a terminal azide group and then clicking this with 2,6-diethynylpyridine [146]. The resulting tridentate ligand was reacted with Ru(II) to give a four-arm metallostar architecture which is estimated to have an absolute molecular weight of about 12,000 based on SEC analyses and UV–vis experiments. O'Reilly and coworker have reported a similar approach to metallostar compounds, using RAFT methodologies to elaborate a 2,6-bis(1-R-1,2,3-triazol-4-yl)pyridine core unit with polystyrene, poly(methyl methacrylate), or poly(*t*-butylacrylate) polymer chains and then synthesizing 2:1 complexes with Ru(II) and 3:1 complexes with Eu(III) [200]. Experiments showed that these aggregated further at high concentration (Scheme 14).

The related 2,6-bis(4-R-1,2,3-triazol-1-ylmethyl)pyridine inverse "click" ligands, **101a–b**, have also been studied [95, 201]. Crowley and coworkers have explored the coordination chemistry of ligands **101** with Cu(II) and Ag(I) [95]. In both cases, products with the expected 1:1 metal–ligand ratio were obtained. However, while ^{1}H NMR spectra of **102** (R = Ph) suggested a simple discrete species, X-ray crystallography showed the Ag(I) compound to in fact be polymeric in the solid state, with tetrahedral Ag(I) centers coordinated by pyridyl and triazolyl donors from one ligand, along with two other triazolyl donors from other ligands, generating a 4,4 net structure. Zhu and coworkers prepared complexes of **101** with Cu(II), Fe(II), and Co(II) (**103**) [201]. X-ray structures of these showed them have the expected octahedral structures. They also reacted **101** with $CuCl_2$ and were able to determine that the Cu(II) center has a trigonal bipyramidal geometry by X-ray analysis (**104**) (Fig. 23).

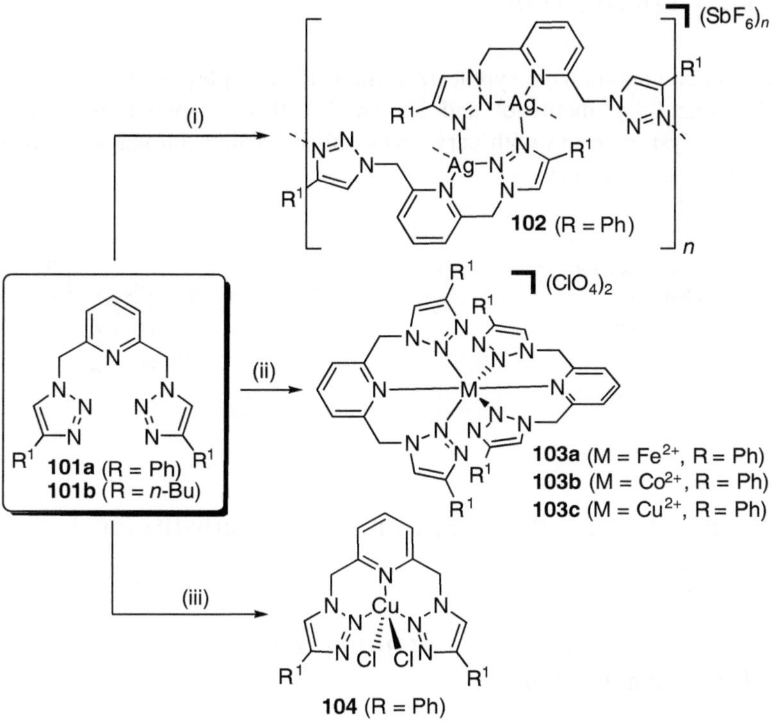

Scheme 15 Synthesis of 2,6-bis(4-R-1,2,3-triazol-1-ylmethyl)pyridine inverse "click" complexes; (*i*) $AgSbF_6$, acetone, CH_2Cl_2, RT, 1 h; (*ii*) $M(ClO_4)_2$, MeOH, RT; (*iii*) $CuCl_2$, EtOH/$CHCl_3$ (1:1), 60°C, 1 h

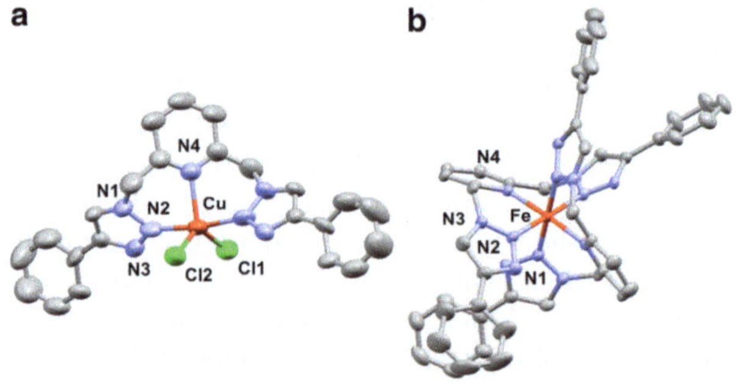

Fig. 23 X-ray structures of **104** (**a**) and **103a** (**b**) showing the different coordination geometries that the 2,6-bis(4-R-1,2,3-triazol-1-ylmethyl)pyridine inverse "click" complexes adopt [201]. Hydrogen atoms and non coordinating counter ions have been omitted for clarity

5.3 Tridentate Ligands Containing Three 1,4-Disubstituted-1,2,3-Triazole Units

Pericas and coworkers have synthesized the CuCl complex, **105**, of tris(1-benzyl-1H-1,2,3-triazol-4-yl)methanol and shown that this compound is an excellent CuAAC catalyst, working with very low catalyst loadings on water or under neat conditions [202] (Fig. 24).

Fig. 24 CuCl-tris(1-benzyl-1H-1,2,3-triazol-4-yl) methanol CuAAC catalyst, **105**

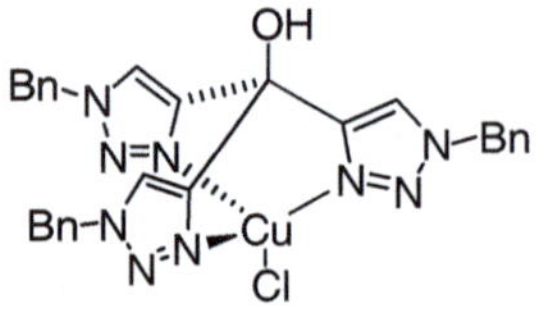

105

6 Polydentate Ligands Containing 1,4-Disubstituted-1,2,3-Triazole Units

6.1 Polydentate Ligands Containing One 1,4-Disubstituted-1,2,3-Triazole Unit

There have been many polydentate ligands reported which contain a single 1,4-disubstituted-1,2-3-triazole unit, but for the most part, these have been synthesized to be metal ion sensors and, as such, will not be discussed further [203–209].

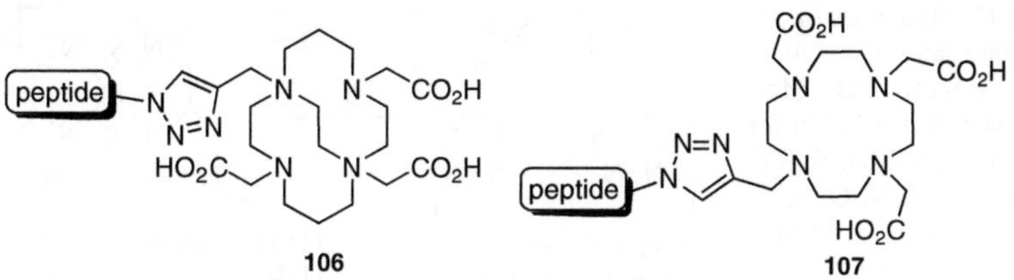

Fig. 25 Lewis' peptide-conjugated ligands for ^{64}Cu radiopharmaceuticals **106–107**

Scheme 16 Synthesis of the Cu(II) macrocycle complex, **109**; (*i*) $Cu(ClO_4)_2$, EtOH/MeCN/H_2O, reflux, 1 h

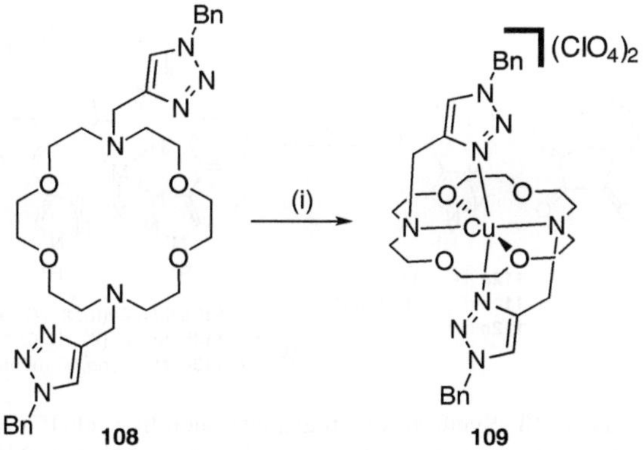

As part of efforts toward new radiopharmaceuticals, Lewis and coworkers [210] have used CuAAC chemistry to attach peptide fragments onto the common tetraazamacrocycle, **106** and **107**, and shown that these bioconjugated ligands can be readily radiolabeled with ^{64}Cu (Fig. 25).

6.2 Polydentate Ligands Containing Two 1,4-Disubstituted-1,2,3-Triazole Units

Joly and coworkers have reported a triazole-appended aza-crown, **108**, which coordinates Cu(II) ions, **109** [211]. The two triazole units, appended to the aza-crown via the ring nitrogens, bind to the Cu(II) ion via the N3 nitrogens, and these, along with the two ring nitrogens and two of the ring oxygens, give a distorted octahedral geometry (Scheme 16).

Hao, Wang et al.[212] have prepared a series of salen-like tetradentate ligands by incorporating two triazole units into an ethylene diamine skeleton (**110a–d**). Reaction of these with Ni(II), Zn(II), and Fe(II) gave complexes (Scheme 17) which, for **110a**, were structurally characterized by X-ray crystallography. In each case, the metal ion is coordinated by the two amine nitrogens and the two

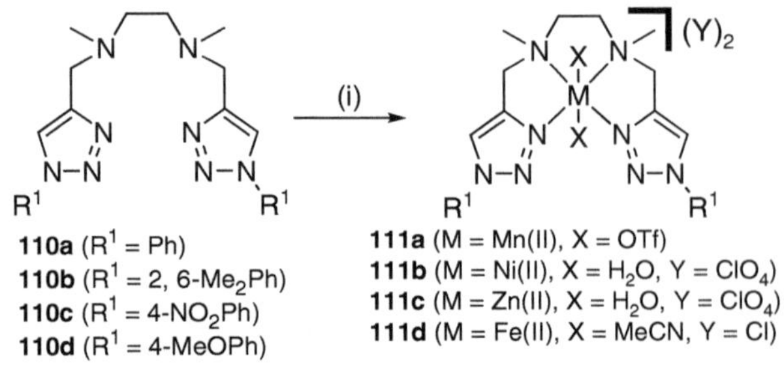

Scheme 17 Synthesis of tetradentate salen-like "click" complexes; (*i*) for **111a** Mn $(OTf)_2$, MeCN, RT, 24 h; for **111b** $Ni(ClO_4)_2$, MeCN, RT 24 h; for **111c** $Zn(ClO_4)_2$, MeCN, RT, 24 h; for **111d** $FeCl_2$, $AgSbF_6$, MeCN, RT 24 h

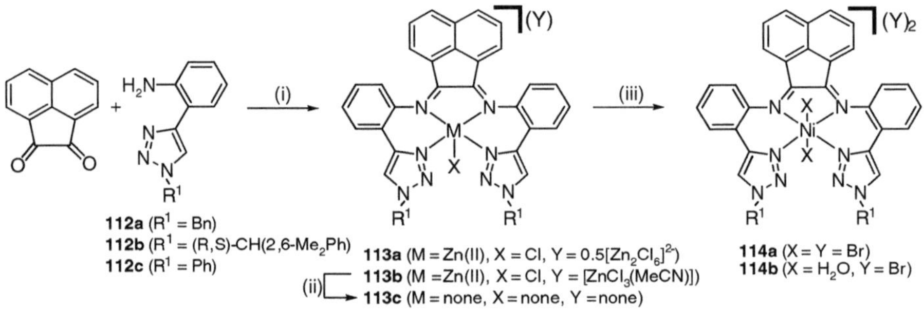

Scheme 18 Synthesis of tetradentate salen-like "click" complexes; (*i*) $ZnCl_2$, glacial acetic acid, reflux, 5 h; (*ii*) $K_2C_2O_4$, CH_2Cl_2, RT, 4 h; (*iii*) $NiBr_2$(DME), CH_2Cl_2, RT, 24 h

triazole N3 donors. The remaining two *cis* coordination sites are occupied by solvent molecules. Complexes of each of the ligands with Mn(II) were structurally characterized, and all showed the manganese ion to be coordinated by the four nitrogen donors of the triazole ligand, as well as two *cis*-coordinated triflate anions. The Mn(II) complexes were found to act as powerful catalysts for the epoxidation of terminal olefins. In particular, **111a** (R^1 = 4-MeOPh), in combination with two equivalents of peracetic acid, was found to achieve 98% conversion of 1-tetradecene at 0.5% loading within 3 min at 0°C.

Gomes and coworkers [213] have reported similar tetradentate ligands, in which they incorporated triazole units into a bis(imino)acenaphthene skeleton. The ligands were prepared by the condensation of aniline-substituted triazoles with acenaphthenequinone in the presence of zinc chloride. X-ray crystallography of **113a–b** reveals a cationic complex with the five-coordinate zinc ion and a chloride ligand in addition to the four nitrogen donors. Treatment of the zinc complexes with potassium oxalate removes the Zn(II) cation and gives the free ligands. Reaction of these with $NiBr_2$(DME) gives Ni(II) complexes **114a** in high yield (Scheme 18). X-ray crystallographic analysis revealed an pseudooctahedral Ni(II) ion, coordinated by the four N donors of the triazole ligand and two bromide ions, which coordinate in the axial positions (Fig. 26). Recrystallization in air leads to the isolation of crystals which, upon X-ray analysis, revealed an distorted octahedral

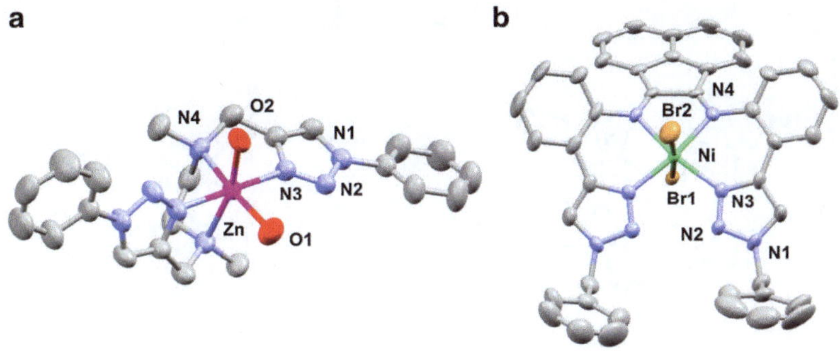

Fig. 26 X-ray structures of salen-like "click" complexes **111c** (**a**) [212] and **114a** (**b**) [213]. Hydrogen atoms and noncoordinating counter ions have been omitted for clarity

Ni(II) complex, with two forms existing the asymmetric unit. One has the *trans* bromide ligands replaced by *cis* water ligands, the bromides being present as counterions only. In this form, the triazole N donors occupy the axial positions. In the second form, the apical positions are occupied by one of the triazole N donors and one of the water molecules. The apparent stereochemical nonrigidity of the complex (and by inference, the [Ni(L)Br_2] complex) is proposed to be due to the lability of the Ni–N(triazole) bonds.

The complexes **114a–b** were tested for their ability to catalyze the addition polymerization of ethene, styrene, and norbornene. Using MAO as a cocatalyst, the complexes were unsuccessful with ethene but gave good results with styrene and norbornene. The different substituents on the ligands were found to have a small effect on polymerization activity and resulting molecular weights. The lability of the Ni–N(triazole) bonds was inferred as being important to the observed activity.

6.3 Polydentate Ligands Containing Three 1,4-Disubstituted-1,2,3-Triazole Units

There have been a number of potentially polydentate ligands that contain three 1,2,3-triazole units reported [92]. The most common is the tris(benzyltriazolylmethyl)amine (TBTA, **115**) which is often used as a coligand in CuAAC reaction. Donnelly, Williams, and coworkers have reacted **115** with Cu(II) and Cu(I) precursors with the aim of understanding the natures of these catalyst species [214]. Reaction with $CuCl_2$ in acetonitrile gave a complex cation with the formula [Cu(**115**)Cl]$^+$. X-ray crystallography showed the cation to possess a trigonal bipyramidal Cu(II) ion. The ligand **115** coordinates in a tetradentate fashion, with the central amine nitrogen occupying an apical position and the chloride ligand coordinated in the other apical site. The structure is similar to that of [Cu(**115**)Br]Br [215]. In DMSO solution, the Cu(II) complex is found to undergo reversible reduction to a

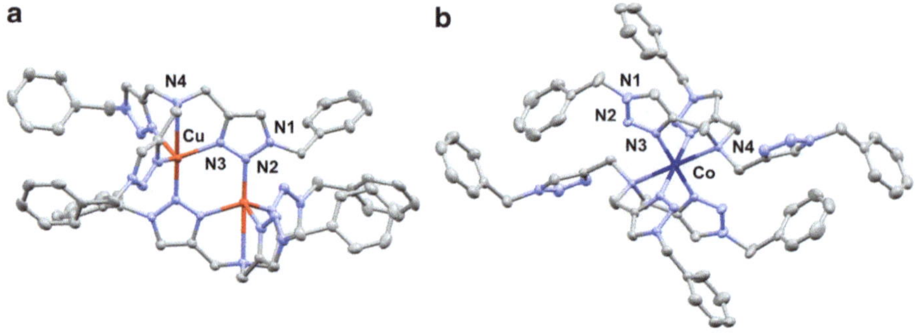

Fig. 27 X-ray structures of tripodal tetradentate "click" complexes **117** (**a**) [214] and **119a** (**b**) [216]. Hydrogen atoms and non-coordinating counter ions have been omitted for clarity

Cu(I) species. However, the cyclic voltammetry experiment had to be done at slow scan rates, indicating that a significant reorganization of the complex ion accompanied the reduction. Reaction of **115** with $[Cu(MeCN)_4]BF_4$ in acetonitrile gave a colorless, air-sensitive complex with the formula $[Cu_2($**115**$)_2](BF_4)_2$. X-ray crystallography revealed a dinuclear cation in which each Cu(I) ion is coordinated by three N3 triazole donors from one ligand and an N2 donor from a second ligand, resulting in a distorted tetrahedral geometry (Fig. 27a). Each of the Cu(II) and Cu(I) complexes was submitted to a model CuAAC reaction and found to successfully catalyze the reaction, suggesting that the Cu(I) species may be related to the actual catalytic species in these reactions (Scheme 19).

Bogani, Sarkar et al. [216] have prepared Co(II) complexes with **115** and related ligands. These were explored for their potential as SCO systems. A 1:1 reaction of TBTA and $Co(ClO_4)_2$ gave **118**, which was shown by X-ray crystallography to possess a high-spin Co(II) ion in an unusual seven-coordinate capped octahedral geometry; the central amine N and three triazole N3 donors coordinate to the Co(II) ion, along with three acetonitrile molecules. Reaction of this 1:1 species with a second equivalent of ligand gives the 1:2 complex, **119a**. X-ray crystallography shows the Co(II) ion to be coordinated by the central amine N donor and two triazole N3 donors of the two ligands, giving an octahedral geometry (Fig. 27b). Magnetic susceptibility and EPR spectroscopy show the Co ion in this case to be low spin. This is proposed to be due to intermolecular π-stacking and T-stacking interactions between the uncoordinated benzyltriazole arms of the TBTA ligands of adjacent molecules. To confirm this, **119b** and **119c**, which are incapable of such interactions due to their cyclohexyl and n-butyl groups, respectively, were prepared and found to contain high-spin Co(II) ions.

Mascal et al. [217] have appended phenyltriazole arms onto the azatriquinacene skeleton to give a potentially tetradentate ligand **120**. Reaction with either Zn $(OAc)_2$ or $Co(OAc)_2$ in methanol, followed by anion exchange with chloride and hexafluorophosphate, gave crystalline complexes **121a–b** (Scheme 20) which could be analyzed by X-ray crystallography. In each case, a trigonal bipyramidal

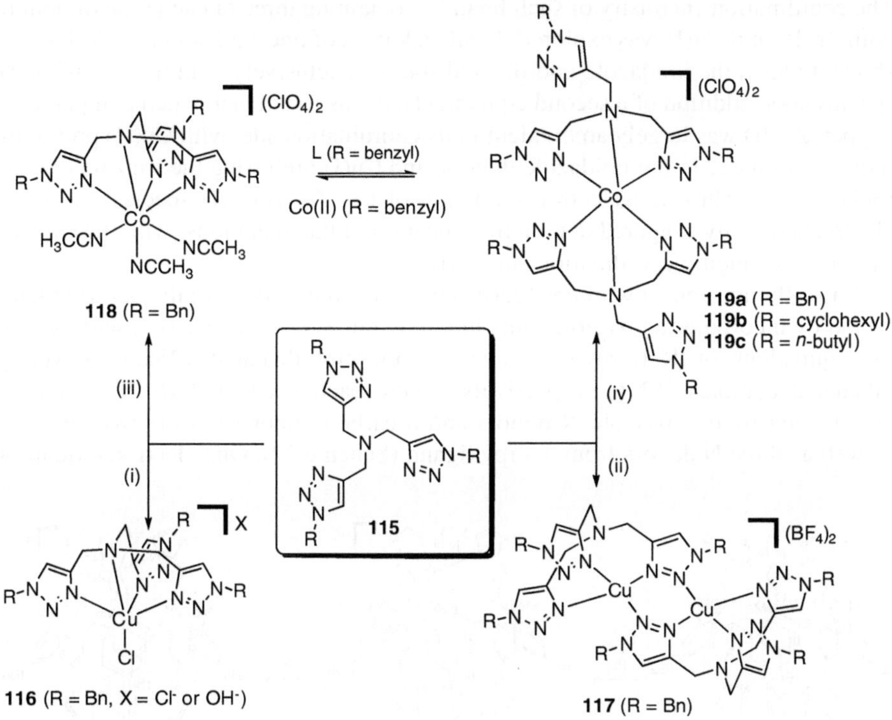

Scheme 19 Synthesis of tripodal tetradentate "click" complexes; (*i*) $CuCl_2$, MeCN; (*ii*) [Cu $(MeCN)_4]BF_4$, MeCN; (*iii*) 1 eq $Co(ClO_4)_2$, MeCN, RT, 2 h; (*iv*) for **119a** 0.5 eq $Co(ClO_4)_2$, MeCN, RT, 2 h; for **119b** 0.5 eq $Co(ClO_4)_2$, MeOH, reflux, 1 h; for **119c** 0.5 eq $Co(ClO_4)_2$, CH_2Cl_2, RT, 2 h

Scheme 20 Synthesis of azatriquinacene-based tripodal "click" complexes; (*i*) $M(OAc)_2$, MeOH, NaCl, KPF_6, RT

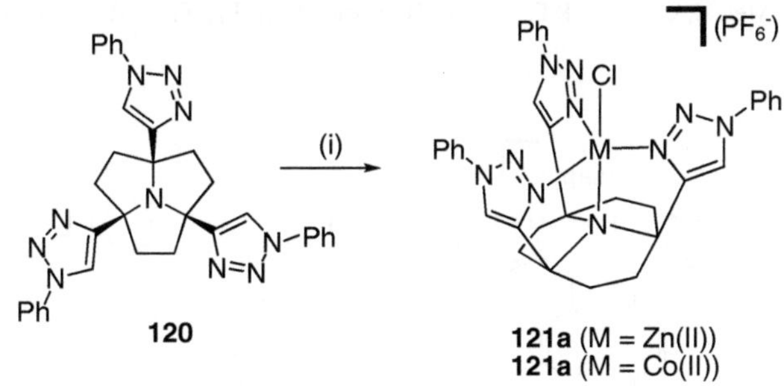

geometry for the metal ion was found, with the central nitrogen of the azatriquinacene ring and the chloride ion occupying the apical positions and the three triazole N3 donors in the equatorial plane.

Reinaud and coworkers have appended triazole donors onto the large rim of calix[6]arenes, along with imidazole donors on the small rim **122** [218, 219].

The coordination chemistry of such ligands, containing three of each type of donor, with Zn(II) and Cu(I) was explored. While addition of one equivalent of Zn(II) was shown to form the imidazole-coordinated species exclusively (and then a di-Zn(II) species upon addition of a second equivalent), it was found that, when compared to copper, Zn(II) was largely ambivalent in its coordination site, with Cu(II) preferentially coordinated to the imidazole donors, and Cu(I) preferring the triazole donors (Scheme 21). Thus, a 1-electron oxidation of the Zn(II)/Cu(I) species led to an electrochemically triggered double translocation of the metal ions, with the Cu and Zn ions exchanging coordination sites [218].

Yao, Zhong, and coworkers [220] have prepared a potentially bis-tridentate ligand by incorporating triazole units into a pyrene skeleton. Reaction with one or two equivalents of [Ru(terpy)Cl_3] gave a monocycloruthenated, **125a**, or dicycloruthenated complex, **125b**, respectively, where each octahedral Ru(II) center is coordinated by two triazole N donors and a carbon donor from the pyrene core, as well as three N donors from a terpy ligand (Scheme 22). One electron oxidation

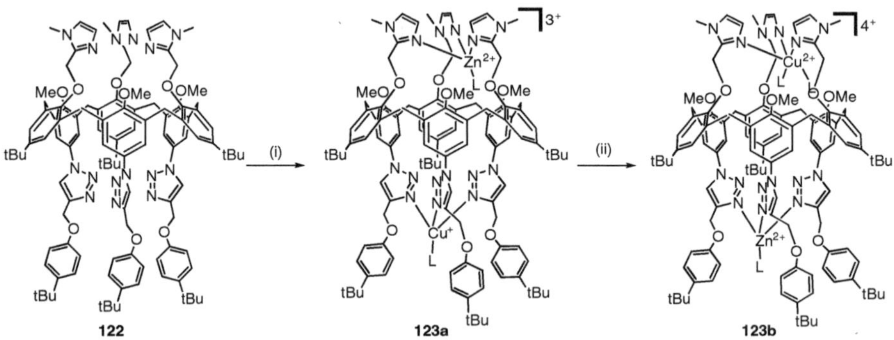

Scheme 21 Synthesis of clix-[4]-arene-based bis-tridentate "click" complexes; (*i*) (*a*) $Zn(OTf)_2$, MeCN, CH_2Cl_2, RT, 1 h; (*b*) $[Cu(MeCN)_4]PF_6$, THF, RT, 1 h; (*ii*) electrochemical oxidation

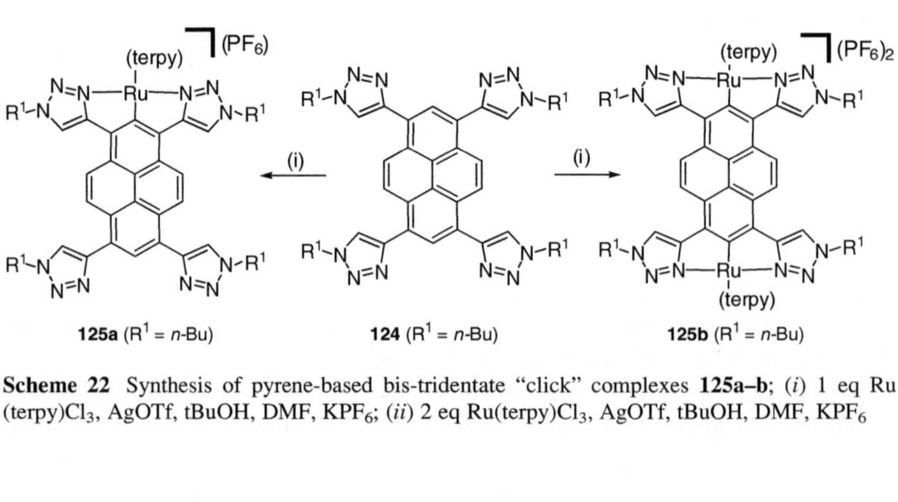

Scheme 22 Synthesis of pyrene-based bis-tridentate "click" complexes **125a–b**; (*i*) 1 eq Ru (terpy)Cl_3, AgOTf, tBuOH, DMF, KPF_6; (*ii*) 2 eq Ru(terpy)Cl_3, AgOTf, tBuOH, DMF, KPF_6

of the diruthenium species gave a stable mixed-valence complex. Electronic coupling of the resulting mixed-valent redox centers was comparable to that observed in complexes with the related pyridine-containing ligand.

6.4 *Polydentate "Click" Ligands Used for the Synthesis of Metallosupramolecular Architectures and Coordination Polymers*

Schubert and coworkers [221] have prepared 3,6-bis(triazolyl)pyridazine ligands, **126a–b**, and reacted these with Cu(I) and Ag(I) salts to synthesize tetrametallic grid species, **127a–b** (Scheme 23). Each metal ion is tetrahedrally coordinated by triazole and pyridazine N donors from two orthogonally disposed ligands. The structures of the compounds were confirmed by one-dimensional and two-dimensional ^{1}H NMR spectroscopy, ESI-QTOF mass spectrometry, and sedimentation velocity experiments.

Several groups have developed the chemistry of poly-2-(1-R-1H-1,2,3-triazol-4-yl) pyridine ligands, **128** [222]. Crowley [94] and Policar [223] have both shown that these ligands will form metallomacrocycles. By using a one-pot synthetic procedure, in which the required potentially explosive polyazide intermediates are generated in situ and then immediately captured in the CuAAC reaction, Crowley and coworkers safely prepared a series of ligands, **128**, where two or three pyridyltriazole groups are joined by various spacer units [94]. Reaction of the ligands, **128**, with $AgSbF_6$ leads to the formation of discrete metallosupramolecular architectures, and ^{1}H NMR and ESI–MS experiments indicate that these are stable in solution (Scheme 24). The structure of the silver(I) metallomacrocycle, **129a**, was confirmed using X-ray crystallography (Fig. 28a).

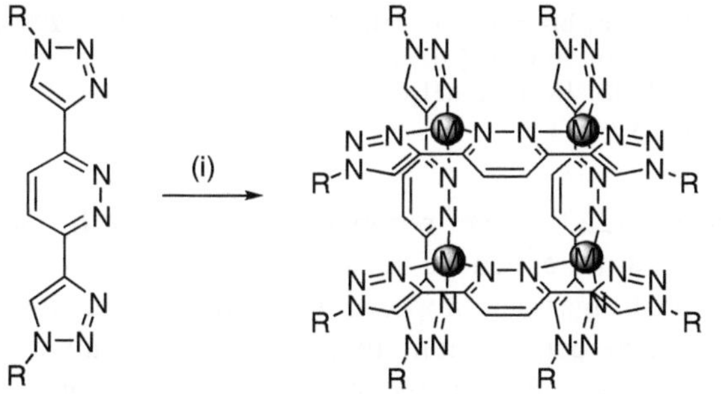

126a (R = mes)
126b (R = $(CH_2CH_2O)_{12}CH_3$)

127a (M = Cu^I, X = PF_6^-)
127b (M = Ag^I, X = SbF_6^-)

Scheme 23 Synthesis of triazole pyridizine based grid complexes; (*i*) for **127a** [Cu(MeCN)$_4$]PF$_6$, CH_2Cl_2, RT, 30 min; for **127b** $AgSbF_6$, CH_2Cl_2, RT, 1 h

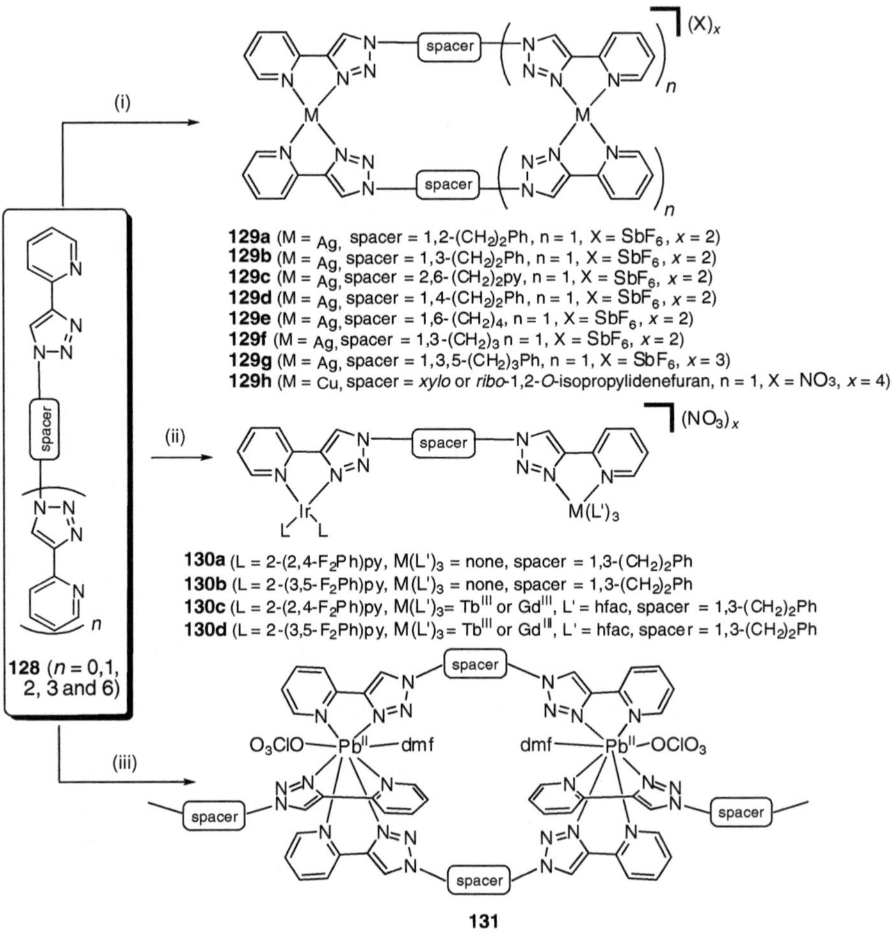

Scheme 24 Synthesis of bis-bidentate "click" complexes; (*i*) for **129a–g** for **129h** $Cu(NO_3)_2$, EtOH, H_2O, RT; (*ii*) for **130a–b** $[Ir(F_2Phpy)_2(\mu\text{-}Cl)]_2$, CH_2Cl_2, MeOH, N_2, 50°C, 12 h,; for **130c–d**, **130a**, or **130b**, $[M(hfac)_3(H_2O)_3]$ (M = Gd(III) or Tb(III)), CH_2Cl_2, RT; (*iii*) Pb $(ClO_4)_2$, MeOH, CH_2Cl_2, RT

Policar et al. [223] have reported bis-bidentate ligands comprising two 2-pyridyltriazole arms appended to a *xylo-* or *ribo*-1,2-*O*-isopropylidenefurano core. Disubstituted sugars are proposed to be attractive but underutilized, structural scaffolds for ligand construction. In particular, the authors suggest that these backbones provide a predetermined stereochemistry to the metal ions. Reactions of each with 1 equiv. of $Cu(NO_3)_2$ gave complexes with a 1:1 metal–ligand ratio (Scheme 24). An X-ray crystal structure of the complex with the *xylo*-backbone showed the complex to be an unsaturated Cu_2L_2 helicate, with approximately square pyramidal Cu(II) cations coordinated by four N donors from two ligands, with a water ligand completing each copper's N_4O coordination sphere (Fig. 28b).

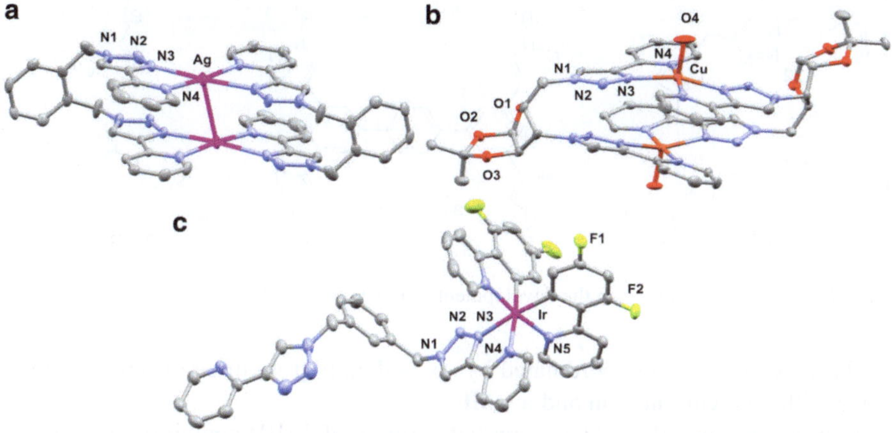

Fig. 28 X-ray structures of bis-bidentate "click" complexes **129a** (**a**) [94] **130j** (**b**) [223], and **131** (**c**) [224] Hydrogen atoms and non-coordinating counter ions have been omitted for clarity

Fluorescence, EPR, CD, and UV–visible spectroscopy studies suggest that both copper complexes share the same structures in the solid state and in solution.

Ward and coworkers have prepared a ditopic pyridyltriazole ligand as part of their studies into visible light sensitization of Tb(III) luminescence by d/f dyad species [224]. Ligand **128**, containing two pyridyltriazole-chelating sites separated by an *m*-phenylene spacer unit, reacts with less than one equivalent of Ir(III) to give a monometallic species which was characterized by X-ray crystallography and found to have the expected octahedral coordination geometry for the Ir(III) ion (Fig. 28c). Detailed spectroscopic studies of solutions of the Ir(III) complex to which Tb(III) ions had been added suggested that there are two conformers of the resulting heterodimetallic complex present in solution, and although both displayed the desired energy transfer from the excited Ir(III) center to the Tb(III) center, one conformer (proposed to have the metal ions closer together than the other) showed faster energy transfer than the other.

Building on their work on ditopic ligands with pyridylpyrazolyl donor units, Ward and coworkers have prepared Pb(II) complexes of the potentially bis-bidentate and tris-bidentate ligands, **128** [225]. X-ray crystallographic analyses revealed these all have polymeric structures in the solid state, with ligands bridging Pd(II) ions. Complex **131** contains eight-coordinate Pb(II) cations, whose coordination sphere includes a DMF ligand, a perchlorate, and nitrogen donors from three different ligands. $[Pb_2(\mu\text{-}L)_2]$ units are assembled into chains by bridging **128** (spacer = 1,3-(CH_2)Ph) ligands to give one-dimensional chains, **131**. The Pb(II) complex with **128** (spacer = 1,4-(CH_2)Ph) is also found to contain eight-coordinate Pb(II) ions, but in this case, all are N donors from four different ligands, giving rise to a two-dimensional 4,4 square net structure. In the case of the potentially tris-bidentate ligands **128** (spacer = 1,3,5-(CH_2)Ph and 1,3,5-(CH_3)-2,4,6-(CH_2)Ph), the Pb(II) complexes were found to have two-dimensional [6,3] net structures, with

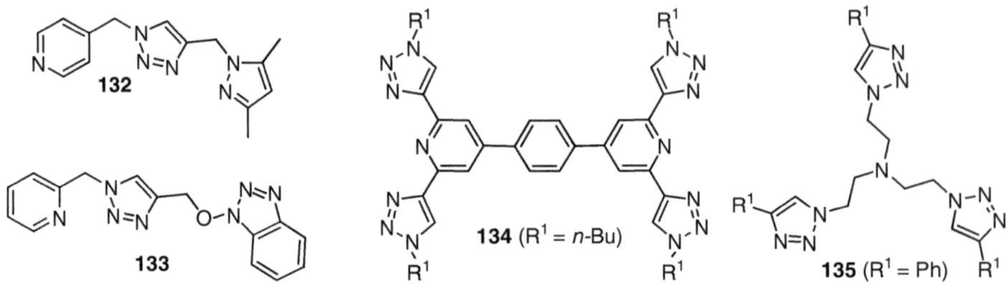

Fig. 29 "Click" ligands used in the development of novel coordination polymers

the Pb(II) cations being coordinated by six N donors from three different ligands, along with a perchlorate ion and a DMF.

Rowan and coworkers have reported a series of Ir(III)-containing complexes constructed from ligands containing pyridyltriazole coordinating units [226]. "Parallel" and "antiparallel" ligands with cyclometallated $[Ir(ppy)_2]^+$ units binding to each of one, two, or three pyridyltriazole units were prepared, and studies of their photophysical properties showed that each Ir center behaves independent of the others. No structural characterization of the complexes was reported.

A number of groups are beginning to exploit "click" ligands (for example see **132–135**) in the formation of coordination polymers [108, 227–229]. In particular, Hor and coworkers have reported an elegant helical Cu(II) coordination polymer formed from **133**. Addition of one equivalent of $CuCl_2$ gave a mixture of green and blue crystals. X-ray analysis showed both the compounds to exists as one-dimensional helical polymers, with each five-coordinate Cu(II) ion chelated by a pyridyltriazole unit, a benzotriazole donor from a second ligand, and two chlorides. In the case of the green compound, these polymers, which have a helical pitch of ca. 45.2 Å, assemble into an overall chiral structure, with all chains having the same helicity, whereas for the blue compound, a racemic mixture of helical chains is found, where the pitch is much shorter at ca. 7.6 Å, and the chains are bridged by a solvent water molecule (Fig. 29).

7 Conclusion

The mild, modular, functional group tolerant and reliable CuAAC reaction has emerged as a highly efficient methodology for the construction of functionalized ligand scaffolds. The 1,4-disubstituted-1,2,3-triazole units that are generated under these "click" conditions display a range of structurally characterized coordination modes including monodentate binding (through either the N3 nitrogen or C5 carbon positions of the 1,2,3-triazole) and bridging (through the N2 and N3 nitrogen atoms). A diverse array of "regular" and "inverse" bi-, tri-, and polydentate "click" chelators incorporating 1,4-disubstituted-1,2,3-triazole units have also been synthesized and characterized. The "regular" click ligands generally form more stable metal complexes presumably because they coordinate through the more electron-rich N3 nitrogen atom of the 1,2,3-triazole unit. "Click" complexes are beginning to find applications in catalysis, metallosupramolecular chemistry, photophysics, and as metallopharmaceuticals and bioimaging agents.

Addendum

This review initially covered the field up to June 30, 2011, and while the manuscript was under review, the area has continued to grow with additional examples of bi- [230–232], tri- [233–237], and polydentate [238, 239] "click" ligands appearing in literature.

References

1. Michael A (1893) J Prakt Chem 48:94
2. Huisgen R (1963) Angew Chem 75:604
3. Huisgen R (1967) Helv Chim Acta 50:2421
4. Huisgen R (1984) In: Padwa A (ed), vol 1. 1, 3-Dipolar Cycloaddition Chemistry Wiley, New York, p 1
5. Huisgen R, Szeimies G, Möbius L (1967) Chem Ber 100:2494
6. Tornoe CW, Christensen C, Meldal M (2002) J Org Chem 67:3057
7. Rostovtsev VV, Green LG, Fokin VV, Sharpless KB (2002) Angew Chem Int Ed 41:2596
8. L'Abbé G (1984) Bull Soc Chim Belg 93:579
9. Wu P, Fokin VV (2007) Aldrichim Acta 40:7
10. Kolb HC, Finn MG, Sharpless BK (2001) Angew Chem Int Ed 40:2004
11. Meldal M, Tornoe CW (2008) Chem Rev 108:2952
12. Juricek M, Kouwer PHJ, Rowan AE (2011) Chem Commun 47:8740
13. Lutz J-F (2007) Angew Chem Int Ed 46:1018
14. Lutz J-F, Zarafshani Z (2008) Adv Drug Deliv Rev 60:958
15. Ganesh V, Sudhir VS, Kundu T, Chandrasekaran S (2011) Chem Asian J 6:2670
16. Li Y, Cai C (2011) Chem Asian J 6:2592
17. Altintas O, Tunca U (2011) Chem Asian J 6:2584
18. Meldal M (2008) Macromol Rapid Commun 29:1016
19. Golas PL, Matyjaszewski K (2010) Chem Soc Rev 39:1338
20. Qin A, Lam JWY, Tang BZ (2010) Chem Soc Rev 39:2522
21. Durot S, Frey J, Sauvage J-P, Tock C (2010) Org Azides 413
22. Aprahamian I, Miljanic OS, Dichtel WR, Isoda K, Yasuda T, Kato T, Stoddart JF (2007) Bull Chem Soc Jpn 80:1856
23. Miljanic OS, Dichtel WR, Aprahamian I, Rohde RD, Agnew HD, Heath JR, Stoddart JF (2007) QSAR Comb Sci 26:1165
24. Hänni KD, Leigh DA (2010) Chem Soc Rev 39:1240
25. Fahrenbach AC, Stoddart JF (2011) Chem Asian J 6:2660
26. Hu J, Lu JR, Ju Y (2011) Chem Asian J 6:2636
27. Li X (2011) Chem Asian J 6:2606
28. Holub JM, Kirshenbaum K (2010) Chem Soc Rev 39:1325
29. Hein JE, Fokin VV (2010) Chem Soc Rev 39:1302

30. Kolb HC, Sharpless KB (2003) Drug Discov Today 8:1128
31. Tornoee CW, Meldal M (2010) Org Azides 285
32. Moses JE, Moorhouse AD (2007) Chem Soc Rev 36:1249
33. Agalave SG, Maujan SR, Pore VS (2011) Chem Asian J 6:2696
34. Zhang L, Chen X, Xue P, Sun HHY, Williams ID, Sharpless KB, Fokin VV, Jia G (2005) J Am Chem Soc 127:15998
35. Rasmussen LK, Boren BC, Fokin VV (2007) Org Lett 9:5337
36. Boren BC, Narayan S, Rasmussen LK, Zhang L, Zhao H, Lin Z, Jia G, Fokin VV (2008) J Am Chem Soc 130:8923
37. Bock VD, Hiemstra H, van Maarseveen JH (2005) Eur J Org Chem 51
38. Pedersen DS, Abell A (2011) Eur J Org Chem 2011:2399
39. McDonald K, Hua Y, Flood A (2010) In: Gale PA, Dehaen W (eds) Anion recognition in supramolecular chemistry, vol 25, Topics in heterocyclic chemistry. Springer, Berlin, p 341
40. Hua Y, Flood AH (2010) Chem Soc Rev 39:1262
41. Struthers H, Mindt TL, Schibli R (2010) Dalton Trans 39:675
42. Lau YH, Rutledge PJ, Watkinson M, Todd MH (2011) Chem Soc Rev 40:2848
43. Constable EC, Housecroft CE, Price JR, Schweighauser L, Zampese JA (2010) Inorg Chem Commun 13:495
44. Imae K, Shimizu K, Ogata K, Fukuzawa S-i (2011) J Org Chem 76:3604
45. Shimizu K, Ogata K, Fukuzawa S-i (2010) Tetrahedron Lett 51:5068
46. Oura I, Shimizu K, Ogata K, Fukuzawa S-i (2010) Org Lett 12:1752
47. Oki H, Oura I, Nakamura T, Ogata K, Fukuzawa S-i (2009) Tetrahedron Asymmetry 20:2185
48. Kato M, Oki H, Ogata K, Fukuzawa S-i (2009) Synlett 1299
49. Fukuzawa S-i, Oki H, Hosaka M, Sugasawa J, Kikuchi S (2007) Org Lett 9:5557
50. Benoist E, Coulais Y, Almant M, Kovensky J, Moreau V, Lesur D, Artigau M, Picard C, Galaup C, Gouin SG (2011) Carbohydr Res 346:26
51. Paek S, Baik C, Kang M-s, Kang H, Ko J (2010) J Organomet Chem 695:821
52. Devic T, David O, Valls M, Marrot J, Couty F, Ferey G (2007) J Am Chem Soc 129:12614
53. Dai Q, Gao W, Liu D, Kapes LM, Zhang X (2006) J Org Chem 71:3928
54. Shen C, Zhang P-F, Chen X-Z (2010) Helv Chim Acta 93:2433
55. Renfrew AK, Juillerat-Jeanneret L, Dyson PJ (2011) J Organomet Chem 696:772
56. Benoist E, Coulais Y, Almant M, Kovensky J, Moreau V, Lesur D, Artigau M, Picard C, Galaup C, Gouin SG (2010) Carbohydr Res 346:26
57. Gonzalez Cabrera D, Koivisto BD, Leigh DA (2007) Chem Commun 4218
58. Constable EC, Housecroft CE, Neuburger M, Roesel P (2010) Chem Commun 46:1628
59. Lucon J, Abedin MJ, Uchida M, Liepold L, Jolley CC, Young M, Douglas T (2010) Chem Commun 46:264
60. Kikkeri R, Liu X, Adibekian A, Tsai Y-H, Seeberger PH (2010) Chem Commun 46:2197
61. Collin J-P, Durot S, Keller M, Sauvage J-P, Trolez Y, Cetina M, Rissanen K (2011) Chem Eur J 17:947
62. Collin J-P, Frey J, Heitz V, Sauvage J-P, Tock C, Allouche L (2009) J Am Chem Soc 131:5609
63. Schmidt F, Rosnizeck IC, Spoerner M, Kalbitzer HR, Koenig B (2011) Inorg Chim Acta 365:38
64. Urankar D, Kosmrlj J (2010) Inorg Chim Acta 363:3817
65. Savonnet M, Bazer-Bachi D, Bats N, Perez-Pellitero J, Jeanneau E, Lecocq V, Pinel C, Farrusseng D (2010) J Am Chem Soc 132:4518
66. Ruther RE, Rigsby ML, Gerken JB, Hogendoorn SR, Landis EC, Stahl SS, Hamers RJ (2011) J Am Chem Soc 133:5692
67. Gadzikwa T, Farha OK, Malliakas CD, Kanatzidis MG, Hupp JT, Nguyen ST (2009) J Am Chem Soc 131:13613
68. Jauregui M, Perry WS, Allain C, Vidler LR, Willis MC, Kenwright AM, Snaith JS, Stasiuk GJ, Lowe MP, Faulkner S (2009) Dalton Trans 6283

69. Moore AL, Bucar D-K, MacGillivray LR, Benny PD (2010) Dalton Trans 39:1926
70. Wieczorek B, Lemcke B, Dijkstra HP, Egmond MR, Klein Gebbink RJM, van Koten G (2010) Eur J Inorg Chem 2010:1929
71. McDonald AR, Dijkstra HP, Suijkerbuijk BMJM, van Klink GPM, van Koten G (2009) Organometallics 28:4689
72. Decreau RA, Collman JP, Hosseini A (2010) Chem Soc Rev 39:1291
73. Mastarone DJ, Harrison VSR, Eckermann AL, Parigi G, Luchinat C, Meade TJ (2011) J Am Chem Soc 133:5329
74. Gauthier S, Weisbach N, Bhuvanesh N, Gladysz JA (2009) Organometallics 28:5597
75. Zhang Q, Takacs JM (2008) Org Lett 10:545
76. Gasser G, Sosniak AM, Leonidova A, Braband H, Metzler-Nolte N (2011) Aust J Chem 64:265
77. Patra M, Gasser G, Bobukhov D, Merz K, Shtemenko AV, Metzler-Nolte N (2010) Dalton Trans 39:5617
78. Metzler-Nolte N (2010) Top Organomet Chem 32:195
79. Husken N, Gasser G, Koster SD, Metzler-Nolte N (2009) Bioconjugate Chem 20:1578
80. Koester SD, Dittrich J, Gasser G, Huesken N, Henao Castaneda IC, Jios JL, Della Vedova CO, Metzler-Nolte N (2008) Organometallics 27:6326
81. Diez-Gonzalez S (2011) Catal Sci Technol 1:166
82. Kuang G-C, Michaels HA, Simmons JT, Clark RJ, Zhu L (2010) J Org Chem 75:6540
83. Brotherton WS, Michaels HA, Simmons JT, Clark RJ, Dalal NS, Zhu L (2009) Org Lett 11:4954
84. Urankar D, Kosmrlj J (2008) J Comb Chem 10:981
85. Urankar D, Steinbuecher M, Kosjek J, Kosmrlj J (2010) Tetrahedron 66:2602
86. Detz RJ, Arevalo Heras S, De Gelder R, Van Leeuwen PWNM, Hiemstra H, Reek JNH, Van Maarseveen JH (2006) Org Lett 8:3227
87. Dolhem F, Johansson MJ, Antonsson T, Kann N (2007) J Comb Chem 9:477
88. Schuster EM, Botoshansky M, Gandelman M (2008) Angew Chem Int Ed 47:4555
89. Schuster EM, Botoshansky M, Gandelman M (2009) Organometallics 28:7001
90. Schuster EM, Nisnevich G, Botoshansky M, Gandelman M (2009) Organometallics 28:5025
91. Matano Y, Nakashima M, Saito A, Imahori H (2009) Org Lett 11:3338
92. van Assema SGA, Tazelaar CGJ, Bas de Jong G, van Maarseveen JH, Schakel M, Lutz M, Spek AL, Slootweg JC, Lammertsma K (2008) Organometallics 27:3210
93. Fletcher JT, Bumgarner BJ, Engels ND, Skoglund DA (2008) Organometallics 27:5430
94. Crowley JD, Bandeen PH (2010) Dalton Trans 39:612
95. Crowley JD, Bandeen PH, Hanton LR (2010) Polyhedron 29:70
96. Schulze B, Friebe C, Hager MD, Winter A, Hoogenboom R, Goerls H, Schubert US (2009) Dalton Trans 787
97. Happ B, Friebe C, Winter A, Hager MD, Hoogenboom R, Schubert US (2009) Chem Asian J 4:154
98. Feldman AK, Colasson B, Fokin VV (2004) Org Lett 6:3897
99. Tao C-Z, Cui X, Li J, Liu A-X, Liu L, Guo Q-X (2007) Tetrahedron Lett 48:3525
100. Andersen J, Bolvig S, Liang X (2005) Synlett 2941
101. Kilpin KJ, Gavey EL, McAdam CJ, Anderson CB, Lind SJ, Keep CC, Gordon KC, Crowley JD (2011) Inorg Chem 50:6334
102. Aromi G, Barrios LA, Roubeau O, Gamez P (2011) Coord Chem Rev 255:485
103. Kitchen JA, Brooker S (2008) Coord Chem Rev 252:2072
104. Klingele MH, Brooker S (2003) Coord Chem Rev 241:119
105. Beckmann U, Brooker S (2003) Coord Chem Rev 245:17
106. Haasnoot JG (2000) Coord Chem Rev 200–202:131
107. Fanni S, Keyes TE, O'Connor CM, Hughes H, Wang R, Vos JG (2000) Coord Chem Rev 208:77
108. Bai S-Q, Young DJ, Hor TSA (2011) Chem Asian J 6:292

109. Suijkerbuijk BMJM, Aerts BNH, Dijkstra HP, Lutz M, Spek AL, van Koten G, Klein Gebbink RJM (2007) Dalton Trans 1273
110. Badeche S, Daran J-C, Ruiz J, Astruc D (2008) Inorg Chem 47:4903
111. Crowley JD, Gavey EL (2010) Dalton Trans 39:4035
112. Uppal BS, Booth RK, Ali N, Lockwood C, Rice CR, Elliott PIP (2011) Dalton Trans 40:7610
113. Maeda C, Yamaguchi S, Ikeda C, Shinokubo H, Osuka A (2008) Org Lett 10:549
114. Wang F, Zhang J, Ding X, Dong S, Liu M, Zheng B, Li S, Wu L, Yu Y, Gibson HW, Huang F (2010) Angew Chem Int Ed 49:1090
115. Aucagne V, Hänni KD, Leigh DA, Lusby PJ, Walker DB (2006) J Am Chem Soc 128:2186
116. Aucagne V, Berna J, Crowley JD, Goldup SM, Hänni KD, Leigh DA, Lusby PJ, Ronaldson VE, Slawin AMZ, Viterisi A, Walker DB (2007) J Am Chem Soc 129:11950
117. Barrell MJ, Leigh DA, Lusby PJ, Slawin AMZ (2008) Angew Chem Int Ed 47:8036
118. Mullen KM, Gunter MJ (2008) J Org Chem 73:3336
119. Collin J-P, Durola F, Heitz V, Reviriego F, Sauvage J-P, Trolez Y (2010) Angew Chem Int Ed 49:10172
120. Bastero A, Font D, Pericas MA (2007) J Org Chem 72:2460
121. Urankar D, Pinter B, Pevec A, De Proft F, Turel I, Kosmrlj J (2010) Inorg Chem 49:4820
122. Gower ML, Crowley JD (2010) Dalton Trans 39:2371
123. Nolte C, Mayer P, Straub BF (2007) Angew Chem Int Ed 46:2101
124. Partyka DV, Gao L, Teets TS, Updegraff JB, Deligonul N, Gray TG (2009) Organometallics 28:6171
125. Fletcher JT, Keeney ME, Walz SE (2010) Synthesis 3339
126. Hanelt S, Liebscher J (2008) Synlett 1058
127. Jeong Y, Ryu J-S (2010) J Org Chem 75:4183
128. Khan SS, Shah J, Liebscher J (2010) Tetrahedron 66:5082
129. Yacob Z, Shah J, Leistner J, Liebscher J (2008) Synlett 2342
130. Kumar A, Pandey PS (2008) Org Lett 10:165
131. Mullen KM, Mercurio J, Serpell CJ, Beer PD (2009) Angew Chem Int Ed 48:4781
132. Mathew P, Neels A, Albrecht M (2008) J Am Chem Soc 130:13534
133. Poulain A, Canseco-Gonzalez D, Hynes-Roche R, Muller-Bunz H, Schuster O, Stoeckli-Evans H, Neels A, Albrecht M (2011) Organometallics 30:1021
134. Nakamura T, Ogata K, Fukuzawa S-i (2010) Chem Lett 39:920
135. Prades A, Peris E, Albrecht M (2011) Organometallics 30:1162
136. Nakamura T, Terashima T, Ogata K, Fukuzawa S-i (2011) Org Lett 13:620
137. Hohloch S, Su C-Y, Sarkar B (2011) Eur J Inorg Chem 3067
138. Kilpin KJ, Paul USD, Lee A-L, Crowley JD (2011) Chem Commun 47:328
139. Bouffard J, Keitz BK, Tonner R, Guisado-Barrios G, Frenking G, Grubbs RH, Bertrand G (2011) Organometallics 30:2617
140. Karthikeyan T, Sankararaman S (2009) Tetrahedron Lett 50:5834
141. Crowley JD, Lee A-L, Kilpin KJ (2011) Aust J Chem 64:1118
142. Mindt TL, Struthers H, Brans L, Anguelov T, Schweinsberg C, Maes V, Tourwe D, Schibli R (2006) J Am Chem Soc 128:15096
143. Boulay A, Seridi A, Zedde C, Ladeira S, Picard C, Maron L, Benoist E (2010) Eur J Inorg Chem 5058
144. Seridi A, Wolff M, Boulay A, Saffon N, Coulais Y, Picard C, Machura B, Benoist E (2011) Inorg Chem Commun 14:238
145. Obata M, Kitamura A, Mori A, Kameyama C, Czaplewska JA, Tanaka R, Kinoshita I, Kusumoto T, Hashimoto H, Harada M, Mikata Y, Funabiki T, Yano S (2008) Dalton Trans 3292
146. Zhang C, Shen X, Sakai R, Gottschaldt M, Schubert US, Hirohara S, Tanihara M, Yano S, Obata M, Xiao N, Satoh T, Kakuchi T (2011) J Polym Sci A Polym Chem 49:746
147. Felici M, Contreras-Carballada P, Vida Y, Smits JMM, Nolte RJM, De Cola L, Williams RM, Feiters MC (2009) Chem Eur J 15:13124

148. Happ B, Escudero D, Hager MD, Friebe C, Winter A, Gorls H, Altuntas E, Gonzalez L, Schubert US (2010) J Org Chem 75:4025
149. Richardson C, Fitchett CM, Keene FR, Steel PJ (2008) Dalton Trans 2534
150. Felici M, Contreras-Carballada P, Smits JMM, Nolte RJM, Williams RM, De Cola L, Feiters MC (2010) Molecules 15:2039
151. Mydlak M, Bizzarri C, Hartmann D, Sarfert W, Schmid G, De Cola L (2010) Adv Funct Mater 20:1812
152. Zanarini S, Felici M, Valenti G, Marcaccio M, Prodi L, Bonacchi S, Contreras-Carballada P, Williams RM, Feiters MC, Nolte RJM, De Cola L, Paolucci F (2011) Chem Eur J 17:4640
153. Fleischel O, Wu N, Petitjean A (2010) Chem Commun 46:8454
154. Kilpin KJ, Crowley JD (2010) Polyhedron 29:3111
155. D'Amora A, Fanfoni L, Cozzula D, Guidolin N, Zangrando E, Felluga F, Gladiali S, Benedetti F, Milani B (2010) Organometallics 29:4472
156. Schweinfurth D, Pattacini R, Strobel S, Sarkar B (2009) Dalton Trans 9291
157. Schweinfurth D, Strobel S, Sarkar B (2011) Inorg Chim Acta 374:253
158. Stengel I, Mishra A, Pootrakulchote N, Moon S-J, Zakeeruddin SM, Graetzel M, Baeuerle P (2011) J Mater Chem 21:3726
159. Bernet L, Lalrempuia R, Ghattas W, Mueller-Bunz H, Vigara L, Llobet A, Albrecht M (2011) Chem Commun 47:8058
160. Lalrempuia R, McDaniel ND, Mueller-Bunz H, Bernhard S, Albrecht M (2010) Angew Chem Int Ed 49:9765
161. Fu Y, Liu Y, Fu X, Zou L, Li H, Li M, Chen X, Qin J (2010) Chin J Chem 28:2226
162. Fu Y, Liu Y, Zhong C, Li H, Chen X-G, Qin J-G (2010) Chin J Inorg Chem 26:1133
163. Bratsos I, Urankar D, Zangrando E, Genova-Kalou P, Kosmrlj J, Alessio E, Turel I (2011) Dalton Trans 40:5188
164. Urankar D, Pevec A, Turel I, Kosmrlj J (2010) Cryst Growth Des 10:4920
165. Urankar D, Pevec A, Kosmrlj J (2011) Eur J Inorg Chem 1921
166. Pinter B, Demšar A, Urankar D, De Proft F, Košmrlj J (2011) Polyhedron 30:2368
167. Amadio E, Bertoldini M, Scrivanti A, Chessa G, Beghetto V, Matteoli U, Bertani R, Dolmella A (2011) Inorg Chim Acta 370:388
168. Maisonial A, Serafin P, Traikia M, Debiton E, Thery V, Aitken DJ, Lemoine P, Viossat B, Gautier A (2008) Eur J Inorg Chem 298
169. Manbeck GF, Brennessel WW, Eisenberg R (2011) Inorg Chem 50:3431
170. Cambeiro XC, Pericàs MA (2011) Adv Synth Catal 353:113
171. Beyer B, Ulbricht C, Escudero D, Friebe C, Winter A, Gonzalez L, Schubert US (2009) Organometallics 28:5478
172. Fernandez-Hernandez JsM, Yang C-H, Beltrán JI, Lemaur V, Polo F, Fröhlich R, Cornil J, De Cola L (2011) J Am Chem Soc 133:10543
173. Boutadla Y, Davies DL, Jones RC, Singh K (2011) Chem Eur J 17:3438
174. Saravanakumar R, Ramkumar V, Sankararaman S (2011) Organometallics 30:1689
175. Warsink S, Drost RM, Lutz M, Spek AL, Elsevier CJ (2010) Organometallics 29:3109
176. Mattiuzzi A, Jabin I, Moucheron C, Kirsch-De Mesmaeker A (2011) Dalton Trans 40:7395
177. Monkowius U, Ritter S, König B, Zabel M, Yersin H (2007) Eur J Inorg Chem 4597
178. Manbeck GF, Brennessel WW, Evans CM, Eisenberg R (2010) Inorg Chem (Washington, DC, U S) 49:2834
179. Scott SØ, Gavey EL, Lind SJ, Gordon KC, Crowley JD (2011) Dalton Trans 40. 12117
180. Brans L, Garcia-Garayoa E, Schweinsberg C, Maes V, Struthers H, Schibli R, Tourwe D (2010) ChemMedChem 5:1717
181. Mindt TL, Mueller C, Melis M, de Jong M, Schibli R (2008) Bioconjugate Chem 19:1689
182. Mindt TL, Muller C, Stuker F, Salazar J-F, Hohn A, Mueggler T, Rudin M, Schibli R (2009) Bioconjugate Chem 20:1940

183. Mindt TL, Schweinsberg C, Brans L, Hagenbach A, Abram U, Tourwe D, Garcia-Garayoa E, Schibli R (2009) ChemMedChem 4:529
184. Ross TL, Honer M, Lam PYH, Mindt TL, Groehn V, Schibli R, Schubiger PA, Ametamey SM (2008) Bioconjugate Chem 19:2462
185. Struthers H, Spingler B, Mindt TL, Schibli R (2008) Chem Eur J 14:6173
186. Struthers H, Viertl D, Kosinski M, Spingler B, Buchegger F, Schibli R (2010) Bioconjugate Chem 21:622
187. Mindt TL, Struthers H, Spingler B, Brans L, Tourwe D, Garcia-Garayoa E, Schibli R (2010) ChemMedChem 5:2026
188. Dhyani MV, Satpati D, Korde A, Sarma HD, Kumar C, Banerjee S (2010) Nucl Med Biol 37:997
189. Kim E-M, Joung M-H, Lee C-M, Jeong H-J, Lim ST, Sohn M-H, Kim DW (2010) Bioorg Med Chem Lett 20:4240
190. Tinnis F, Adolfsson H (2010) Org Biomol Chem 8:4536
191. Gu S, Xu H, Zhang N, Chen W (2010) Chem Asian J 5:1677
192. Schuster EM, Botoshansky M, Gandelman M (2011) Dalton Trans 40:8764
193. Chandrasekhar N, Chandrasekar R (2010) Dalton Trans 39:9872
194. Li Y, Huffman JC, Flood AH (2007) Chem Commun 2692
195. Meudtner RM, Ostermeier M, Goddard R, Limberg C, Hecht S (2007) Chem Eur J 13:9834
196. Ostermeier M, Berlin M-A, Meudtner RM, Demeshko S, Meyer F, Limberg C, Hecht S (2010) Chem Eur J 16:10202
197. Danielraj P, Varghese B, Sankararaman S (2010) Acta Crystallogr Sect C Cryst Struct Commun C66:m366
198. Schulze B, Escudero D, Friebe C, Siebert R, Goerls H, Koehn U, Altuntas E, Baumgaertel A, Hager MD, Winter A, Dietzek B, Popp J, Gonzalez L, Schubert US (2011) Chem Eur J 17:5494
199. Yang W-W, Wang L, Zhong Y-W, Yao J (2011) Organometallics 30:2236
200. Munuera L, O'Reilly RK (2010) Dalton Trans 39:388
201. Brotherton WS, Guha PM, Phan H, Clark RJ, Shatruk M, Zhu L (2011) Dalton Trans 40:3655
202. Özçubukçu S, Ozkal E, Jimeno C, Pericàs MA (2009) Org Lett 11:4680
203. Huang S, Clark RJ, Zhu L (2007) Org Lett 9:4999
204. Michaels HA, Murphy CS, Clark RJ, Davidson MW, Zhu L (2010) Inorg Chem 49:4278
205. Maity D, Govindaraju T (2010) Inorg Chem 49:7229
206. Maity D, Govindaraju T (2010) Chem Commun 46:4499
207. Jobe K, Brennan CH, Motevalli M, Goldup SM, Watkinson M (2011) Chem Commun 47:6036
208. Tamanini E, Flavin K, Motevalli M, Piperno S, Gheber LA, Todd MH, Watkinson M (2010) Inorg Chem 49:3789
209. Tamanini E, Katewa A, Sedger LM, Todd MH, Watkinson M (2008) Inorg Chem 48:319
210. Lebedev AY, Holland JP, Lewis JS (2010) Chem Commun 46:1706
211. Joly J-P, Beley M, Selmeczi K, Wenger E (2009) Inorg Chem Commun 12:382
212. Hao E, Wang Z, Jiao L, Wang S (2010) Dalton Trans 39:2660
213. Li L, Gomes CSB, Gomes PT, Duarte MT, Fan Z (2011) Dalton Trans 40:3365
214. Donnelly PS, Zanatta SD, Zammit SC, White JM, Williams SJ (2008) Chem Commun 2459
215. Geng J, Lindqvist J, Mantovani G, Chen G, Sayers CT, Clarkson GJ, Haddleton DM (2007) QSAR Comb Sci 26:1220
216. Schweinfurth D, Weisser F, Bubrin D, Bogani L, Sarkar B (2011) Inorg Chem 50:6114
217. Jevric M, Zheng T, Meher NK, Fettinger JC, Mascal M (2011) Angew Chem Int Ed 50:717
218. Colasson B, Le Poul N, Le Mest Y, Reinaud O (2010) J Am Chem Soc 132:4393
219. Colasson B, Save M, Milko P, Roithová J, Schröder D, Reinaud O (2007) Org Lett 9:4987
220. Wang L, Yang W-W, Zheng R-H, Shi Q, Zhong Y-W, Yao J (2011) Inorg Chem 50:7074
221. Happ B, Pavlov GM, Altuntas E, Friebe C, Hager MD, Winter A, Goerls H, Guenther W, Schubert US (2011) Chem Asian J 6:873

222. Souchon V, Maisonneuve S, David O, Leray I, Xie J, Valeur B (2008) Photochem Photobiol Sci 7:1323
223. Garcia L, Maisonneuve S, Xie J, Guillot R, Dorlet P, Riviere E, Desmadril M, Lambert F, Policar C (2010) Inorg Chem 49:7282
224. Sykes D, Ward MD (2011) Chem Commun 47:2279
225. Najar AM, Tidmarsh IS, Ward MD (2010) CrystEngComm 12:3642
226. Juricek M, Felici M, Contreras-Carballada P, Lauko J, Bou SR, Kouwer PHJ, Brouwer AM, Rowan AE (2011) J Mater Chem 21:2104
227. Bai S-Q, Leelasubcharoen S-Y, Chen X, Koh L-L, Zuo J-L, Hor TSA (2010) Cryst Growth Des 10:1715
228. Bai S-Q, Kwang JY, Koh LL, Young DJ, Hor TSA (2010) Dalton Trans 39:2631
229. Ohi H, Shimizu M, Obata M, Funabiki T, Yano S (2008) Acta Crystallogr, Sect E Struct Rep Online E 64:m1256
230. Romero T, Orenes RA, Espinosa A, Tárraga A, Molina P (2011) Inorg Chem 50:8214
231. Kuang G-C, Guha PM, Brotherton WS, Simmons JT, Stankee LA, Nguyen BT, Clark RJ, Zhu L (2011) J Am Chem Soc 133:13984
232. Liu S, Muller P, Takase MK, Swager TM (2011) Inorg Chem 50:7598
233. Müller C, Schibli R (2011) J Nucl Med 52:1
234. Dhyani MV, Satpati D, Korde A, Banerjee S (2011) Cancer Biother Radiopharm 26:539
235. Michaels HA, Zhu L (2011) Chem Asian J 6:2825
236. Chopin N, Decamps S, Gouger A, Médebielle M, Picot S, Bienvenu A-L, Pilet G (2011) J Fluorine Chem 132:850
237. Wang W, Hong S, Tran A, Jiang H, Triano R, Liu Y, Chen X, Wu P (2011) Chem Asian J 6:2796
238. Cai J, Yang X, Arumugam K, Bielawski CW, Sessler JL (2011) Organometallics 30:5033
239. Happ B, Schäfer J, Menzel R, Hager MD, Winter A, Popp Jr, Beckert R, Dietzek B, Schubert US (2011) Macromolecules 44:6277

Top Heterocycl Chem (2012) 28: 85–108
DOI: 10.1007/7081_2011_70

Published online: 8 February 2012

Binding Anions in Rigid and Reconfigurable Triazole Receptors

Semin Lee and Amar H. Flood

Abstract The use of triazole CH•••anion hydrogen bonds, strong and easy to install, has expanded dramatically since 2008. Various aryl-triazole derivatives have been synthesized and investigated to obtain fundamental understandings of anion stabilization as well as to develop new receptors for applications. Receptors have now been created to make use of triazole, triazolium, or iodotriazolium. The triazole CH•••anion binding motif has also been utilized in sophisticated structures such as interlocked molecules and toward fluorescent sensors and ion-selective electrodes. Furthermore, demonstrations on the transport of anions across membranes and studies of light-induced anion regulation have broadened the scope of application for this new anion binding motif. This chapter will focus on these recent developments.

Keywords Anion receptor · CH hydrogen bond · Click chemistry · Foldamers · Preorganization

Contents

S. Lee and A.H. Flood (✉)
Department of Chemistry, Indiana University, Bloomington, IN 47405, USA
e-mail: aflood@indiana.edu

Abbreviations

Bn	Benzyl
CH_2Cl_2	Dichloromethane
$CHCl_3$	Chloroform
CuAAC	Copper(I)-catalyzed azide-alkyne cycloaddition
ESP	Electrostatic potential
ISE	Ion-selective electrode
K_a	Association constant
NMR	Nuclear magnetic resonance
POPC	1-Palmitoyl-2-oleoyl-*sn*-glycero-3-phosphocholine
TBA^+	Tetrabutylammonium cation

1 Introduction

Anion supramolecular chemistry [1] has traditionally created anion receptors with NH (e.g., amide, pyrrole, and urea) and OH hydrogen bond donors or positively charged molecular components [2]. These representative hydrogen bond donors have polarized X–H bonds to generate an electropositive hydrogen for mediating the XH•••anion hydrogen bond. By contrast, CH groups are not very polar and were correspondingly described as weak hydrogen bond donors [3]. Yet this classification is no longer strictly valid. The fact that these hydrogen bonds have enjoyed increasing deployment using click chemistry [4, 5] has allowed practitioners to explore a range of themes. A common thread running through these studies is the shape matching and shape changing that attends anion receptor binding, offering insight and opportunities alike.

Contributing to the growing recognition of CH•••anion hydrogen bonds were early examples involving CH hydrogens that exhibited downfield shifts in their NMR peak positions upon anion titration [6–8]; hallmarks of hydrogen bonding. These and other observations suggested that some CH groups do have supportive interaction with anions yet they do not play determining roles. Overturning this typical position, a crystal structure of a fluorinated macrocycle was found to have a F^- ion bound inside that was held in place solely by CH•••F^- interactions [9]. Nevertheless, CH donors remained as weak "second-class" interactions that were thought to stabilize anions only in the gaseous and solid states. Perhaps, this situation arose historically as described by Desiraju and Steiner in a section of their book on "*The Weak Hydrogen Bond*" [3] that describes early work by Sutor [10] in the 1960s.

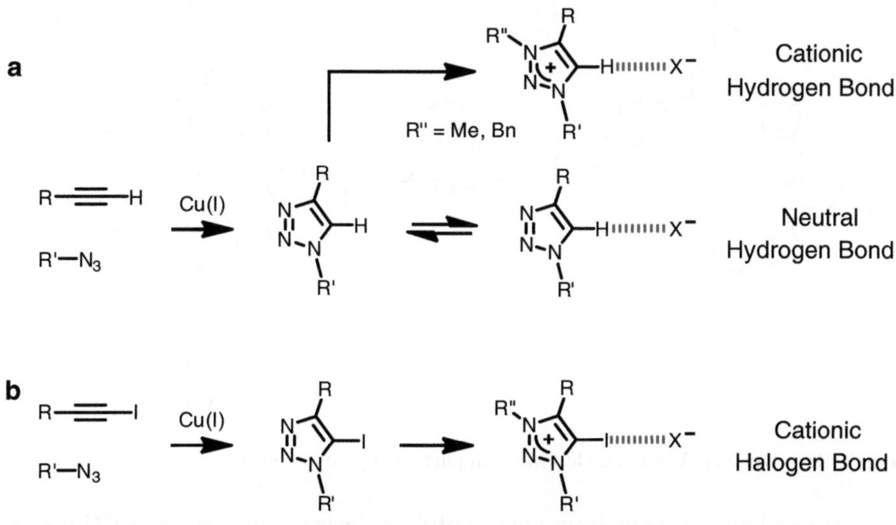

Fig. 1 (**a**) Synthesis of 1,2,3-triazole by copper(I)-catalyzed azide–alkyne cycloaddition (CuAAC) and the CH•••X^- hydrogen bonds that form with neutral triazole and positively charged triazolium moieties. (**b**) Synthesis of iodotriazolium with CI•••X^- halogen bonds

Consequently, studies of this hydrogen bond in solution languished. Fortunately, Benjamin Hay's theoretical studies on aromatic CH•••anion interactions [11], beginning in 2005, reinvigorated interest. Hay showed that C–H bonds could be extrinsically polarized by electron-withdrawing substituents such that they could approach the strength of NH donors. With theoretical evidence at hand, several groups synthesized receptors in 2008 with polarized CH groups to deliberately investigate their ability to bind anions [12–14]. For instance, Johnson and Hay teamed up to test the prediction that dinitro-substituted benzenes within tripodal receptors could bind halides with polarized aromatic CH groups [12].

With the importance of extrinsic polarization in mind, the ability of 1,2,3-triazoles to bind anions with a CH hydrogen bond (Fig. 1) seems logical with hindsight. Triazoles have three electronegative nitrogen atoms connected together within a five-membered ring. This arrangement gives rise to a large dipole moment (~5 D) [15] and an electron-withdrawing framework that polarizes the C^5–H bond to generate an electropositive hydrogen. This outcome meets the requirement for hydrogen bonding.

In 2008, three groups independently observed strong 1,2,3-triazole CH•••anion interactions within aryl-triazole-based receptors. Li and Flood synthesized a shape-persistent macrocyclic receptor, triazolophane [16, 17] (Fig. 2), with an unexpectedly high affinity for chloride that rivaled many synthetic receptors with NH and OH hydrogen bond donors. At the time, the binding constant was evaluated at $K_a = 130{,}000\ M^{-1}$ (CH_2Cl_2), but it has since been reevaluated: first to $11{,}000{,}000\ M^{-1}$ after Hiroses's rule [18] and second to $5{,}500{,}000\ M^{-1}$ by taking 2:1 sandwiches [19], ion-pair formation, and ion-pair competition [20] into account (*vide infra*). Experimental [16, 17, 19, 20] and theoretical [21] studies revealed that

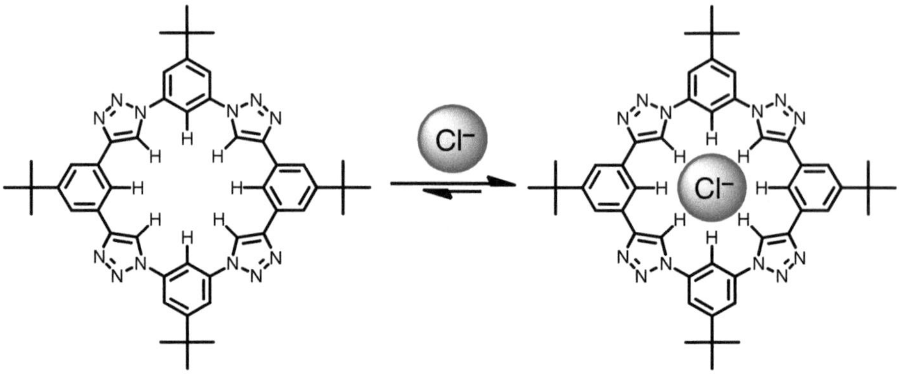

Fig. 2 Triazolophane **1** binding chloride with pure CH hydrogen bonds

such strong binding stems from the use of four electropositive triazole CH donors and four phenylene CH donors generating eight donors that are rigidly preorganized toward the center of the binding pocket. After Hay [11], the role of the phenylene CH group was implicated using electronic substituents [17]. Craig and coworkers synthesized a foldameric receptor **7** and observed anion-directed folding [22]. Hecht and coworkers synthesized an aryl-triazole foldamer **8** with chiral side chains and observed that the helix inverted its handedness in response to the addition of anions [23]. While the shape of Flood's macrocycles [16, 17, 19, 20] changed only a little, Craig's [22] and Hecht's [23] receptors were completely reconfigured from oligomers to folded helices (Fig. 3).

Owing to the popularity and vast modularity of the copper(I)-catalyzed azide-alkyne cycloaddition (CuAAC) [4, 5], so-called click chemistry [24] (Fig. 1), many more 1,2,3-triazole-based anion receptors have been reported during the past 3 years (2008–2011). Reviews [25, 26] have been published to cover this new moiety in anion receptor chemistry. Therefore, this chapter will focus on triazole-based anion receptors that have not been reviewed to date. In addition, applications including sensors, ion-selective electrodes, catalysis, anion transport, and anion regulation, as well as their use in interlocked molecules, will be discussed.

2 Anion Receptors 2009–2011

2.1 Macrocyclic Receptors

Macrocycles are a traditional favorite of supramolecular chemistry and molecular recognition. They offer size selectivity and added strength to binding affinities by the action of preorganization, captured most succinctly by the macrocyclic

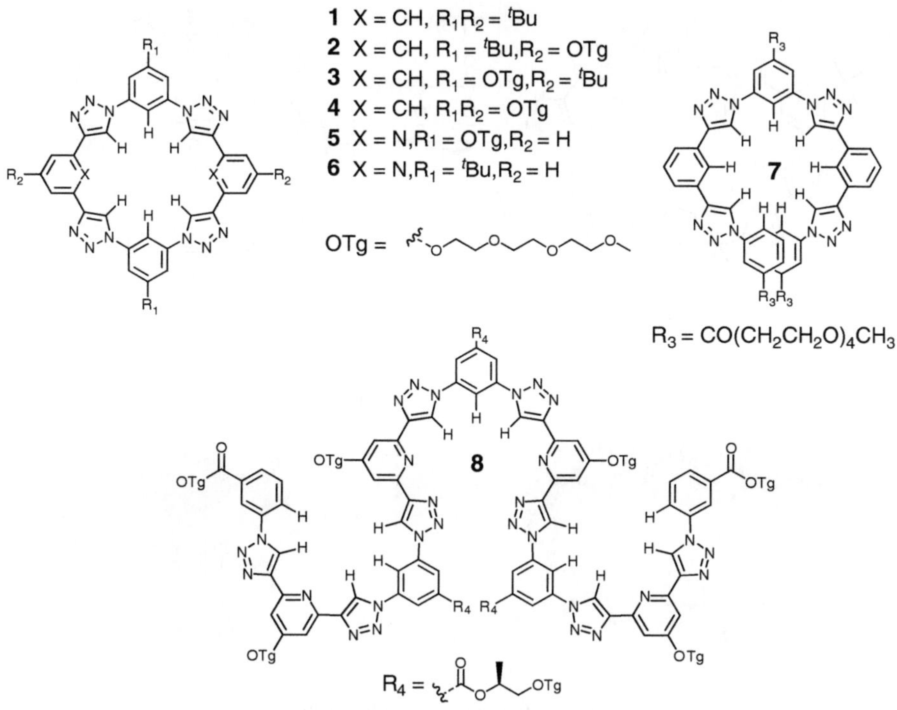

Fig. 3 Seminal aryl-triazole-based anion receptors from 2008

effect [27]. As elaborated later, triazolophanes are unique in anion-triazole chemistry for their high degree of preorganization. By contrast, we report on systems that, while enjoying the benefits of a macrocycle, display some degree of conformational reorganization upon anion binding.

Sessler's group synthesized a macrocyclic receptor **9** (Fig. 4a) that contains four pyrroles and four triazoles [28]. The receptor was found to bind pyrophosphate ($HP_2O_7^{3-}$) with high affinity, $K_a = 2{,}300{,}000\ M^{-1}$, in $CHCl_3$. The X-ray crystal structure of the free receptor (Fig. 4b) shows that two of the pyrroles are facing outside the cavity while the remaining donors are directed inward. Upon pyrophosphate binding, all pyrrole and triazole donors are directed toward the central cavity (Fig. 4c). The authors believed that the pyrroles contribute more to the anion binding than the triazoles based on NMR titration data and the shorter NH•••O distances.

Zhu and coworkers synthesized flexible macrocycles containing indolocarbazoles and triazoles [29] (Fig. 5). On account of the large π-surface of the indolocarbazole moiety and the flexible ethylene glycol–based backbone, both macrocycles **10** and **11** form intramolecular π–π stacked structures as confirmed by NMR ($CDCl_3$) and X-ray structure studies. Upon chloride binding, the UV-vis absorption spectra (CH_2Cl_2) show blue shifts, suggesting that the π–π stacking becomes disrupted and the receptor refolds when the anions are bound.

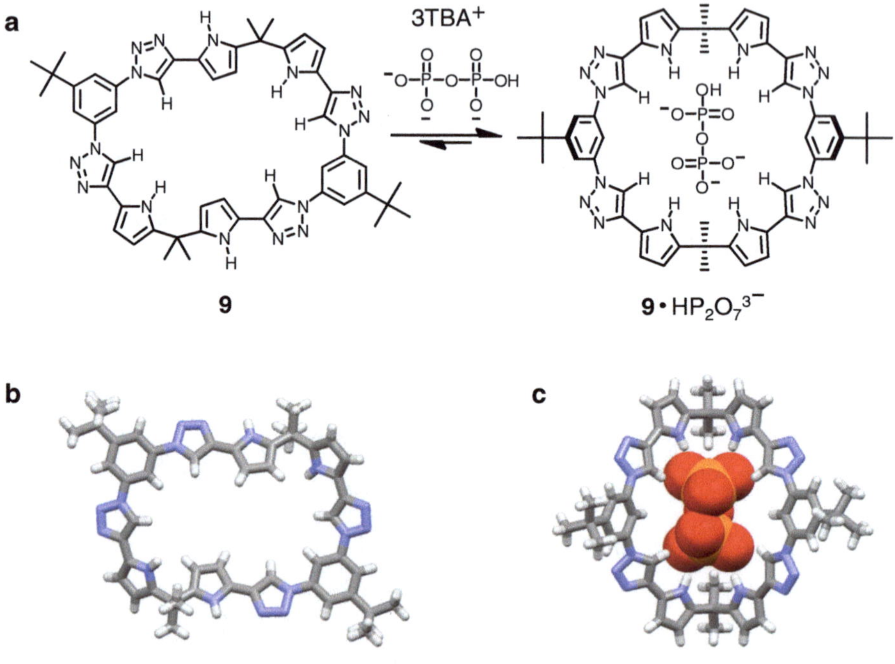

Fig. 4 A pyrrole-based triazolophane, **9**, binding pyrophosphate

Flood and coworkers have prepared a propylene-based triazolophane **12** [20] (Fig. 6), in order to compare the anion binding strength of an aliphatic CH group with the phenylene one in triazolophane **1**. In order to make a faithful comparison between the two receptors, the 1:1 binding constants need to be evaluated. To achieve this end, various equilibria (Fig. 7) were identified to exist in the solution, and they were employed to determine the concentration-independent-binding constants for all equilibria.

Such a detailed account of the processes occurring in solution has never before been described in anion recognition, and so they deserve a restatement here. The titration starts with an empty macrocycle in CH_2Cl_2. When tetrabutylammonium chloride (TBACl) is added as a chloride source, the TBA^+ cation competes with the macrocycle for chloride binding. While ion-pair equilibrium has been acknowledged [30], this is one of the first times it has been included quantitatively in the data analysis. During the addition of up to 0.5 equivalents of the TBACl salt to the solution, a 2:1 sandwich complex is formed, almost exclusively at NMR concentrations (~1 mM). Beyond 0.5 equivalents, the 1:1 complex is progressively formed. Unexpectedly, the authors also observed that the anionic 1:1 complex **1**•Cl^- has sufficient charge density to pair with the TBA^+ cation to generate the 1:1:1 ion-pair complex, **1**•Cl^-•TBA^+. Quantifying the role of such ion-pair complexes are gaining interest in the community for fundamental and applied research as well as for the challenge of designing effective receptors.

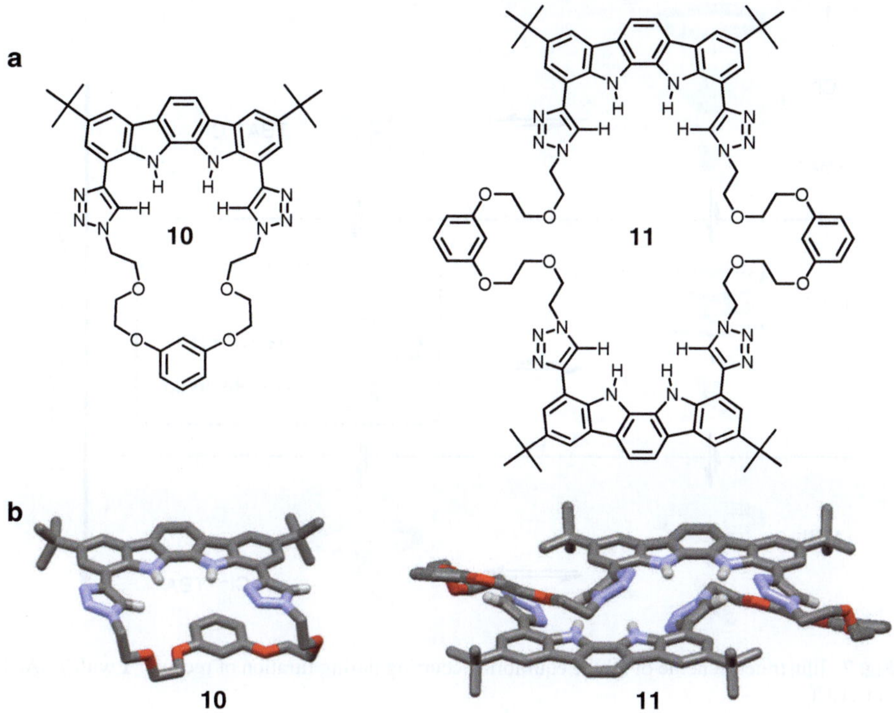

Fig. 5 (**a**) Triazole-derived indolocarbazole macrocycles **10** and **11** and their (**b**) X-ray crystal structures

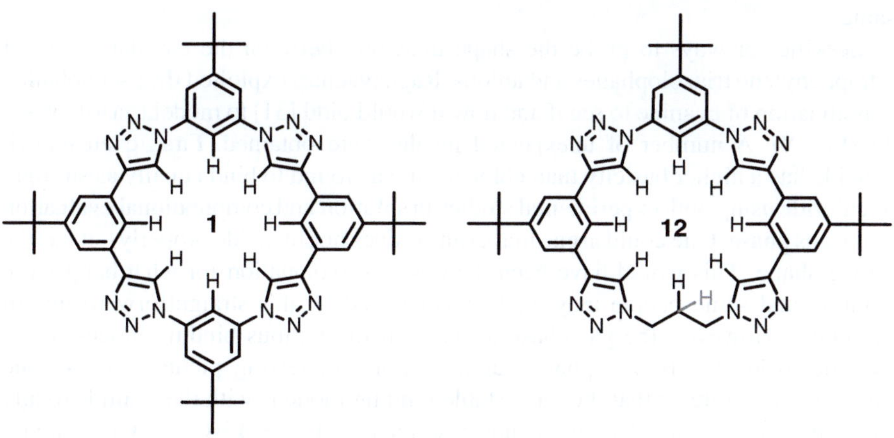

Fig. 6 Triazolophane **1** and propylene-based triazolophane **12**

Careful analysis of the 1:1 chloride binding constants of **1** and **12** showed that they have the same value to within error. The authors employed ^{1}H NMR peak shifts and computed structures to verify that the aliphatic CH donor is weaker than

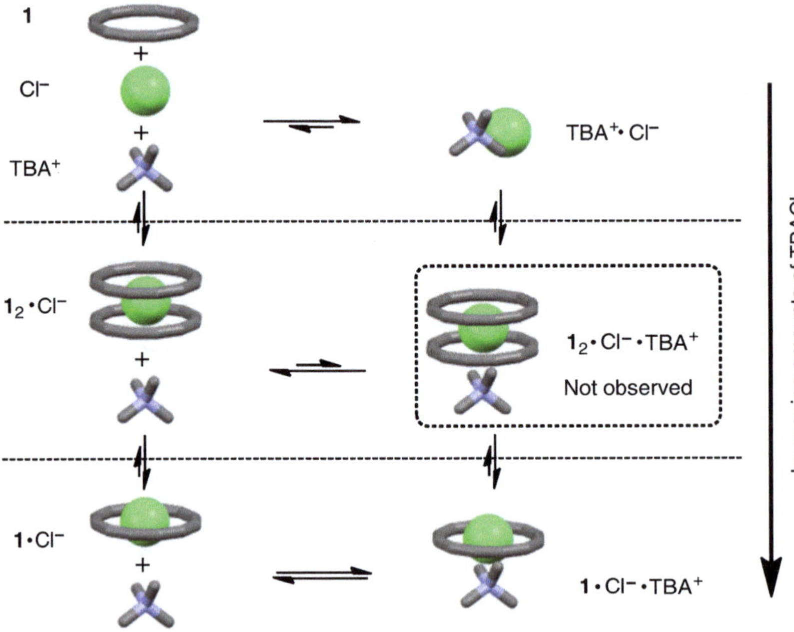

Fig. 7 Illustrated scheme of all the equilibria occurring during titration of receptor **1** with TBACl in CH_2Cl_2

the one from the phenylene. Nevertheless, this factor is compensated inside the receptor on entropic grounds to make the binding free energies of **1** and **12** the same.

Seeking for ways to probe the shape matching between the circular cavity of tetraphenylene triazolophanes and anions, Raghavachari exploited the pseudohalide classification of cyanide to see if and how it would bind [31] to model triazolophane **13** (Fig. 8). A number of unexpected results were obtained. First, even though cyanide has a higher basicity than chloride, it was found to bind equally as strongly as chloride using both experimental studies in solution and computational evaluation in the gas phase. One could argue that cyanide does not fit inside properly – it has the wrong shape. This would have been the obvious explanation for what happens in solution and some people may well be persuaded by this straightforward line of reasoning. However, the gas-phase analysis of the various binding modes of the cyanide inside the triazolophane paints a more interesting picture. The second outcome, therefore, is that the most stable binding mode is with the cyanide totally inside the cavity pointed along the north–south axis (Fig. 8, **13•**CN^-). Other modes (east–west, diagonal, and perpendicular) were all within energetic reach either as slightly higher minima or as nonstationary points. Commensurate with cyanide's compact structure and its propensity to display averaged properties [32], one might even expect all these binding modes to be represented in solution earning cyanide's moniker as a pseudohalide. This brings the discussion back to the equal binding

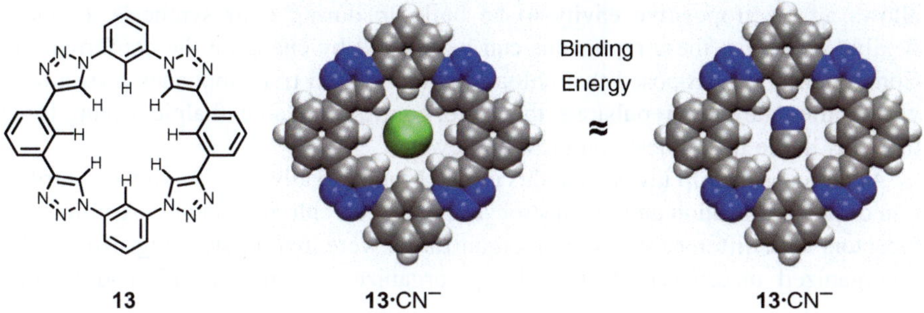

Fig. 8 Calculated binding modes of triazolophane **13** with Cl^- and CN^-

strengths between chloride and cyanide with less weight now placed on the shape mismatch.

A third unexpected observation arose from the computed distribution of charges on the bare and complexed cyanide ions. It is accepted that carbon has a slightly greater share of the negative charge, 55%, with the other 45% on the nitrogen end. It was a surprise to see, therefore, that inside the cavity, the carbon donated a lot of its electron density to the macrocycle whereas the nitrogen held onto most of its charge. The outcome of this charge redistribution was shorter hydrogen bonds to the nitrogen end of the cyanide, presumably arising from its now larger negative charge. Whereas the carbon end presumably enjoys a larger charge-transfer component [33] to its noncovalent interaction consistent with its greater basicity. All of these interesting features are under further investigation.

2.2 *Lessons on Anion Binding Using Macrocyclic and Acyclic Receptors*

On account of their recent discovery, the factors that determine the triazole's ability to stabilize anions are still being elucidated. Craig and coworkers studied compounds **7** and **14** (Fig. 9) to investigate various fundamental aspects of aryl-triazole receptors [34]. Anions with different sizes ($X^- = Cl^-, Br^-, I^-$) and shapes ($X^-, PhCO_2^-, HSO_4^-, NO_3^-, PF_6^-$) were tested and show that benzoate binds to **14** with relatively high affinity compared to the other anions. This observation was rationalized from the shape complementarity between the triatomic –COO^- functionality of the anionic carboxylate and receptor **14**. Considering the effect of solvation, anion titrations were conducted in solvents with a range of Gutmann acceptor numbers, which reflect a solvent's ability to accept lone pairs. The results show weakened anion binding when using solvents with higher acceptor numbers on account of the fact that solvation of the anion becomes more competitive.

During Craig's appraisal of the factors that impact the stabilization of anions in aryl-triazole receptors, he suggested that the rigid structure of triazolophanes (**1–6**) allows an electropositive cavity to be built up during their synthesis. In more flexible receptors, these repulsions can be relaxed by changing the conformation. Consequently, he proposed that anion binding to rigid triazolophanes goes a long way to alleviate these repulsions, therefore, contributing an enthalpic driving force that aids in the host–guest complexation event.

As a means to help investigate this effect, Flood broadly considered the contribution of preorganization and the macrocyclic effect on chloride binding. Aryl-triazole receptors with different levels of preorganization were investigated (Fig. 10): rigidly preorganized macrocycle **1**, partially preorganized macrocycle **15**, and flexible

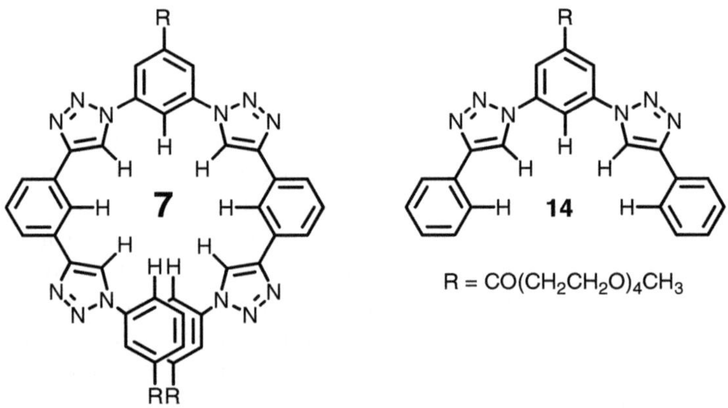

Fig. 9 Oligomeric anion receptors **7** and **14**

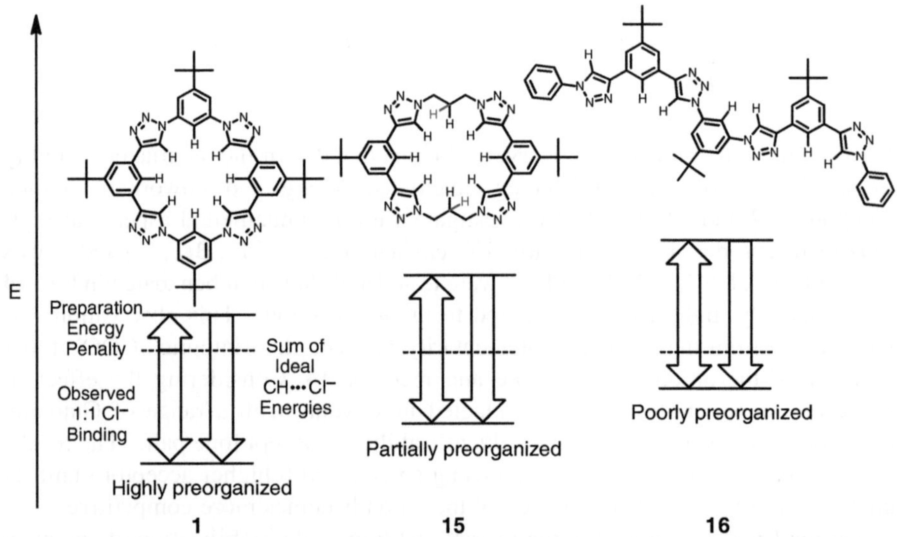

Fig. 10 Aryl-triazole anion receptors with various levels of preorganization

oligomer **16** [35]. The analysis relies upon an estimate of the "preparation energies," which are defined as the free energy cost of generating the ideal geometry for the complex from the initially empty receptor. These energies were estimated using a combination of insight from computations and the experimental solution-phase data. For the rigid tetraphenylene receptors like **1**, the preparation energy that must be paid upon chloride binding is about 10 kJ mol^{-1}. Based on modest entropy and solvation changes, this penalty has been attributed to large changes in the electrostatic potentials (ESP) observed between the two geometries of 327 and 287 kJ mol^{-1} for the ideal and empty triazolophanes, respectively. Using the same analysis for macrocycle **15**, which starts with an ESP of 253 kJ mol^{-1} and ends in the ideal geometry with an ESP of 296 kJ mol^{-1}, gives a preparation cost of 21 kJ mol^{-1}. Therefore, the smaller preparation energy for rigid triazolophane **1** is attributed to its greater rigidity. That is, this rigidity prevents the electropositive hydrogens from moving away from each other to relax the ESP. For oligomer **16**, the analysis shows a larger 41 kJ mol^{-1} preparation energy arising again from ESP factors (236 kJ mol^{-1} and 302 kJ mol^{-1} for the unfolded and folded oligomer, respectively) but also entropy and solvation. In this "disorganized" case, the electropositive hydrogens are fully divergent, and even greater costs are required to bring them together to converge upon the chloride.

2.3 Acyclic Receptors

Acyclic receptors are usually easier to make but may not always provide as great a benefit from preorganization as macrocycles. Nevertheless, they offer a richer set of conformational effects that have various applications. Exploring further the interplay between and binding, Lee and Flood prepared an aryl-triazole oligomer that utilizes OH•••N intramolecular hydrogen bonds to organize the receptor into a crescent shape that is in the correct conformation for binding a chloride [36]. As a result, the preorganized receptor **17** (Fig. 11a) had a 47-fold increase in chloride binding constant (CH_2Cl_2) when compared to the nonpreorganized receptor **18** (Fig. 11b). The crystal structure of a related derivative, **19**, confirmed the existence of the intramolecular hydrogen bond and the crescent shape. This observation emphasizes the fact that triazoles can act as hydrogen bond acceptors as well as donors.

Haridas and coworkers have synthesized a variety of receptors based on isophthalamides [37] to look at the effect of different hydrogen bonding units. Receptors **20**, **21**, and **22** (Fig. 12) showed noticeable anion binding properties upon titrations with TBA^+ salts of various halides at NMR concentrations ($CDCl_3$). Compared to receptor **20** without triazoles, both **21** and **22** had two- to threefold increases in the Cl^- binding constants. However, related receptors bearing ester groups in place of the amides or two flanking triazole units without the amides did not bind anions. Therefore, the triazole CH donors play a supporting role for anion binding in this series of oligomeric anion receptors.

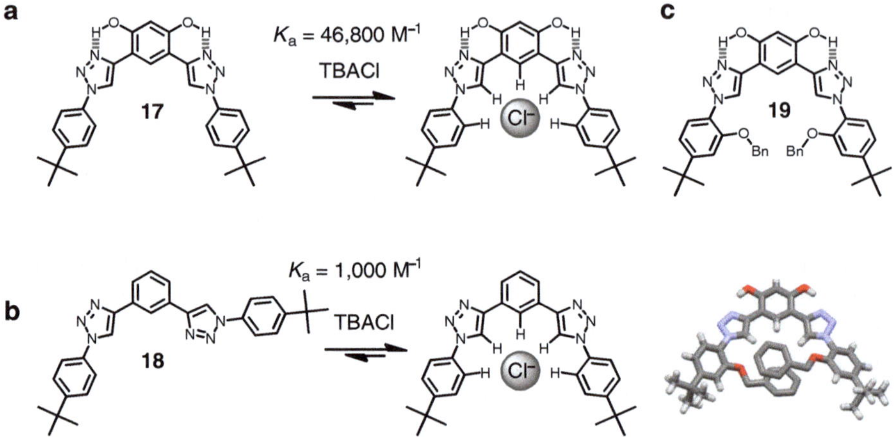

Fig. 11 (**a**) Chloride binding is stronger with receptor (**a**) **17** than (**b**) **18**. (**c**) X-ray crystal structure of **19** (Bn = benzyl)

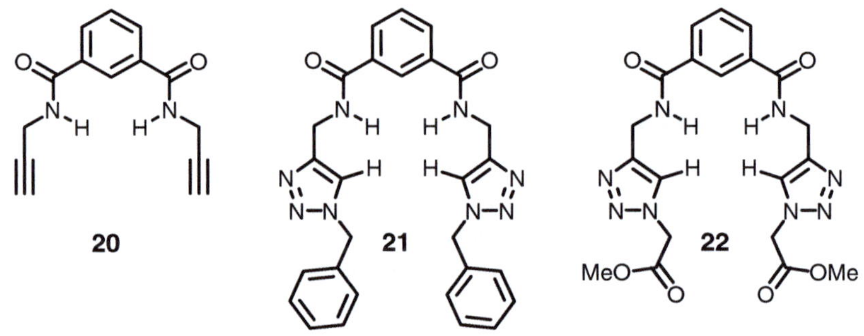

Fig. 12 Isophthalamide anion receptors with and without triazoles

Jiang and coworkers synthesized aryl-triazole-amide oligomers **23**, **24**, and **25** (Fig. 13) and investigated their chloride binding properties [38]. It was observed that binding chloride induced the folding of the oligomers. Oligomer **23** with six rings defining six CH groups and an amide contributing one NH donor creates a binding cleft (Fig. 13a) favoring 1:1 binding with Cl^-. However, the elongated oligomers **24** and **25** with 12 and 18 aromatic rings, respectively, were able to fold into a helix with overlapping π-systems. Once folded, they could accommodate two chlorides by changing into either a loose helix (as shown in Fig. 13b) or an S-shaped conformation. On account of the fact that the longer oligomer **25** has more hydrogen bond donors than **24**, it had a greater tendency to bind two chlorides.

Sánchez and coworkers synthesized amphiphilic aryl-triazole oligomers **26** and **27** (Fig. 14a, c) and observed their resulting morphology upon aggregation in acetonitrile (Fig. 14b, d) [39]. Scanning electron microscope images of the

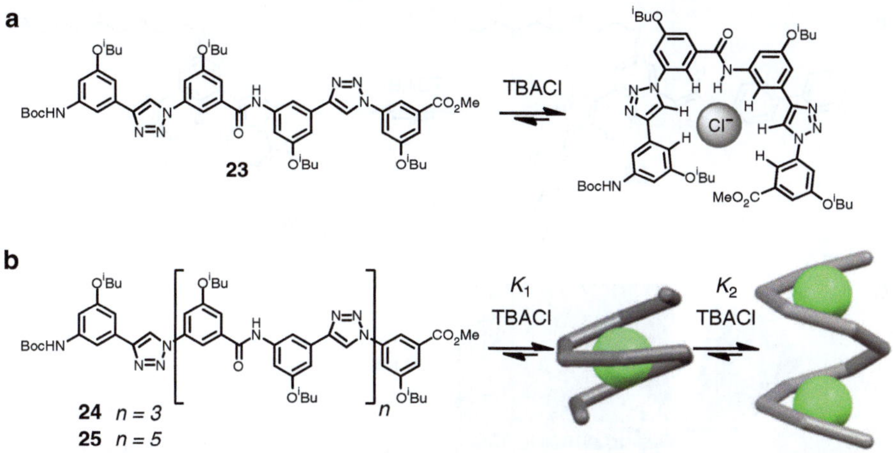

Fig. 13 (**a**) Scheme of an aryl-triazole oligomer (**23**) binding chloride. (**b**) Foldamers **24** and **25** and cartoon representations of their conformations upon binding up to two chlorides

receptors show that **26** aggregates into flat lamellae structures while **27** generates spherical structures. The authors consider that these aggregate morphologies stem from the greater planarity of **26** relative to **27**, which has out-of-plane naphthyl groups. In an anion-responsive manner, both molecules lose their aggregated structures upon binding bromide.

Schubert and coworkers synthesized oligomeric receptors containing triazolium units [40]. Receptor **28** with two triazolium units is shown to have a zigzag conformation where one triazolium CH donor is facing out of the empty anion binding pocket (Fig. 15a). On the other hand, receptor **29** with one triazolium and one triazole unit is believed to direct all CH donors into the binding site. A crystal structure of an iodide ion in the 1:1 complex (Fig. 15b) certainly shows this conformation exists in the complexed structure. Both receptors had strong affinities with sulfate (**28** in CD_3CN/CD_3OD 4:1, **29** in CD_3CN). Furthermore, **29** has a strong tendency to form a 2:1 complex with sulfate ions. Such an anion binding motif shows promises for use as a template for making interlocked molecules.

3 Applications

As a testament to the growing interest in anion receptor chemistry [2] and the novel chemical profile of triazoles, not to mention their simple preparation, many receptors are being developed to test various applications.

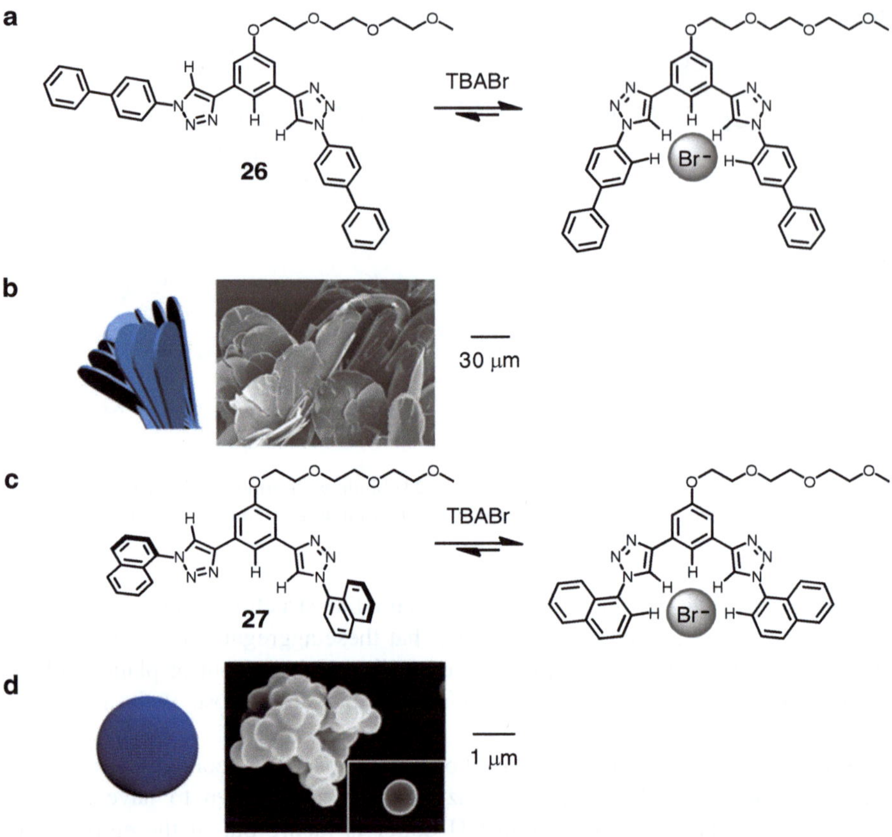

Fig. 14 Amphiphilic oligomers (**a**) **26** and (**c**) **27** can aggregate into different morphologies, (**b**) and (**d**), or bind anions in a nonaggregated manner (Reprinted with permission from Chem Commun 47:5016. Copyright 2011 Royal Society of Chemistry)

3.1 Interlocked Molecules

The triazole unit in its various forms (Fig. 1) has been used to prepare rotaxanes and enable their switching. Li and coworkers have synthesized a rotaxane (Fig. 16) that utilizes a triazole unit as a hydrogen bonding station for a isophthalamide-based macrocycle [41]. As synthesized, rotaxane **30**$^{+}$**-H** has the macrocycle situated at the ammonium station where the polyether linkage can be strongly hydrogen-bonded. Upon deprotonation to **30**, the macrocycle prefers to sit at the triazole station where two isophthalamide NH protons are bound to the triazole nitrogen and the triazole CH proton is bound to the polyether oxygens of the macrocycle. The triazole proton shows an upfield shift from a combination of effects, the aromatic shielding by the macrocycle's phenylenes, and the hydrogen bonding with the polyether group. When TBA-halide (Cl^{-}, Br^{-}, I^{-}) salts are titrated with **30**, the cooperative anion

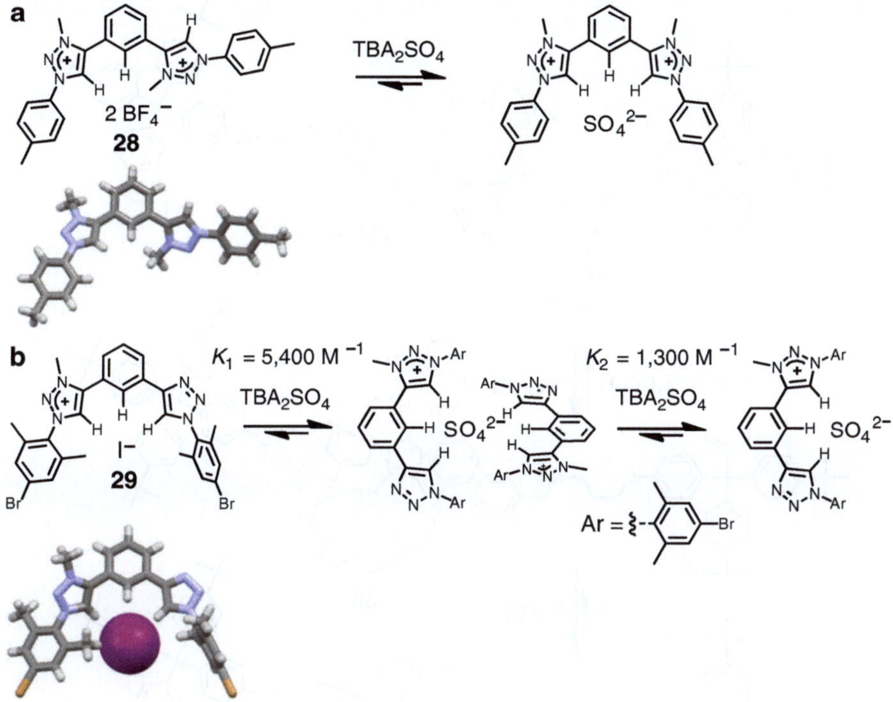

Fig. 15 X-ray crystal structure of receptors (**a**) **28** and (**b**) **29** and their sulfate binding equilibria

binding of the triazole CH and the isophthalamide NH causes the macrocycle to pirouette about the triazole station.

After their use of methylthiazolium moieties [42], Beer and coworkers synthesized anion-templated interlocked molecules by utilizing *halogen bonds* [43] (Fig. 17) derived from iodotriazolium moieties. The R^+–I•••X^- halogen bonds of iodotriazolium-halide ion pairs (**31**•X^-) were identified by X-ray crystal structures (Fig. 17c). These triazolium compounds formed pseudorotaxanes with isophthalamide-based macrocycle **32** (Fig. 17a). It was determined that **31**•Br^- had the strongest association constant with **32** in $CDCl_3$. Therefore, the authors utilized the R^+–I•••Br^-ion pair as the thread and the template to synthesize a rotaxane using a clipping strategy. Metathesis of premacrocycle **34** in the presence of **33**•Br^- provided the rotaxane **33**•Br^-•**34'**.

3.2 Sensors

Kim and coworkers synthesized an anion sensor (**35**, Fig. 18a) that utilizes calix[4] arene as the backbone structure, two pyrene moieties as chromophores and two triazoles as anion binding units [44]. Receptor **35** has a strong excimer band at

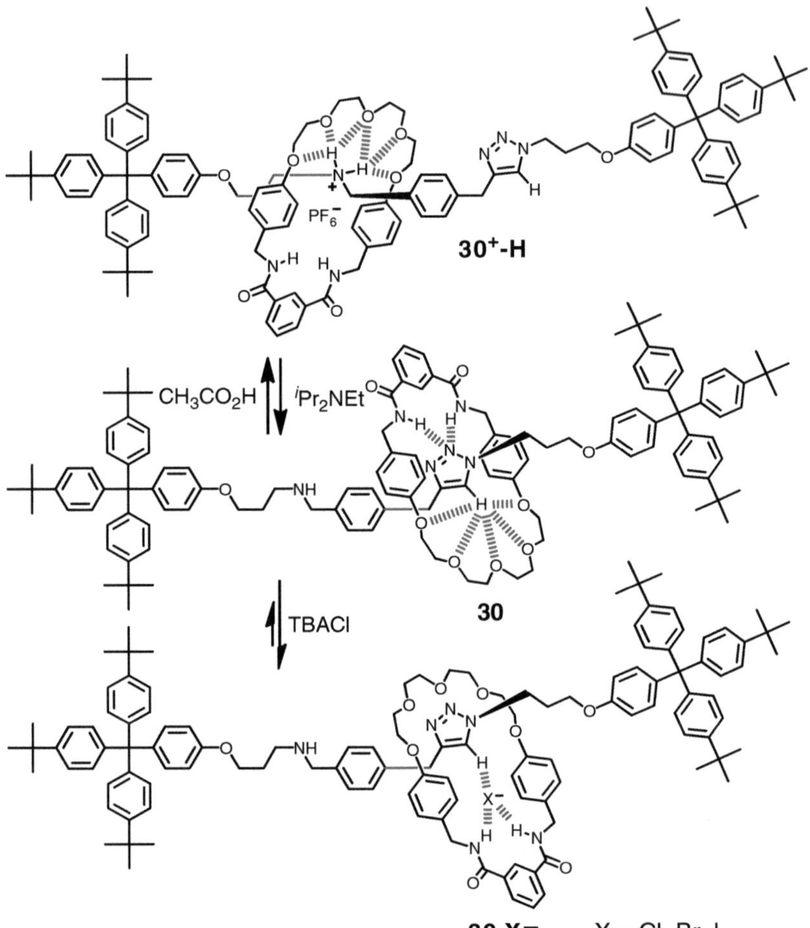

Fig. 16 Switching of a rotaxane **30** upon acid/base and chloride addition

475 nm originating from the pyrene dimer. Upon addition of TBAI, the receptor binds iodide and undergoes a conformational change that dissociates the pyrene excimer, decreasing the intensity of the emission band at 475 nm in CH_3CN (Fig. 18b).

Potentiometric sensing of anions using triazole CH•••anion interactions have also been investigated. Bachas and Flood have incorporated triazolophanes **1** [45] and **5** [46] into poly(vinyl chloride) (PVC) membrane electrodes to demonstrate their efficacy as ion-selective electrodes (ISE). The ISE with **1** exhibited anti-Hofmeister selectivity toward bromide. In particular, the electrode showed high selectivity and accuracy toward 10–50 mM of bromide using horse blood serum containing interfering background anions such as chloride (~100 mM) and bicarbonate (~30 mM). The ISE incorporated with **5** was highly selective toward iodide.

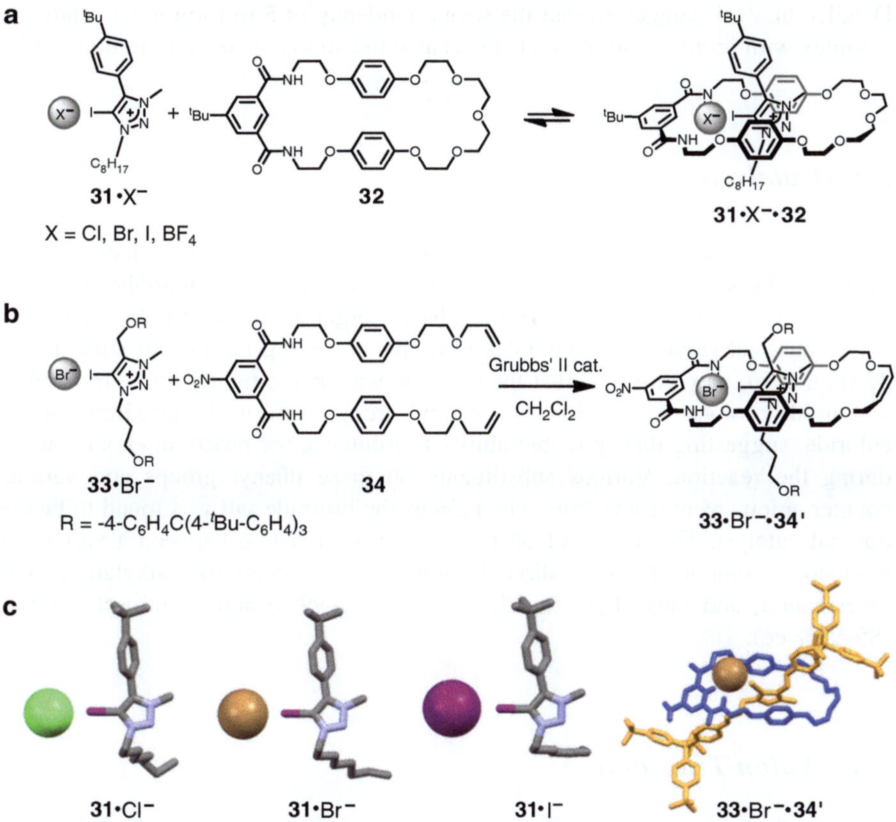

Fig. 17 (**a**) Pseudorotaxane and (**b**) rotaxane formation by halogen bonding. (**c**) Crystal structures of **31**•Cl^-, **31**•Br^-, **31**•I^-, and **33**•Br^-•**34**′

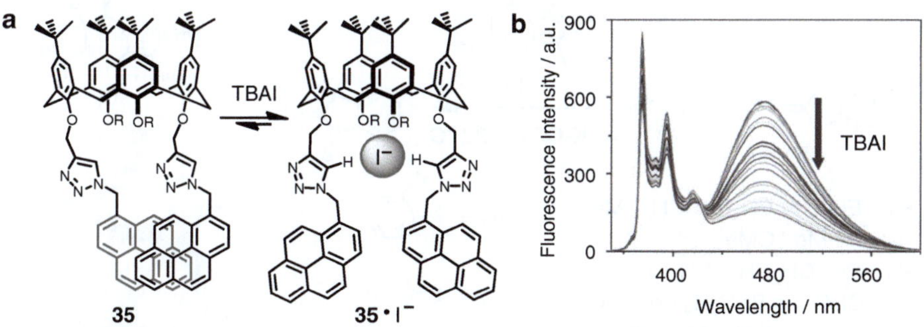

Fig. 18 (**a**) Scheme of **35** binding with I^-. (**b**) Decreasing excimer band upon addition of TBAI to **35** (6 μM in CH_3CN) (Reprinted with permission from Bull Korean Chem Soc 31:624. Copyright 2010 Korean Chemical Society)

Detailed analysis suggested that the strong tendency of **5** to form a 2:1 sandwich complex with iodide in solution [19] was also the source of selectivity in the ISE.

3.3 *Catalysis*

Hydrogen-bonded organocatalysis is a growing field that typically makes use of various NH donor moieties [47]. Contributing to this new field of application, Ooi and coworkers developed a triazolium-based organocatalyst (**36**, Fig. 19a) for asymmetric alkylation of oxyindoles [48]. By analyzing the crystal structure of **37** (Fig. 19b), they observed that the chloride was bound tightly to the triazolium and amide protons. In addition, two phenyl groups were in the proximity of the chloride, suggesting their possible ability to influence the prochiral enolate anion during the reaction. Various substituents on these phenyl groups and various counter anions were tested from which **36** as the bromide salt was found to be the optimal catalyst. The ability of **36** to catalyze the reaction between a variety of oxyindole compounds and alkyl bromides for asymmetric alkylation was investigated, and they show excellent yields (82–99%) and enantioselectivities (85–98% ee).

3.4 *Anion Transport*

Another growth area for anion receptor chemistry involves the transport of ions across lipid membranes [49], where practitioners seek both fundamental

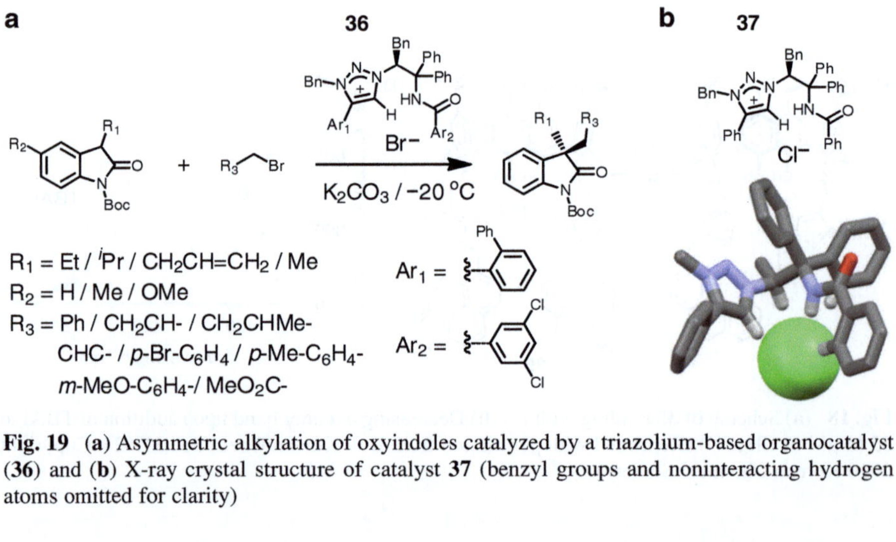

Fig. 19 (**a**) Asymmetric alkylation of oxyindoles catalyzed by a triazolium-based organocatalyst (**36**) and (**b**) X-ray crystal structure of catalyst **37** (benzyl groups and noninteracting hydrogen atoms omitted for clarity)

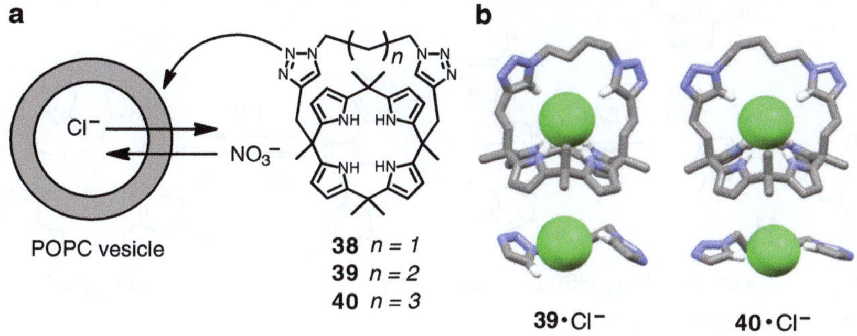

Fig. 20 (**a**) Scheme of triazole-strapped calix[4]pyrroles-transporting anions through a POPC vesicle membrane. (**b**) X-ray crystal structure of **39**•Cl^- and **40**•Cl^-, side view (*top*), partial structure focused on triazole-Cl^- interaction (*bottom*)

understanding and potential benefits for human health. Gale and coworkers have prepared triazole-strapped calix[4]pyrroles **38**, **39**, and **40** with three, four, and five methylene (small, medium, and large) spacers between two triazoles, respectively (Fig. 20a) [50]. Anion binding constants were measured by isothermal titration calorimetry in acetonitrile. The chloride binding constant was the largest for medium-sized cavity, **39** (**39** > **40** > **38**). However, X-ray crystal structures of **39**•Cl^- and **40**•Cl^- show that both triazoles in **40** are hydrogen-bonded to the chloride while only one triazole is bonded in **39** (Fig. 20b). The chloride transport abilities were tested with POPC (1-palmitoyl-2-oleoyl-*sn*-glycero-3-phosphocholine) vesicles loaded with various chloride salts suspended in an outer solution of $NaNO_3$. The chloride efflux rate was highest for the largest transporter **40** (**40** > **39** > **38**). The results indicated that in this system the chloride transport ability does not readily follow the trend of the anion binding strength.

3.5 Anion Regulation

Taking inspiration from Shinkai's early work with light-driven binding and release of alkali cations [51], new studies are emerging to control anions in a similar manner. Jiang and coworkers have synthesized an aryl-triazole oligomer (**41**, Fig. 21) with a photoswitchable azobenzene unit in the center of the chain [52] after related work by Hecht [53]. The oligomer was able to switch between *trans* and *cis* conformations upon UV or Vis light irradiation. ^{1}H NMR titrations with various TBA anion salts in d_6-acetone showed that the *cis* form, $\mathbf{41}_{cis}$, had higher anion binding affinities than the *trans* form, $\mathbf{41}_{trans}$, by factors as large as four: $K_{cis}/K_{trans} = 4.1$ (Cl^-), 3.9 (Br^-), 2.8 (I^-), 2.1 (NO_3^-), 1.7 (HSO_4^-). These changes in the anion binding affinity were speculated to come from the spatial complementarity between the receptor and anion, since $\mathbf{41}_{cis}$ has a smaller anion binding cavity than $\mathbf{41}_{trans}$.

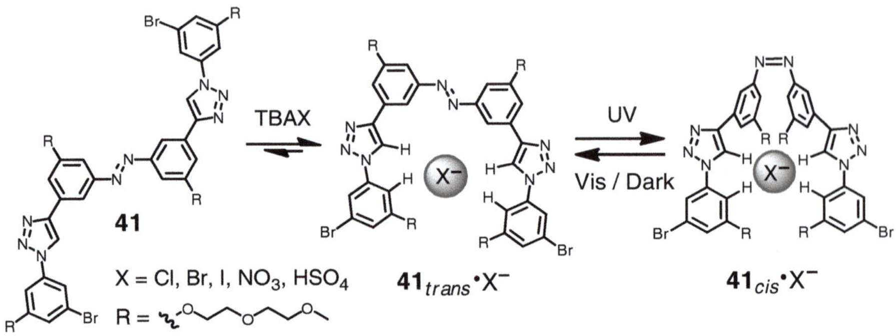

Fig. 21 Photoswitchable oligomeric anion receptor

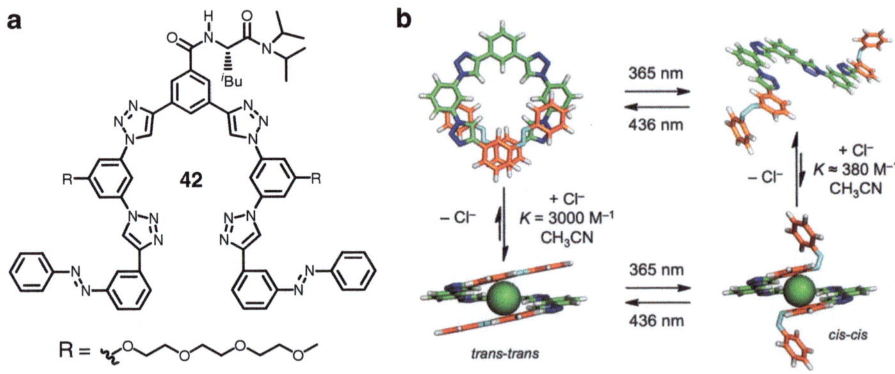

Fig. 22 (**a**) A foldameric light-active anion receptor **42** and (**b**) its cycle of binding and releasing chloride upon UV or visible light irradiation (Reprinted with permission from J Am Chem Soc 132:12838. Copyright 2010 American Chemical Society)

Independently, Hua and Flood synthesized a light-active aryl-triazole foldamer (**42**, Fig. 22) which has two azobenzene units on each end [54]. When the azobenzene units are photoisomerized from *trans* to *cis* with UV (365 nm) irradiation, the foldamer loses two of its three aromatic π–π interactions that were being employed to stabilize the helically folded form. Therefore, UV irradiation destabilizes the folded helix in favor of the unfolded random-coil conformation. The folded conformation could be reinstated by irradiation with visible light (436 nm). In solution, the foldamer binds with chloride in its *trans* isomer (K_a = 3,000 M^{-1}, MeCN) and releases chloride in the *cis* isomer (K_a = 380 M^{-1}). On account of the fact that free chloride ions conduct electricity through solution better than receptor-bound chloride, the anion binding-release cycle was observed by conductivity measurements. Consistently, a 1:1 solution of **42** and TBACl (MeCN) showed increased conductivity under UV light and decreased conductivity under Vis light.

4 Perspectives

4.1 Broad Utilization of 1,2,3-Triazoles

Recent years have seen a creative flourishing of receptors that are designed to take advantage of some aspect of the click-generated triazole's noncovalent chemistry. From neutral triazole [16] to cationic triazolium [40] CH hydrogen bonding and now cationic halogen bonding [43], all are available for stabilizing anions. Sometimes the triazoles have been used with other CH donors [20, 22, 23, 31, 34–36, 39, 40, 52, 54], and in other instances, they have acted alongside NH donors [28, 29, 37, 38, 49]. Starting from these creations, practitioners have also shaped interlocked molecules [43] and pH-driven molecular switches [41], sensors for anions [44–46] and catalysts [48] for organic transformations as well as transporters [50] for ferrying anions across lipid membranes. While these forays are but proof-of-principle demonstrations, they do hint at a growing platform that may one day enable a commercially viable molecular product.

These receptors represent a veritable smorgasbord of ways for stabilizing anions, yet a common thread running through the majority of them is structural change – or the lack thereof [35] – whether it is driven by anion binding or some other stimuli (pH [41] and light [52, 54]). These changes sometimes illuminate a fundamental aspect of the anion receptor binding event, and at other times, they are used to generate another favorable outcome.

4.2 Structural Effects: Preorganized, Stimuli Induced, and Stimuli Driven

Preorganization has been verified [34, 35] as a crucial factor in enhancing binding strengths; or when stated more appropriately in reverse: Preorganization reduces the magnitude of the energetic penalties that need to be paid to prepare the "perfect" structure for complexation. This feature was shown in macrocycles [35] and acyclic receptors [36] alike. It is believed that the enthalpic costs that get prepaid when triazole-derived receptors are prepared in a preorganized format are associated with bringing the electropositive hydrogens into close proximity with each other. Thus, it is rewarding to spot systems that are preorganized for anion complexation. Of course, it is also enjoyable to consider how the shape and structure of the organic receptor changes upon anion binding. Some systems are designed to make some good use of the energy provided by the complexation event. These types of anion stimulation events have driven folding of long and short oligomers [38–40], as well as promoting sensor-type readouts [44]. Finally, we are seeing the use of orthogonal stimuli, in the form of light-driven azobenzene isomerization, to make and break the receptor's structure [52] for the binding and releasing of anions [54].

5 Concluding Remarks

The click-generated 1,2,3-triazoles and its derivatives are enjoying favor among the supramolecular community. One sees classic NH donors being replaced with triazoles and a plethora of different receptor chemistries being trialed. Forays into various applications, be they switches, sensors, catalysts, transporters, or regulators, are being investigated at a level commensurate with their representation in traditional amide, urea, and pyrrole receptors. A rewarding theme has been the use of covalent chemistry to enforce preorganization, as well as the opposite, where structural changes that are caused by anion binding have enabled new understanding and applications. Finally, as easy to install as they are, their similarity in strength (and geometry) to NH donors bodes well for the triazoles continued use in anion receptor chemistry.

References

1. Bianchi A, Bowman-James K, García-España E (1997) Supramolecular chemistry of anions. Wiley-VCH, New York
2. Sessler JL, Gale PA, Cho W-S (2006) Anion receptor chemistry. RSC Publishing, Cambridge, UK
3. Desiraju GR, Steiner T (1999) The weak hydrogen bond. Oxford University Press Inc, New York
4. Rostovtsev VV, Green LG, Fokin VV, Sharpless KB (2002) Angew Chem Int Ed 41:2596
5. Tornøe CW, Christensen C, Meldal M (2002) J Org Chem 67:3057
6. Kwon JY, Jang YJ, Kim SK, Lee K-H, Kim JS, Yoon J (2004) J Org Chem 69:5155
7. Ilioudis CA, Tocher DA, Steed JW (2004) J Am Chem Soc 126:12395
8. Chmielewski MJ, Charon M, Jurczak J (2004) Org Lett 6:3501
9. Farnham WB, Roe DC, Dixon DA, Calabrese JC, Harlow RL (1990) J Am Chem Soc 112:7707
10. Sutor DJ (1963) J Chem Soc 1105
11. Bryantsev VS, Hay BP (2005) J Am Chem Soc 127:8282
12. Berryman OB, Sather AC, Hay BP, Meisner JS, Johnson DW (2008) J Am Chem Soc 130:10895
13. Zhu SS, Staats H, Brandhorst K, Grunenberg J, Gruppi F, Dalcanale E, Lützen A, Rissanen K, Schalley CA (2008) Angew Chem Int Ed 47:788
14. Yoon D-W, Gross DE, Lynch VM, Sessler JL, Hay BP, Lee C-H (2008) Angew Chem Int Ed 47:5038
15. Palmer MH, Findlay RH, Gaskell AJ (1974) J Chem Soc Perkin Trans 2:420
16. Li Y, Flood AH (2008) Angew Chem Int Ed 47:2649
17. Li Y, Flood AH (2008) J Am Chem Soc 130:12111
18. Hirose K (2001) J Inclusion Phenom Macrocyclic Chem 39:193
19. Li Y, Pink M, Karty JA, Flood AH (2008) J Am Chem Soc 130:17293
20. Hua Y, Ramabhadran RO, Uduehi EO, Karty JA, Raghavachari K, Flood AH (2011) Chem Eur J 17:312
21. Bandyopadhyay I, Raghavachari K, Flood AH (2009) ChemPhysChem 10:2535
22. Juwarker H, Lenhardt JM, Pham DM, Craig SL (2008) Angew Chem Int Ed 47:3740
23. Meudtner RM, Hecht S (2008) Angew Chem Int Ed 47:4926

24. Kolb HC, Finn MG, Sharpless KB (2001) Angew Chem Int Ed 40:2004
25. Hua Y, Flood AH (2010) Chem Soc Rev 39:1262
26. McDonald KP, Hua Y, Flood AH (2010) Top Heterocycl Chem 24:341
27. Cabbines DK, Margerum DW (1969) J Am Chem Soc 91:6540
28. Sessler JL, Cai J, Gong H-Y, Yang X, Arambula JF, Hay BP (2010) J Am Chem Soc 132:14058
29. Zhao Y, Li Y, Li Y, Huang C, Liu H, Lai S-W, Che C-M, Zhu D (2010) Org Biomol Chem 8:3923
30. Alunni S, Pero A, Reichenbach G (1998) J Chem Soc Perkin Trans 2:1747
31. Ramabhadran RO, Hua Y, Li Y, Flood AH, Raghavachari K (2011) Chem Eur J 17:9123
32. Meot-Ner M, Cybulski SM, Scheiner S, Liebman JF (1988) J Phys Chem 92:2138
33. Alabugin IV, Zeidan TA (2002) J Am Chem Soc 124:3175
34. Juwarker H, Lenhardt JM, Castillo JC, Zhao E, Krishnamurthy S, Jamiolkowski RM, Kim K-H, Craig SL (2009) J Org Chem 74:8924
35. Hua Y, Ramabhadran RO, Karty JA, Raghavachari K, Flood AH (2011) Chem Commun 47:5979
36. Lee S, Hua Y, Park H, Flood AH (2010) Org Lett 12:2100
37. Haridas V, Sahu S, Venugopalan P (2011) Tetrahedron 67:727
38. Wang Y, Xiang J, Jiang H (2011) Chem Eur J 17:613
39. García F, Torres MR, Matesanz E, Sánchez L (2011) Chem Commun 47:5016
40. Schulze B, Friebe C, Hager HD, Günther W, Köhn U, Jahn BO, Görls H, Schubert US (2010) Org Lett 12:2710
41. Zheng H, Zhou W, Lv J, Yin X, Li Y, Liu H, Li Y (2009) Chem Eur J 15:13253
42. Mullen KM, Mercurio J, Serpell CJ, Beer PD (2009) Angew Chem Int Ed 48:4781
43. Kilah NL, Wise MD, Serpell CJ, Thompson AL, White NG, Christensen KE, Beer PD (2010) J Am Chem Soc 132:11893
44. Kim JS, Park SY, Kim SH, Thuéry P, Souane R, Matthews SE, Vicens J (2010) Bull Korean Chem Soc 31:624
45. Zahran EM, Hua Y, Li Y, Flood AH, Bachas LG (2010) Anal Chem 82:368
46. Zahran EM, Hua Y, Lee S, Flood AH, Bachas LG (2011) Anal Chem 83:3455
47. Doyle AG, Jacobsen EN (2007) Chem Rev 107:5713
48. Ohmatsu K, Kiyokawa M, Ooi T (2011) J Am Chem Soc 133:1307
49. Davis AP, Sheppard DN, Smith BD (2007) Chem Soc Rev 36:348
50. Yano M, Tong CC, Light ME, Schmidtchen FP, Gale PA (2010) Org Biomol Chem 8:4356
51. Shinkai S, Nakaji T, Ogawa T, Shigematsu K, Manabe O (1981) J Am Chem Soc 103:111
52. Wang Y, Bie F, Jiang H (2010) Org Lett 12:3630
53. Khan A, Kaiser C, Hecht S (2006) Angew Chem Int Ed 45:1878
54. Hua Y, Flood AH (2010) J Am Chem Soc 132:12838

Top Heterocycl Chem (2012) 28: 109–136
DOI: 10.1007/7081_2011_69

Published online: 8 February 2012

Click Triazoles as Chemosensors

Michael Watkinson

Abstract The "click" generation of a 1,4-disubstituted triazole from a terminal alkyne and a terminal azide has been widely used in recent years to prepare a plethora of sensor systems for the detection of cations, anions, small molecules and biomolecules. The focus of the chapter is on sensors in which the triazole plays an intimate role in the sensing event, although some examples of its use as a means of ligation are also presented.

Keywords Click chemistry · Fluorescence · Sensors · Triazole

Contents

M. Watkinson (✉)
The Joseph Priestley Building, School of Biological and Chemical Sciences, Queen Mary University of London, Mile End Road, London E1 4NS, UK
e-mail: m.watkinson@qmul.ac.uk

1 Introduction

The detection of the interaction between a target analyte and a receptor forms the basis of chemical sensing, and a wide range of sensors has been reported [1–6]. The majority of sensors rely on the effect analyte binding has on a fluorophore [7], although a number of other sensing modalities exist. In recent years, the concept of click chemistry has been introduced by Sharpless and the archetypal click reaction to generate a 1,4-disubstituted triazole, the Cu(I)-catalysed azide–alkyne cycloaddition [8–10], widely used [11–13] not least in the field of chemical sensors, which will be reviewed in this chapter. The popularity of this particular "click" reaction lies in its modular nature as well as its synthetic simplicity and reliability. Whilst these features make it an attractive strategy for the ligation of complex analytes to reporter units in chemical sensors, the main focus of this chapter lies in applications in which the 1,2,3-triazole plays a crucial role in analyte detection.

2 Cation Sensing

Due to the fundamental roles cations play in biological, chemical and environmental processes considerable efforts have been directed towards the synthesis of a range of systems capable of their detection with high selectivity and sensitivity. Of these methods, fluorescence-based sensors prepared using click chemistry have undoubtedly received most attention, although there are a range of other methods which merit discussion.

2.1 *Solution-Based Switch-On Fluorescent Sensors for the Al^{3+} Cation*

Aluminium is widely used in many industrial fields as a result of it being the most abundant metallic element by weight and the third most abundant of all elements after oxygen and silicon. Although a modest weekly dietary intake of aluminium can be tolerated (ca. 7 mg/kg body weight), excessive intake results in a range of associated pathologies. Moreover, 40% of the world's acidic soils are caused by aluminium toxicity. There is thus a need for effective Al^{3+} chemosensors for application in medical and environmental research. To date, there have been few such sensors reported, but amongst these click-generated systems are prominent. The anthraquinone-based sensor **1** showed a significant fluorescence emission in the presence of Al^{3+} at 557 nm, with insignificant changes to the emission spectrum in the presence of a range of other cations including Cu^{2+} and Hg^{2+}. Analysis of binding by Job plot and Benesi–Hildebrand plot experiments revealed a 2:1 stoichiometry between **1** and Al^{3+}, which in addition to ^1H NMR titration data lead to the proposed formation of a sandwich complex in which carbonyl and triazole coordinate to the Al^{3+} centre [14]. Bipyridine-containing chemosensors **2** and **3** were prepared by the same group, with click chemistry being utilised to link the metal-binding unit to two different fluorophores. Both sensors proved effective in the switch-on detection of Al^{3+}. Sensor **2** showed a clear blueshift in the presence of Al^{3+} with the emission maximum at 446 nm cf. 532 nm supporting the coordination of the $–N(CH_3)_2$ moiety and an ICT mechanism [15], whilst **3** showed a slightly redshifted emission at 443 nm [16]. Analysis of the binding of Al^{3+} to both sensors by ^{1}H NMR spectroscopy indicated the involvement of both the click triazole and the bipyridine unit in the metal's coordination sphere. Both sensors were quenched by Cu^{2+} and In^{3+}. Unfortunately, all three sensors were only reported to operate in CH_3CN solution.

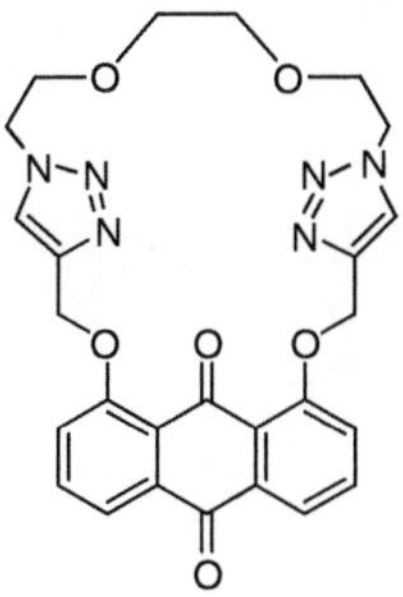
1

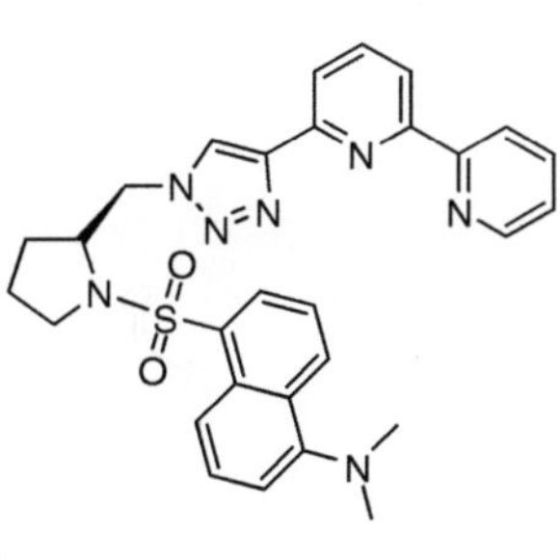
2

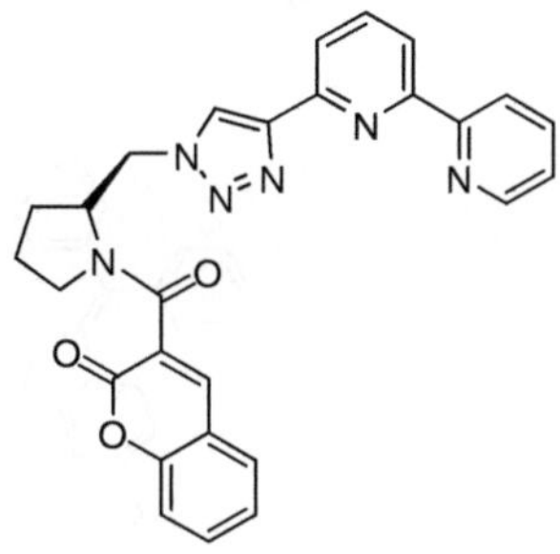
3

2.2 *Solution-Based Triazole-Containing Fluorescence Switch-On Sensors for the Cu^{2+} Cation*

Copper is an essential trace element in living systems, but the most common means of its detection is through fluorescence quenching mode because of the paramagnetic nature of the Cu^{2+} cation. In recent years, however, a number of fluorescence enhancement sensors have been reported, some utilising a click-derived triazole. Fluorescence of the anthracene moiety in sensor **4** is quenched by both Cu^{2+} and Hg^{2+}, but intriguingly, the closely related system **5** exhibits a switch-on response exclusively for Cu^{2+} [17, 18]. The diazo-containing system **6** proved to be an effective colorimetric Cu^{2+}-specific sensor with detection possible with the naked eye. Binding of the metal ion induced in CH_3CN/H_2O (4:1 v/v) solution induced a blueshift in the emission maximum from 459 nm to 342 nm consistent with an ICT mechanism, with the formation of a 1:1 metal–ligand complex supported by Job's method [19]. Recently, aggregation-induced emission materials have been applied to the sensing of Cu^{2+} with a detection limit of ca. 1 μM. The Cu^+ catalysed click reaction of azide **7** and alkyne **8** resulted in THF/H_2O (1:1 v/v) solution showing a maximum fluorescence emission at 490 nm after standing for 48 h. Increasing the concentration of Cu^{2+} (>20 mM) quenched fluorescence due to the precipitation of highly cross-linked materials; however, heating the mixture at 50°C with stirring resulted in a faster response (12 h) and no decrease in fluorescence at higher Cu^{2+} concentrations. High selectivity for Cu^{2+} was also confirmed through competition experiments with a number of other cations [20].

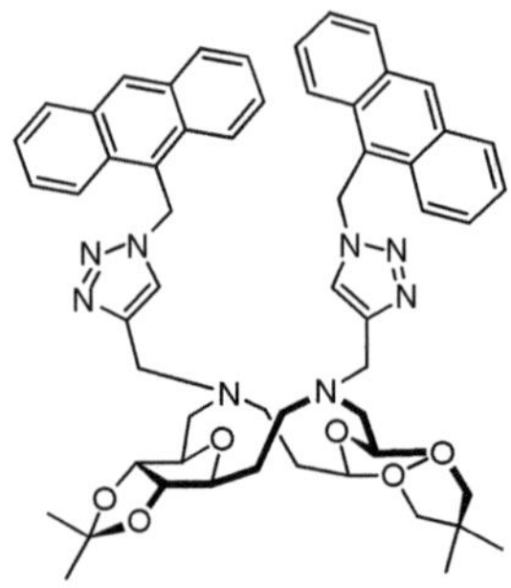

4

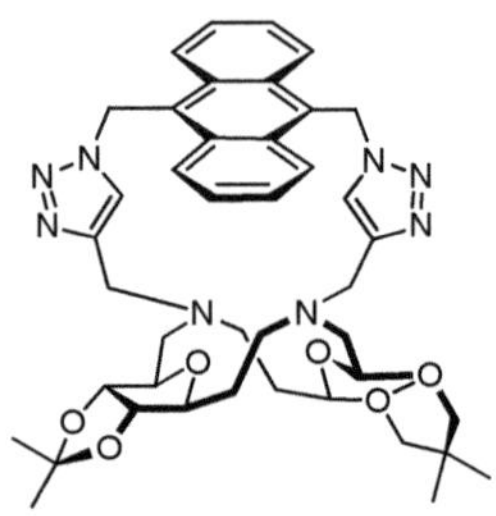

5

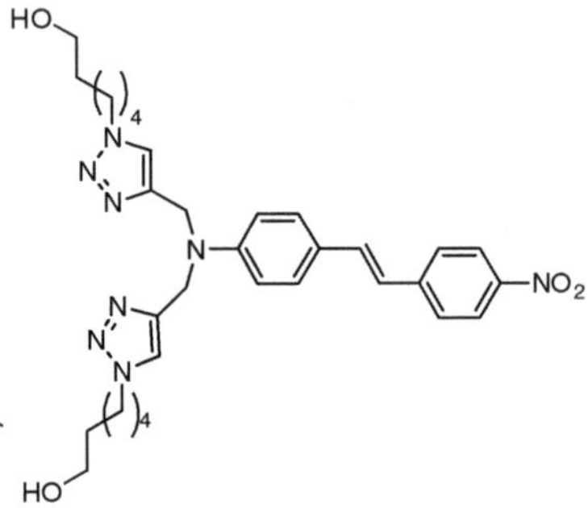

6

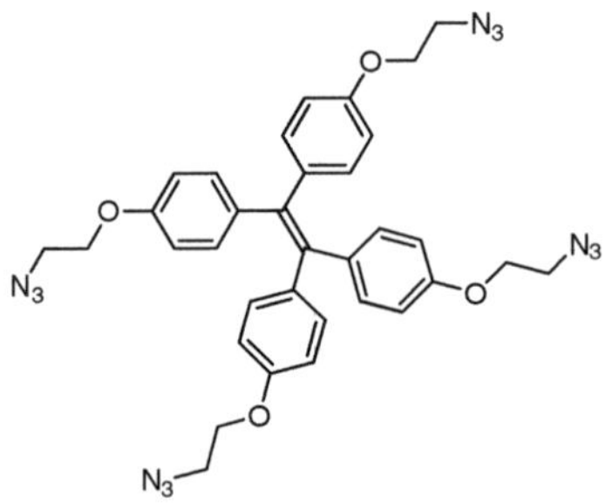

7

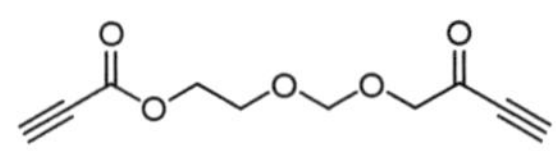

8

2.3 Solution-Based Switch-On Fluorescent Sensors for Zn^{2+}, Cd^{2+} and Hg^{2+} Cations

Zinc is increasingly recognised as a key player in a host of biological processes and disease states. This is not surprising given that it is the second most abundant transition metal in the human body. Whilst much of this zinc is tightly associated with the active sites in enzymes and with DNA-binding proteins, a number of tissues contain pools of mobile zinc with a wide range of concentrations including the pancreas, retina and brain. As a result, there has been, and continues to be, a drive to develop a deep understanding of the biology of zinc [21]. Due to its spectroscopically silent d^{10} electronic configuration, most efforts have focussed on the development of fluorescent chemosensors, which has been the subject of an excellent recent review [22], and it is no surprise that these have also included click-based sensors in which the triazole plays an intimate role in zinc coordination and detection. Frequently, these sensor systems show similar switch-on sensitivity for the stereoelectronic isostere Cd^{2+}, whilst Hg^{2+} quenches fluorescence. Nonetheless, such molecular devices are of significant importance due to the potential toxicity of these ions to the environment and biological systems.

The metal ion chemosensor **9** was synthesised independently by two groups by clicking two fluorescent pyrene subunits to a calix[4]arene framework [23, 24]. As a result of the different sensor responses to Zn^{2+} and Cu^{2+}, which quenches fluorescence (as does Hg^{2+}), it was suggested that it could be used as an INHIBITION NOR logic gate [23]. The Zn^{2+}-free species exhibits a strong pyrene excimer emission at a shorter wavelength than would be biologically desirable ($\lambda_{em} = 482$ nm, $\lambda_{ex} = 343$ nm) in CH_3CN. Upon addition of Zn^{2+} or Cd^{2+}, **9** exhibits a ratiometric response, with increased pyrene monomer emission and quenching of excimer emission. ^{1}H NMR experiments suggest that the pyrene subunits of **9** are initially involved in π–π stacking, but their separation is forced by the introduction of the metal ion into the triazole binding pocket. To test the effectiveness of **9** as a chemosensor, the fluorescence response was recorded in the presence of other metal ions. Only Cu^{2+}, Hg^{2+} and Pb^{2+} were found to disrupt the sensor by quenching fluorescence completely [24]. The photophysical properties of pyridin-2′-yl-1,2,3-triazole groups have been investigated through installation of these fluorophores on a calix[4]arene framework in sensor **10** [25]. An overall decrease in fluorescence and the appearance of a new emission band were observed upon addition of Zn^{2+} and Cd^{2+} in MeOH, changes attributed to excimer formation. Whilst the exact binding mode was not determined, UV–vis absorbance titrations indicated the formation of both 1:1 and 1:2 ligand–metal complexes. The effect of competing metal ions was not investigated. A calix[4]arene sensor was synthesised which contained two potential binding sites attached to the lower rim: one bis-triazole binding pocket and another consisting of two Schiff base groups functionalised with hydroxymethyl groups [26]. Although extremely sensitive and selective sensing of Zn^{2+} over a range of divalent metal ions was observed, including Cd^{2+}, ^{1}H NMR studies revealed that Zn^{2+} only binds at the Schiff base site. Furthermore, an analogue without the hydroxyl and imino groups did not show any absorbance or fluorescence changes upon the addition of Zn^{2+}, confirming that, perhaps surprisingly, the triazoles play an exclusively structural role in this sensor. In an elegant extension to this work, the sensing of Zn^{2+} was demonstrated in blood serum as well as related albumin proteins [27]. Although sensor **11** was reported to have a highly selective affinity for Pb^{2+} through enhancement of the monomer emission of pyrene in an organic/aqueous solution, a significant output was also observed in the presence of excess Zn^{2+} in organic solvents [28]. This feature was utilised to develop a highly selective $H_2PO_4^-$ anion sensor utilising the zinc complex of **11** vide infra [29].

Like calixarenes, cyclodextrins are a widely used scaffold in host–guest chemistry [30], although fewer examples of click-based cyclodextrin sensors have been reported. Liu and co-workers created a fluorescent Cd^{2+} sensor **12** by attaching 8-hydroxyquinoline to a β-cyclodextrin core via a triazole linker [31]. The hydroxyquinoline and triazole groups chelate Cd^{2+} in aqueous solution at physiological pH, resulting in enhanced fluorescence of the hydroxyquinoline. The binding mode of Cd^{2+} to the sensor was confirmed by a combination of 1D and 2D NMR techniques and by testing analogues lacking the triazole or the quinoline nitrogen. The chemosensor exhibits reasonable selectivity for Cd^{2+}, with Zn^{2+} the only other metal ion able to cause a similar fluorescence response. However, fluorescence quenching occurred in the presence of Cu^{2+} and Fe^{3+} and to a lesser extent Ag^+, Hg^{2+} and Pb^{2+}. Introducing an adamantine carboxylic acid guest into the cyclodextrin significantly enhanced the fluorescence response and selectivity for Cd^{2+} and Zn^{2+}, due to cooperative binding by the guest carboxylate group. Xie's group attached pyridin-2′-yl-1,2,3-triazole fluorophores to β-cyclodextrin to provide a similar sensor to their calixarene-based system **10** [32]. Fluorescence titrations in MeOH showed that compound **13** binds Zn^{2+} in a 1:1 stoichiometry with a sevenfold increase in fluorescence intensity. Initial investigations found that of a range of metal ions, only Mg^{2+} triggered a similar, though smaller, response, whilst various transition metal ions quench fluorescence. However, it was later found that Cd^{2+} also caused a fluorescence increase, forming 1:1, 1:2 and 1:3 host–guest complexes [25]. A smaller model compound representing a one sugar subunit of cyclodextrin **13** was also synthesised and shown to form responsive but more weakly bound 1:1 complexes with Zn^{2+} and Cd^{2+}. The importance of the pyridinyl-triazole group for sensing was demonstrated by creating phenyl- and hydroxymethyl-triazole model analogues that showed no observable fluorescence emission.

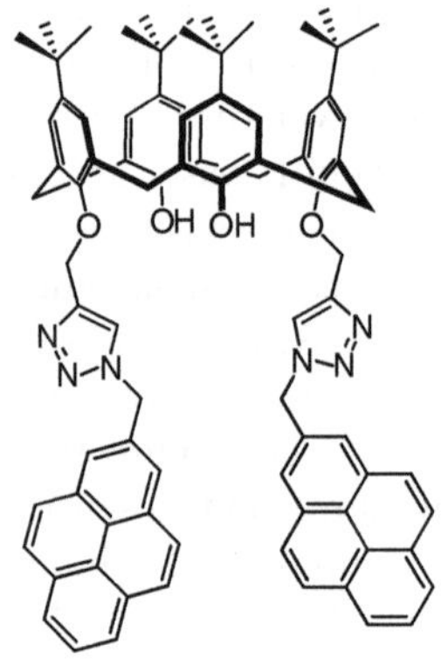

9

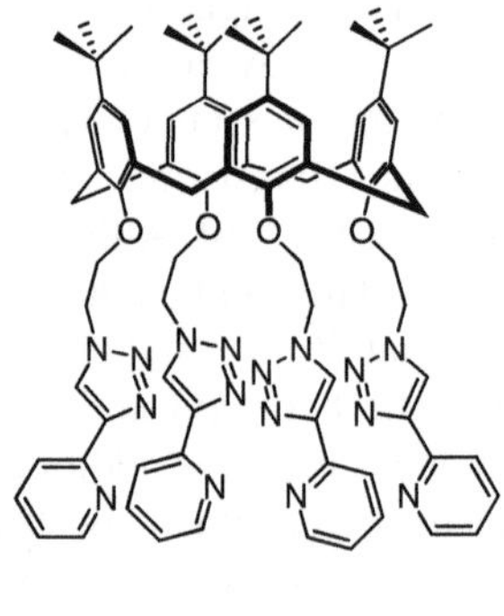

10

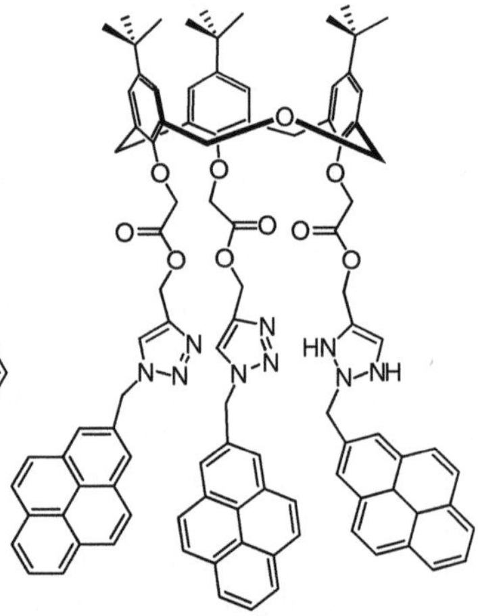

11

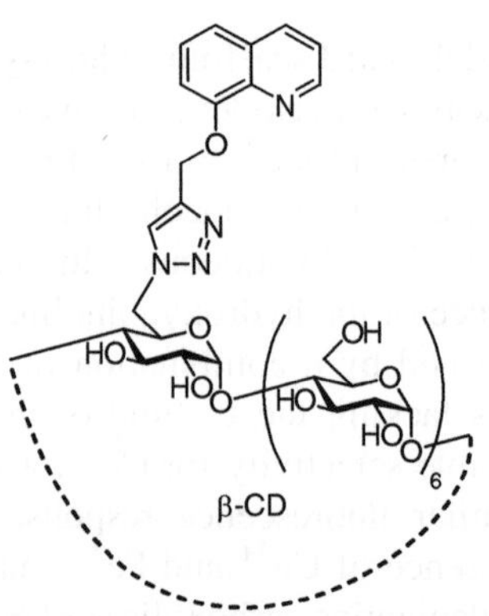

12

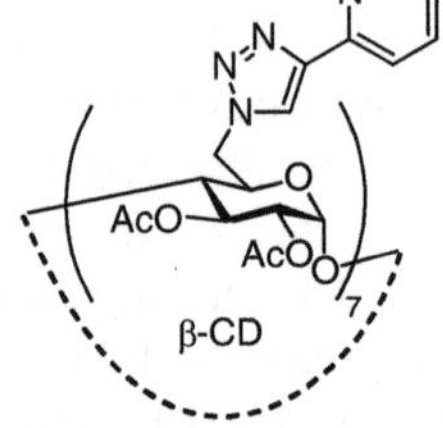

13

A number of small molecule sensors have also been reported. The simple pyridine-based system **14** displayed a significant enhancement in the fluorescence emission as well as a hypsochromic shift upon the addition of Zn^{2+}. When a number of anions were added to the **14**/Zn^{2+}, complex fluorescence was quenched [33]. The commonly used DPA ligand has been linked to two fluorescent aromatic groups using click chemistry to give the highly Zn^{2+} selective and water-soluble sensors **15** and **16** [34]. It was found that **15** also provides a ratiometric response with Cd^{2+}, whilst fluorescence was quenched for both sensors by Co^{2+} and Cu^{2+}. We have developed a number of macrocyclic-based sensors for zinc using both the cyclam and cyclen metal-binding motif. Our initial system appended a naphthalimide fluorophore to the cyclam centre to give **17**, which displayed excellent sixfold switch-on selectivity for Zn^{2+} with a $K_d = 43$ nM over a range of competing metal ions and physiologically relevant pH range [35]. The sensor was also shown to be able to detect a Zn^{2+} flux in apoptotic murine thymocytes without the need to add extracellular zinc. Heterogeneous sol–gel analogues of this system were also shown to maintain the switch-on response for zinc, whilst the analogue **18** incorporating two triazole-linked fluorophores showed double the response of **17** whilst upholding similar selectivity for Zn^{2+} [36]. The related cyclen-based system **19** displayed similar, but subtly different, behaviour to **16** in vitro, e.g. Fe^{3+} quenches fluorescence for **19** but not **17**. Both **17** and **19** were shown to be able to sense Zn^{2+} in *Danio rerio* again with subtle differences being observed in the organelle localisation of the two sensors [37]. A number of bis(1,2,3-triazolyl)fluorenyl probes **20** have been developed and shown to have switch-on selectivity for Zn^{2+} and Hg^{2+} in a range of organic media. As expected for an ICT-based sensor, the emission maxima undergo a blueshift upon metal binding, and probe **20d** was shown to detect Zn^{2+} in HeLa cells [38, 39]. A triazole-linked 8-hydroxyquinoline dimer has been reported to be an effective sensor for Cd^{2+}, showing selective turn-on fluorescence selectivity for Cd^{2+} over Zn^{2+} over a broad pH range in aqueous solution (CH_3CN:H_2O = 5:95) [40]. Utilising a similar approach to other sensors, particularly pyrene-functionalised calixarene **9**, two pyrene fluorophores have been joined by alkyl and polyoxyethylene spacers. For example, enhanced monomer and reduced excimer emission were observed for **21** in the presence of Zn^{2+} and Cd^{2+}, whilst Ni^{2+}, Pb^{2+}, Cu^{2+}, Hg^{2+} and Cr^{3+} showed non-specific quenching of both monomer and excimer emission in CH_3CN [41, 42].

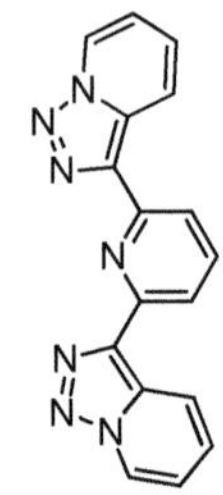

14

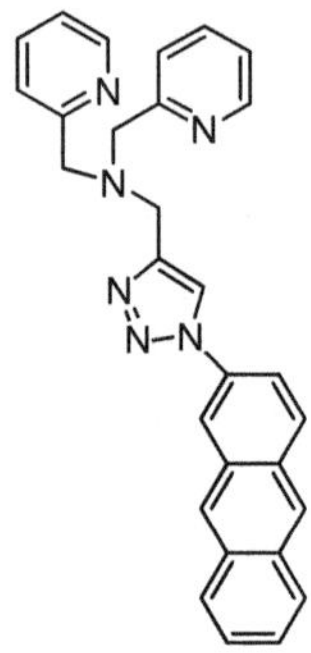

15

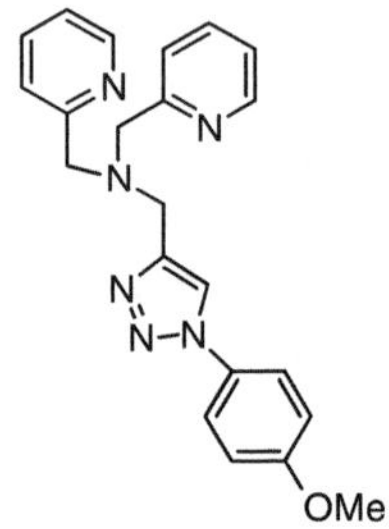

16

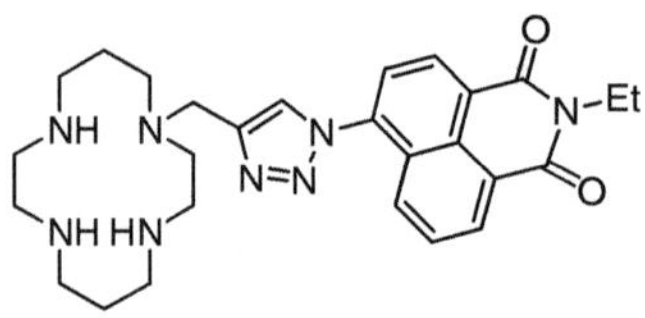

17

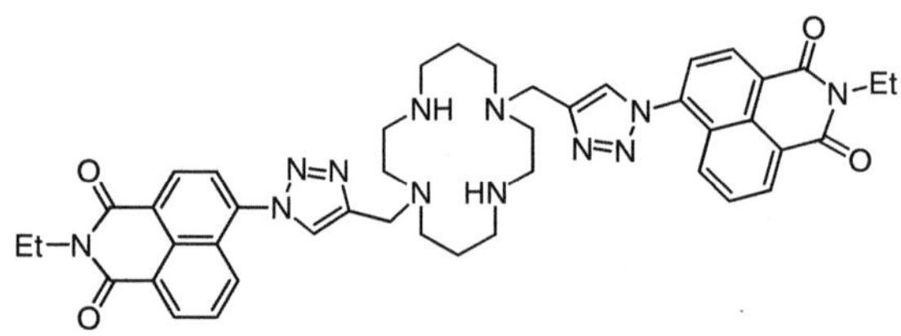

18

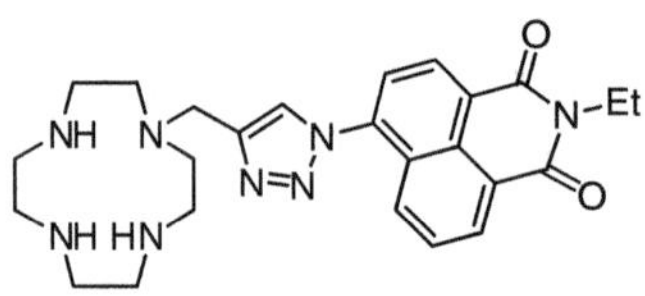

19

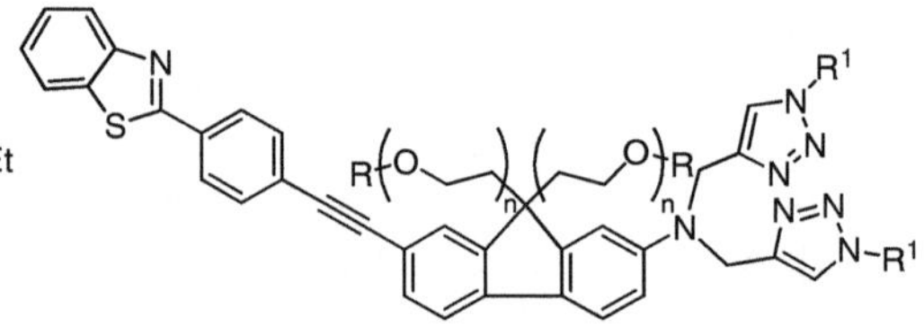

20a R = C_2H_5, R^1 = n-C_4H_9, n=2
20b R = C_2H_5, R^1 = n-$C_{11}H_{22}SH$, n= 2
20c R = CH_3, R^1 = n-$C_{11}H_{22}SH$, n= 6-10
20d R = C_2H_5, R^1 = $CH_2CO_2C_2H_5$, n=2

21

In a different strategy, Aslyn and co-workers elegantly employed the click reaction as a means of joining an alkynyl pyrene and an azide-functionalised coumarin to produce **22** in which the close proximity of the two fluorophores allowed Förster resonance energy transfer (FRET) to occur. It was shown that addition of EDTA to the two monomers prevented the click reaction occurring due to the chelation of Cu^{2+}, but that the addition of exogenous Zn^{2+}, and to a lesser extent Pb^{2+}, allowed the reaction to proceed as a result of competitive displacement of Cu^{2+} from the EDTA. The FRET response was shown to have a dependence on the competing metal ion concentration, allowing an allosteric method of Zn^{2+} sensing in the concentration range 3.5×10^{-5}–1.7×10^{-3} M [43]. A similar approach has also been used by Wang and co-workers in which the synthesis of the fluorogenic click product **23** signalled the presence of Cu^{+} or Cu^{2+}. Again, addition of EDTA resulted in chelation of Cu^{2+} which terminated the click reaction. Subsequent addition of Ca^{2+}, Zn^{2+} and Cd^{2+} again resulted in displacement of Cu^{2+} and increased fluorescence [44].

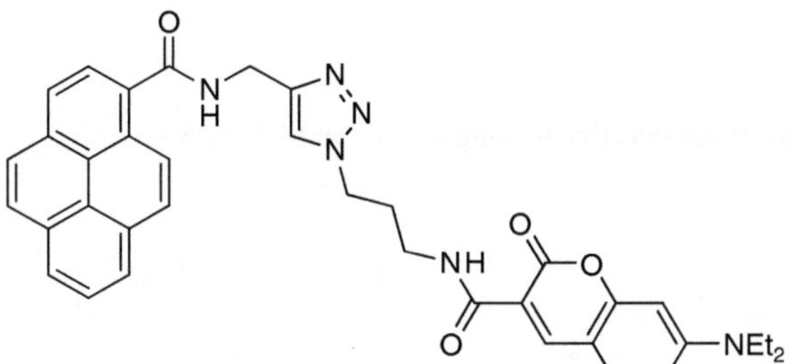

22

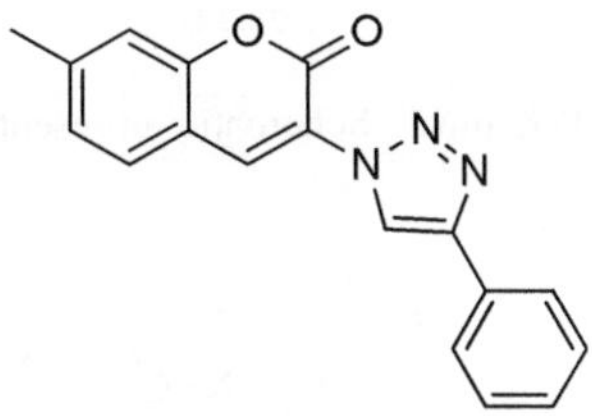

23

Solid-supported Hg^{2+} sensor incorporating a piperazine–pyridine functionalised naphthalimide fluorophore has been attached to mesoporous SBA-15 silica and shown to have a selective fluorescence response for Hg^{2+}. Interestingly, the selectivity for Hg^{2+} was found to be higher than for the precursor alkynyl dye; however, the click triazole was not proposed to be involved in coordination to the metal ion in the active system [45].

2.4 *Solution-Based Switch-On Fluorescent Sensors for Miscellaneous Cations*

In an extension of previous studies on functionalised calixarenes with on–off fluorescence selectivity for Pb^{2+} and K^{+}, Chung and co-workers prepared **24**. It was shown that fluorescence quenching occurred in the presence of Pb^{2+}, but in the presence of excess K^{+}, strong fluorescence could be recovered. Using model studies, the authors proposed that binding of Pb^{2+} necessitated the presence of two triazoles (Scheme 1) [46].

A rare example of a Pt^{2+} selective fluorescent probe **25** was reported by Tae and co-workers in which it was hypothesised that Pt^{2+} coordination would induce ring opening of the rhodamine spirolactam. Excellent selectivity for the detection of Pt^{2+} over a wide range of other metal ions was observed by fluorescence emission.

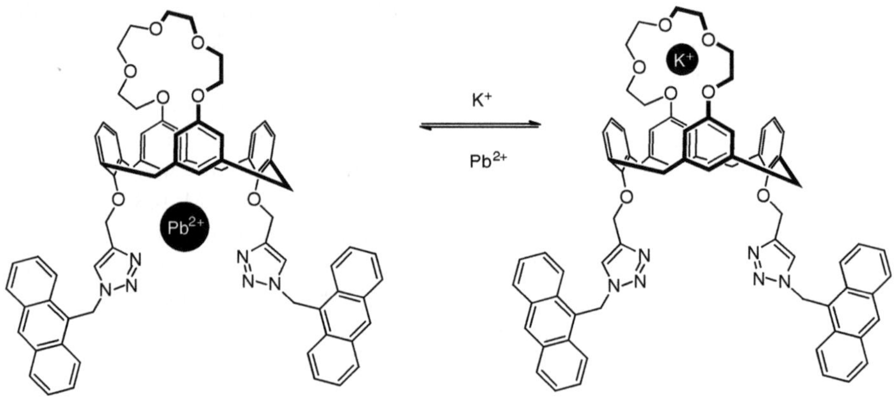

Scheme 1 Schematic representation of the selective binding of Pb^{2+} and K^+ to **24**

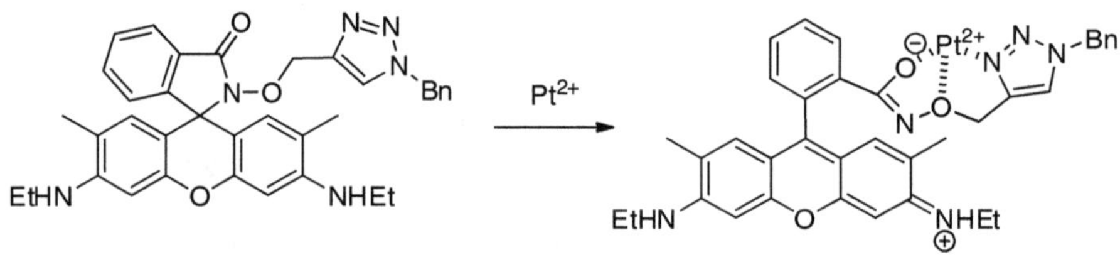

Scheme 2 Ring opening of rhodamine spirolactam **25** in the presence of Pt^{2+}

The response was linearly proportional to the concentration of Pt^{2+} up to the detection limit of 125 nM. Moreover, a colorimetric change was also observed with a solution of the sensor changing from colourless to pink upon the addition of Pt^{2+}. Interestingly, the colorimetric response was also observed for Pd^{2+}. It was also demonstrated that the system was capable of detecting the presence of cisplatin in aqueous solutions [47] (Scheme 2).

Finally, a number of 1,4-diaryltriazoles were reported to have turn-on fluorescence upon addition of a number of metal cations, although their selectivity was insufficient to render them useful chemosensors [48]. Guo and co-workers have also investigated the potential of these compounds for pH sensing [49].

2.5 *Cation Selective Switch-Off Fluorescent Sensors*

A number of the sensors already reported to have selective fluorescence switch-on sensitivity for particular cations also exhibit fluorescence quenching in the presence of certain other cations. Whilst this is a potentially significant disadvantage if either/both the kinetics or thermodynamics of complex formation of the competing quenching cation render the sensor ineffective under the conditions of use, selective quenching of fluorescence can also have attendant advantages and provide a means of on–off detection, provided that selectivity is high. For example, Zhu and co-workers have elegantly shown that derivatisation of a glycerol backbone with cholic acid units provides a supramolecular assembly which binds pyrene. Addition of a number of metal ions results in quenching of the fluorescence of the bound pyrene, but the lack of selectivity renders this system ineffectual as a specific chemosensor despite the elegant system design [50]. Further design iterations resulted in a related system with the triazoles situated at the extremity of the supramolecular pyrene binding unit, but no improved cation selectivity was reported [51].

In contrast, Wang and co-workers reported that anthracene-functionalised fluorophore **26** was highly selective for the detection of Cu^{2+} with a five times greater relative quenching observed for this cation over the range of other cations reported to also cause quenching, which included Cd^{2+} and Pb^{2+}, but not Hg^{2+} [52]. A slightly less selective Cu^{2+} quenched fluorophore based on the click-functionalised anthraquinone scaffold **27** was reported by Chen and co-workers. Although Cr^{3+}, Pb^{2+} and Al^{3+} also quenched fluorescence to some extent, complete quenching of fluorescence was achieved by the addition of only one equivalent of Cu^{2+} to fluorescent solutions containing ten equivalents of the competing ions, indicating the potential of the system as a copper selective chemosensor [53]. In the same vein, Xie and co-workers have reported a benzothiadiazoyl click-functionalised β-cyclodextrin to be a highly selective fluorescence switch-off sensor for Ni^{2+}. Addition of fifty equivalents of a variety of cations to the sensor resulted in complete quenching by Ni^{2+}. Although significant quenching was also observed for Co^{2+}, Cu^{2+} and Hg^{2+}, it was shown that Ni^{2+} could be detected in the presence of these cations [54]. Li, Chen, and co-workers reported two-click derivatised pyranoglycosides to display selective fluorescence quenching in the presence of Ni^{2+}, whilst a number of other cations showed no or little quenching (Na^{+}, K^{+}, Mn^{2+}, Ca^{2+}, Co^{2+}, Ag^{+}, Mg^{2+}, Cd^{2+} and Cu^{2+}). Although both sensors required considerable molar excesses of Ni^{2+} to achieve significant quenching, sensor **28** was the more sensitive of the two. Both sensors also showed favourable inhibitory activity of protein tyrosine phosphatase 1B [55]. Bunz and co-workers have recently reported two related thiadiazole-based triazoles. Of these, **29** showed fluorescence quenching in aqueous solutions in the presence of Hg^{2+}, Cu^{2+} and Ni^{2+}; however, there was also a redshift of ca. 20–30 nm in the absorption band for the latter two ions, enabling them to be distinguished from Hg^{2+} [56]. Zhu and co-workers reported the synthesis of polymer **30** which showed some selectivity for the detection of Hg^{2+} over Cd^{2+}, Co^{2+}, Ni^{2+}, Zn^{2+} and Cu^{2+} as a result of quenching of fluorescence. Some quenching was also observed for Ag^{+}, but the efficiency of quenching was reported to be 2.8 times higher for Hg^{2+} [57]. A similar polymer containing a BINOL ether also showed Hg^{2+} quenching of fluorescence with good selectivity except for Pd^{2+} [58]. Cheng and co-workers introduced an (*S*)-BINOL-based sensor which showed only quenching in the presence of Hg^{2+} and a modest switch-on fluorescence in the presence of Ag^{+}. None of the other metals tested showed any effect on the sensor. Interestingly, an analogue of the original sensor bearing only one-click triazole unit and a free BINOL-OH group did not show such behaviour, indicating the importance of both triazoles in the sensing behaviour [59, 60]. In contrast, Chen recently reported an Ag^{+} selective switch-off sensors based on coumarin–sugar conjugates. Excellent switch-off selectivity was reported for **31**, although a significant excess of the cation was required [61]. A calix[4]arene appended with 8-hydroxyquinoline groups via triazoles gave a Hg^{2+}-selective fluorescence quenching sensor in acetonitrile, with only Fe^{3+} exhibiting quenching to a small extent [62]. Finally, Xie and co-workers have recently reported **32** to be a highly selective Hg^{2+} sensor. In EtOH/HEPES (v/v = 9:1), a significant quenching was observed for only Hg^{2+} which was accompanied by a redshift of 40 nm in the emission. In addition, a colour change from light yellow to light orange in the absorption spectrum allowed a dual modality of sensing this cation [63].

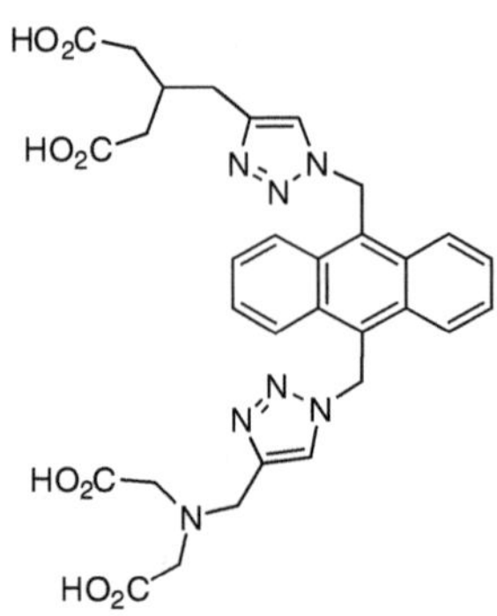

26

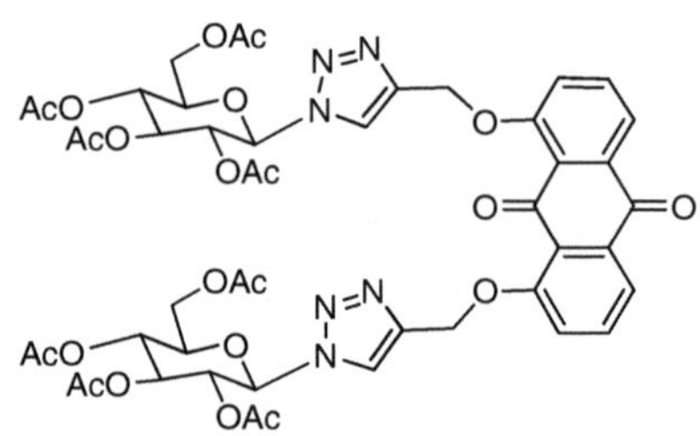

27

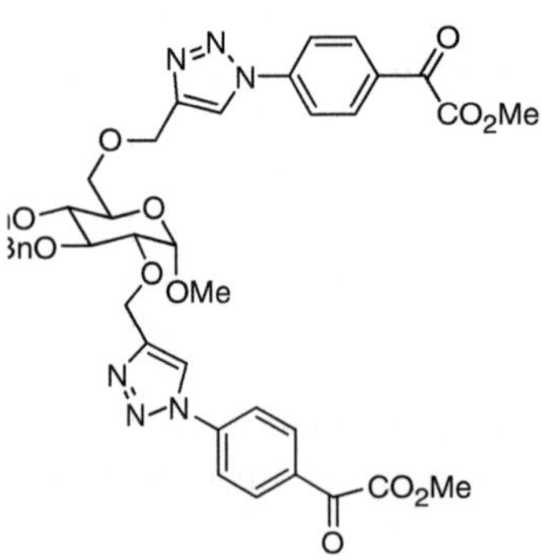

28

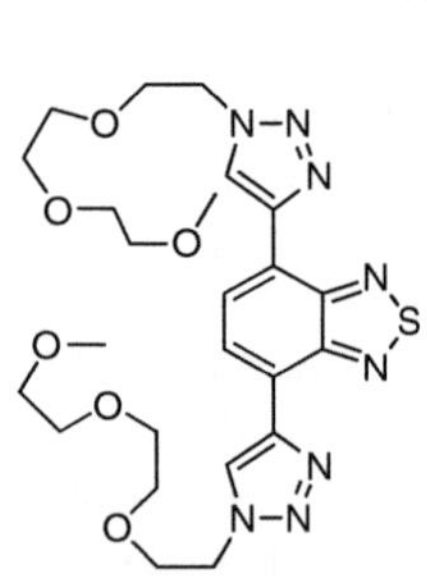

29

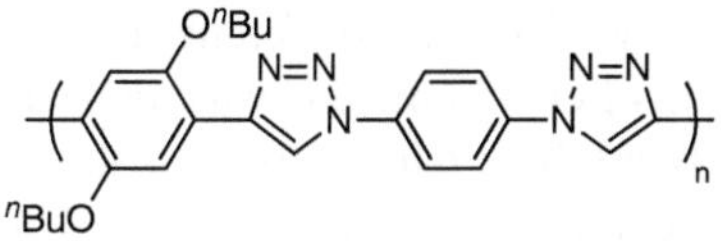

30

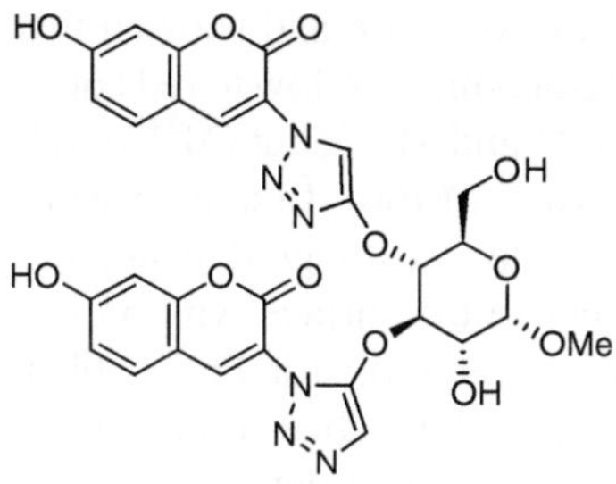

31

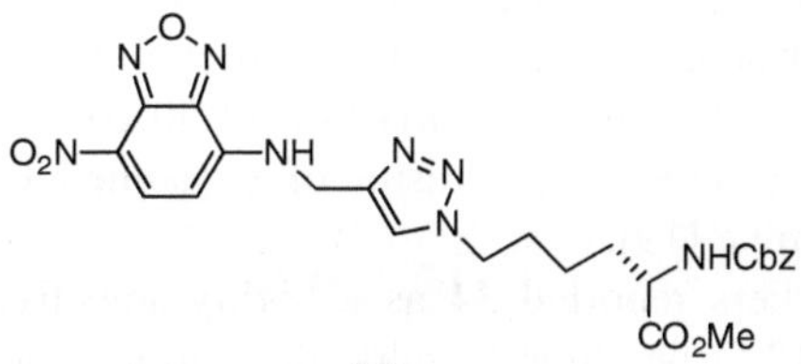

32

2.6 *Cation Sensing by Other Methods*

In addition to fluorescence switch-on and switch-off cation sensors, a number of other sensing modalities have also been reported. Of these, by far the most common are colorimetric sensors. Indeed, a number of the fluorophore-based sensors already discussed (vide supra) also display colorimetric sensing. In this section, however, sensors only displaying colorimetric properties are discussed. Li and co-workers reported that bifunctionalisation of silver nanoparticles with 2-mercaptoacetic acid and 4-(prop-2-ynyloxy)pyridine, which was then clicked to azidoethanol, furnished a highly selective colorimetric sensor for Co^{2+}. A clear shift from yellow (405 nm) absorption to red (550 nm) occurred exclusively for Co^{2+}, and the click triazole was shown to be an integral part of the sensor. Detection of Co^{2+} was possible in the presence of a range of competing cations down to a concentration of ca. 7×10^{-6} M [64]. In an alternative strategy, Jiang and co-workers used the Cu^{+} catalysed click reaction as a means of detecting Cu^{2+} through the aggregation of gold nanoparticles. The nanoparticles were functionalised with terminal azide and alkyne-functionalised thiols in addition to a commercial PEG thiol to prevent nanoparticle aggregation. Addition of Cu^{2+} and ascorbate to the nanoparticles resulted in a gradual change in the solution colour from pink to colourless with the concomitant precipitation of the nanoparticles due to aggregation. Although this process was relatively slow, it was shown to be selective for Cu^{2+} with a detection limit of 50 μM. Of a number of other cations tested, only Ag^{+} gave a colour change, which was attributed to surface electron effects of silver on gold nanoparticles [65]. Using the same alkyne,

Li and co-workers used click chemistry to functionalise silver nanoparticles with ethyl-2-azidoacetate and showed that they were selective colorimetric sensors for Cd^{2+}, displaying a yellow to red colour change. Some changes were also reported for Pb^{2+}, Hg^{2+} and Mn^{2+}, but Cd^{2+} could be clearly distinguished with a detection limit of 2×10^{-5} M [66]. Li and co-workers subsequently adopted a different strategy to detect Pb^{2+} with gold nanoparticles in which click chemistry was used to functionalise the surface with an oligo-ethyleneglycol. The nanoparticles showed excellent selectivity for Pb^{2+} which induced a clear colour change from red to navy blue, and the click triazole was shown to be necessary to maintain selectivity over other cations. A detection limit of 7 mM was established, and the nanoparticles were shown to have practical applications in the sensing of Pb^{2+} in both drinking water and paint [67]. Kannan and co-workers reported **33** to be a highly selective colorimetric sensor for Cu^{2+} in CH_3CN/H_2O (v/v 80:20). A clear blueshift upon binding of Cu^{2+} was observed as a result of weakened D–π–A ICT from the coordinated tertiary amine [19].

Chung and co-workers reported **34** as a highly selective colorimetric cation sensor for Ca^{2+} and Pb^{2+} perchlorate salts, the cations causing a bathochromic shift from 365 nm to 527 nm and 541 nm, respectively. The click triazole was shown to be integral to sensing of the cations, although the analysis of the cation binding properties of the derivatised calixarene was restricted to organic solvents and was hampered by the poor of solubility of the Pb^{2+} complex [68]. In an extension of this work, the MeO substituent was replaced with an $-NO_2$ group which produced a system able to act as an INHIBIT logic gate with a YES logic function for Ca^{2+} and F^-, although it was not able to act as a ditopic receptor for the ion pair [69].

A "click" derivatised analogue of glycyrrhetinic acid was reported to be a highly selective receptor for Hg^{2+} by Ju and co-workers. Addition of Hg^{2+} was shown to result in changes in the UV region of the absorbance spectrum, with the loss of bands at 253 and 317 nm, and with the appearance of a new absorption at 279 nm and of the other cations tested, only Cu^{2+} showed appreciable affinity [70]. Similarly, Kumar and Pandey synthesised two bile-acid-based receptors containing 1,2,3-triazoles and showed them to induce changes in the UV region in $CHCl_3$/MeOH (v/v 7:3) solution upon the addition of Hg^{2+}, with the loss of the bands at 254 and 317 nm and the appearance of a weak absorption at 279 nm. Again, Cu^{2+} showed some affinity for the receptor, as did Pb^{2+}. The receptors were shown to have NOR and OR type logic gating properties using Hg^{2+} and H^+ [71].

Li and co-workers reported **35** to be capable of extracting both group 1 and divalent metal cations due to the presence of both hard and soft donor ligands. Amongst a number of analogues prepared, this sensor showed the greatest efficacy in the extraction of group 1 picrate salts from organic media, although with no selectivity, as well as higher affinity for Pb^{2+} over the other cations tested [72]. Li and co-workers have also reported that **36** and its methyl ester analogue show good selectivity for the extraction of Cs^+ from the organic phase over other group 1 picrate salts with a Cs^+/Na^+ extraction ratio of 10.52 [73].

Viguier and Hulme reported a new approach to the possible in vivo detection of Cu^{+} through the formation of the luminescent lanthanide chelate **37** in which the dansyl fluorophore acts as an antenna allowing energy transfer to the europium centre. Although the click triazole was not directly involved in coordination to the europium centre, it was an essential part of the sensor as it prevented dansyl coordination to europium and quenching of the lanthanide luminescence, as had been observed in previously reported analogues. It was hypothesised that Cu^{+} chelates to glutathione in vivo, and that if this complex could catalyse the click reaction, the system could detect Cu^{+}. It was shown that addition of a Cu^{+} complex of glutathione at micromolar concentration to the alkynyl and azide precursors of **37** induced a tenfold increase in europium luminescence emission and confirmed its formation with control experiments, demonstrating its potential efficacy in vivo [74]. Collman has elegantly shown that local electrochemical generation of Cu^{+} can be used to pattern electrodes using click chemistry, allowing the selective functionalisation of identical but individually addressable electrodes with analytical scaffolds [75]. Finally, Lin, Chen and co-workers demonstrated that functionalisation of self-assembled azide-terminated monolayers could be used as a means of detecting Cu^{2+} using electrochemiluminescence (ECL). Silica nanoparticles containing $Ru(bipy)_3^{2+}$ and derivatised with an alkyne-terminated chain. The [3+2] cycloaddition was then induced by local generation of Cu^{+} electrochemically which resulted in a strong ECL signal being detected [76].

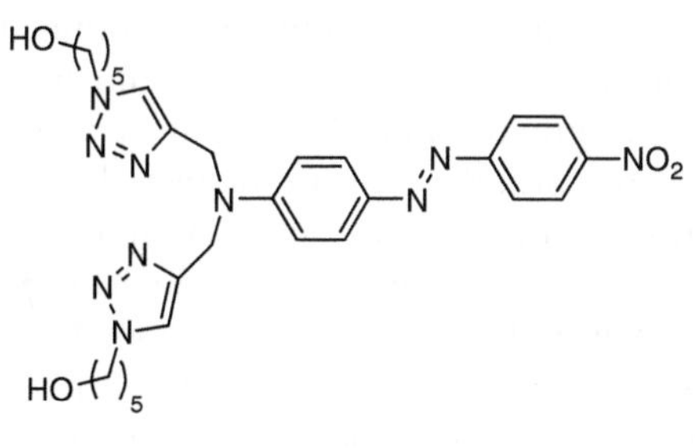

33

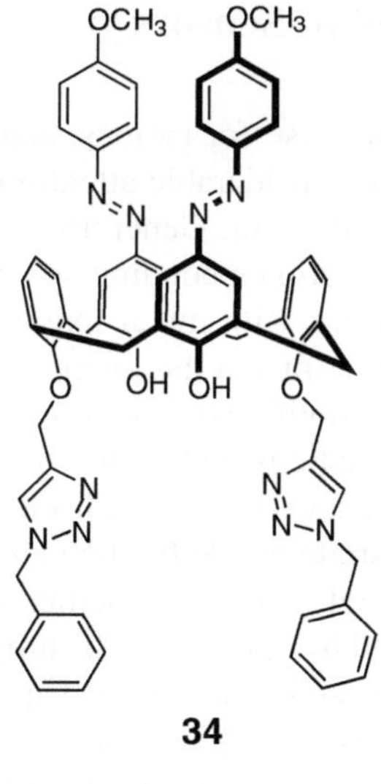

34

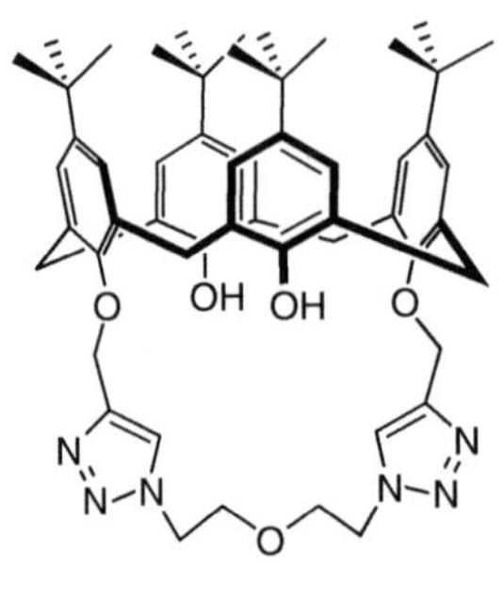

35

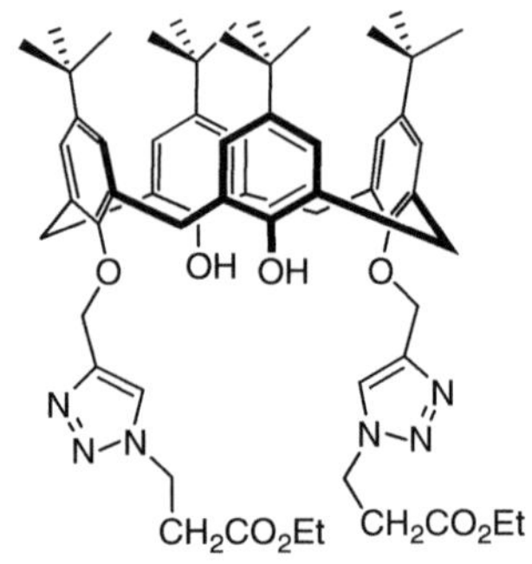

36

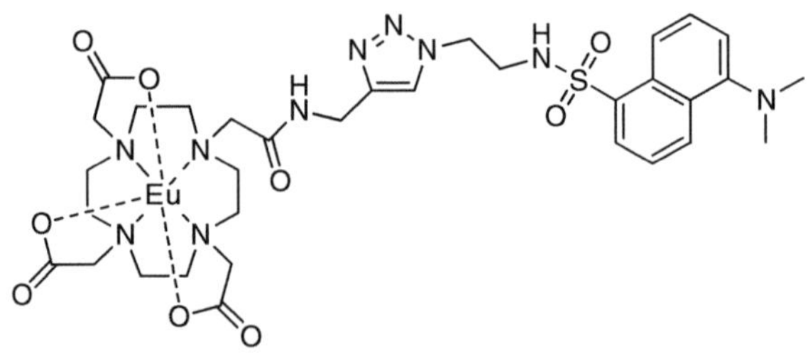

37

3 Anion Sensors

As is the case for cations, anion recognition by supramolecular architectures has received considerable attention due to the important roles they play in biology and the environment. Generally, receptor interactions with anions rely on Coulombic forces, hydrogen bonding, ion pairs, π-stacking interactions or through coordinative interactions with metal ions incorporated into the structure. Click chemistry has recently attracted the interest of a number of groups, and whilst an array of anion binding motifs have been reported [4, 6] and form the basis of another chapter in this thematic issue by Flood, sensors, however, are more restricted. Bachas and co-workers have exploited the eight C–H hydrogen bond donors in triazolophane **38** to demonstrate halide binding and recently showed their application as ionophores in poly(vinyl chloride) membrane electrodes. Selectivity for Cl^{-} and Br^{-} was achieved by optimising the proportions of triazolophane, plasticiser and a lipophilic additive, and their practical application was demonstrated in the detection of Br^{-} in horse serum [77]. Hua and Flood also demonstrated that photoinduced release and binding of Cl^{-} by **39** could be detected using conductivity measurements. The *trans–trans* stereoisomer forms a helical arrangement which optimises hydrogen bonding with the halide. Upon photoisomerisation to either the *cis–trans* or *cis–cis* form, the effectiveness of hydrogen bonding was reduced due to the conformational change that the isomerisation induced. As a result, the binding constant for Cl^- was lowered. It was shown using TBACl as the chloride source that photochemically induced changes in conductivity could be achieved over eight photochemical cycles due to sequential binding and release of Cl^-[78]. An electrochemical ferrocene reported and pyrene fluorophore were combined in **40** which was shown to be capable of detecting pyrophosphate both electrochemically and by an increase in fluorescence. Of a number of anions tested, only pyrophosphate induced a cathodic shift of the ferrocene/ferrocenium redox couple, and it was shown that a 2:1 receptor–anion complex formed. In addition, whilst the monomer showed a weak fluorescence emission typical of a pyrene monomer, addition of pyrophosphate induced a significant increase in the excimer emission at 448 nm, demonstrating the dual modality of anion sensing [79]. Triazoles have been utilised in a squaraine rotaxanes. In one case, the squaraine fluorescence is diminished on macrocyclic encapsulation but can be recovered by dye displacement with anions, although the recovery was not specific to a particular anion [80]. In another application, the rotaxane system acts as an optical chloride sensor; however, in both cases, the triazole only plays a structural role [81]. Pandey and co-workers prepared a series of bile acid-containing polymers **41** using click chemistry, which stabilised silver nanoparticles formed in their presence from the reduction of $AgNO_3$. The triazoles were essential to the stability of the nanoparticles. The effect of a range of anions on the nanoparticles was tested, and only I^- was observed to result in a colorimetric change as a result of nanoparticle aggregation with a detection limit of 250 μM [82]. A slight modification of **9** to incorporate methyl esters containing arms at the phenolic positions resulted in a system which exhibited enhanced pyrene emission in the presence of I^-. A 1:1 complex was formed, but the binding mechanism was not elucidated [83]. Another chemosensor containing the bis-triazole motif was prepared during efforts to create cavitands and nanotubes from calix[4]arenes using click chemistry. Incorporation of an anthracene fluorophore as a lower rim cap resulted in a cavitand that bound Br^- and I^- in methanol, which was detected by a small increase in fluorescence emission [84]. Meudtner and Hecht have reported oligomers containing alternating triazole and aryl units which adopt helical conformations in aqueous acetonitrile solution which are stabilised by π–π interactions. Addition of Cl^- and Br^- resulted in a reversal of helicity which could be detected by CD spectroscopy. In contrast, F^- showed no change [85]. Jiang and co-workers have employed CD spectroscopy to investigate the behaviour of cationic aryl triazole oligomers **42** which adopt helical conformations in aqueous methanol. In the absence of Cl^- and F^-, the individual foldamers form aggregates with opposite helicity; however, in their presence, aggregation is prevented as a result of halide binding, resulting in a means of their detection [86]. Finally, Wang and co-workers have elegantly demonstrated the efficacy of the copper-free strain promoted double [3+2] cycloaddition of dialkyne **43** with aqueous azide as a means of detection by LC–MS. Quantitation of aqueous azide was shown to be possible in a dynamic range 0.5–100 μM with a detection limit of at least 21 ppb, and a wide range of other anions did not show any interference [87].

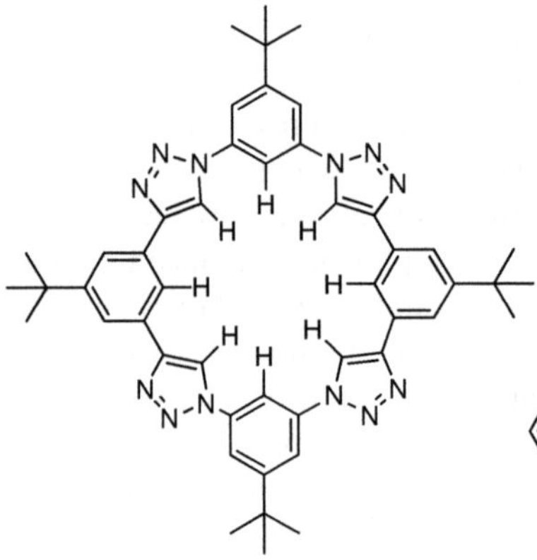

38

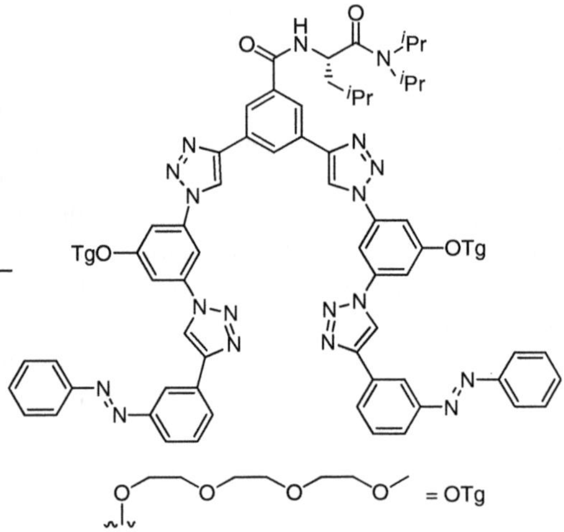

39

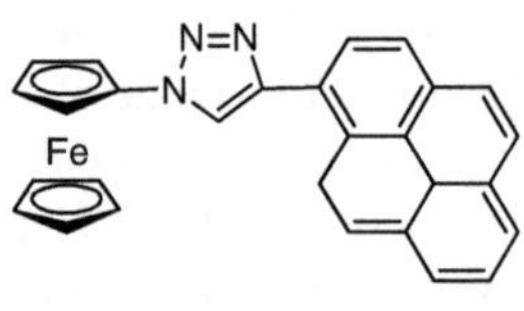

40

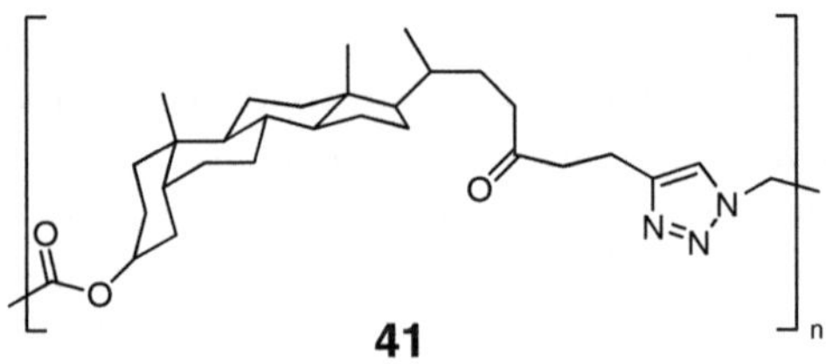

41

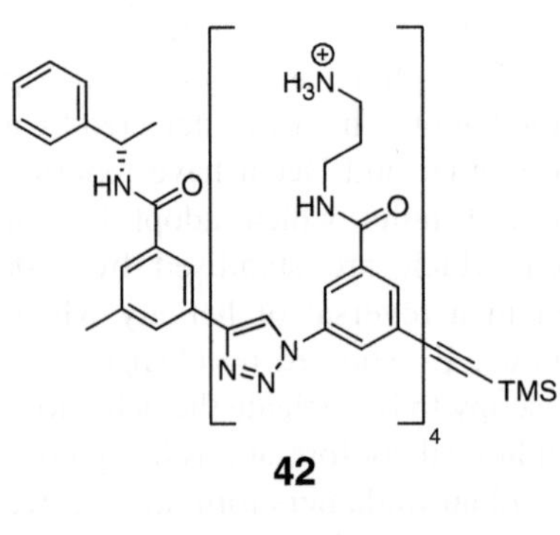

42

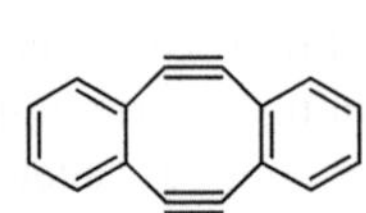

43

4 Combined Anion and Cation Sensors

Reports of triazole-containing anion and cation sensors are very limited in number. Yamato and co-workers recently reported that addition of the $H_2PO_4^-$ anion (40 equivalents) to the Zn^{2+} complex of **11** resulted in a 67% increase in the excimer band at 485 nm in the presence of small amounts of water in organic media. The system was also demonstrated to be an efficient logic circuit for a molecular traffic signal with the help of binary logic based on an INHIBIT and OR gate [29]. In an extension of previous work on cyclam-based zinc sensors, Todd, Rutledge and co-workers prepared **44**, a rare example of a "click" triazole which coordinates to metals through the electron-deficient N2 triazole atom. They showed that there was no enhancement in fluorescence in the presence of a number of cations, including Zn^{2+}, but that it was quenched by both Cu^{2+} and Zn^{2+}. Addition of a range of anions to the complexes of these metals resulted in recovery of fluorescence for Hg^{2+} as a result of demetallation but not for Cu^{2+}. I^- and $S_2O_3{}^{2-}$ were most effective with less than five equivalents of the anion required. This provided a means of distinguishing between the two cations which was demonstrated using cation-doped water from Sydney harbour [88].

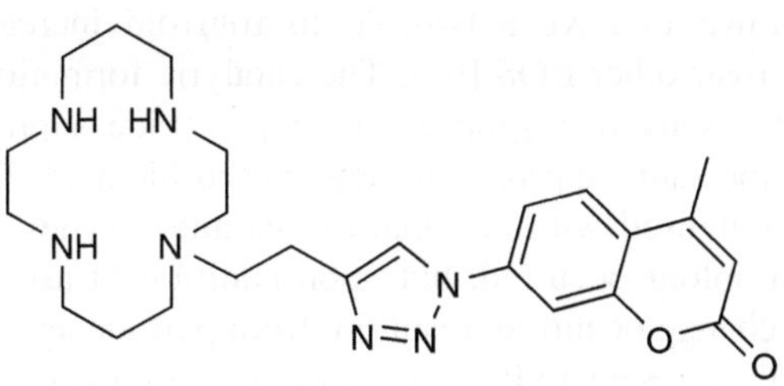

44

Astruc and co-workers have reported a number of dendrimers for sensing applications in which ferrocenyl groups were appended at the termini of dendrons using click chemistry. The electrochemical properties of the ferrocene units were affected by ion binding to the triazoles. Binding of cations shifted the Fc/Fc^+ couple to more positive potentials, whereas anion binding moved it to more negative potential. By tracking these changes in potential by cyclic voltammetry, it proved possible to non-specifically detect Pd^{2+}, Pt^{2+}, Cu^+ and Cu^{2+} as well as $H_2PO_4{}^{2-}$ and ATP^{2-} [89–91].

5 Sensing of Small Molecules

Probes for small molecules have a wealth of potential applications, e.g. in monitoring health, biological systems and homeland security. As a result of the widespread application of such sensors, it is unsurprising that click triazole sensors have been developed in this area for an array of applications which will be presented in this section.

There is increasing interest in the development of effective probes for nitroxyl (HNO) due to the recognition that it plays a number of important biological roles. Rosenthal and Lippard recently reported **45** to be an effective nitroxyl sensor. Fluorescence of the free ligand was quenched by Cu^{2+}. Reduction of Cu^{2+} to Cu^{+} with excess cysteine resulted in recovery of fluorescence. Treatment of **45** with Angeli's salt which generates equimolar HNO and nitrite under physiological conditions was shown to result in a greater than fourfold increase in fluorescence with visualisation possible with 50 μM Angeli's salt. The probe was shown to be selective for HNO over a number of other reactive oxygen and nitrogen species. Further, the efficacy of the probe for the detection of HNO in HeLa cells was demonstrated [92]. Wang and co-workers reported **46** to be an effective probe for the reactive oxygen species (ROS) hydrogen peroxide. In the presence of hydrogen peroxide, boronate ester conversion to phenol resulted in a significant enhancement in fluorescence at 475 nm as well as a 70 nm redshift in excitation wavelength. The probe was also shown to have a twofold to fourfold increase in response to hydrogen peroxide over other ROS [93]. The catalytic formation of a triazole has been used to elicit a sensing response for copper (vide supra), and Li and co-workers have used the same approach to detect ascorbic acid. They prepared gold nanoparticles functionalised with terminal azides and alkynes. In the presence of Cu^{2+}, no change in colour occurred, but upon addition of ascorbic acid, a red to pink/purple colour change occurred, resulting from particle aggregation which was demonstrated by TEM. Selectivity for ascorbic acid over a number of other biological reductants was established, and the optimised system was shown to have a detection limit of 3 nM. The new sensor system was used to detect ascorbic acid in supermarket orange and grapefruit juice and gave comparable results with standard assays [94]. Fossey, James and co-workers have reported **47** to be an effective switch-on fluorescence sensor for the detection of saccharides. The triazole plays a crucial role in the sensor as in the absence of saccharide fluorescence is quenched by the triazole. Formation of the "ate" complex at boron upon addition of a range of saccharides prevents quenching and resulted in a significant increase in fluorescence for D-fructose, D-galactose and D-mannose [95]. Zhao and co-workers have reported cruciform boronic acid fluorophore **48** to be an effective switch-on sensor for D-fructose, D-ribose, D-galactose and D-glucose. Although the response of the system was broadly similar for all of the saccharides tested, the affinity for fructose and ribose was higher. The triazole was proposed to play a crucial role as a hydrogen bond donor to stabilise the boron–saccharide complex at neutral pH. A linear analogue showed no sensitivity to the sugars [96]. A poly(vinyl alcohol) based glucose sensor has been reported in which the polymer chain was first decorated with pyrene units using click chemistry, and then the remaining alcohol units were functionalised with glucose oxidase (GOx) using standard coupling methodology. Quenching of pyrene fluorescence by dioxygen was shown to decrease in the presence of glucose due to dioxygen consumption by GOx. Glucose concentrations could be quantified between 0.25 and 3.0 mM, and although it is not clear whether the triazole plays any role in sensing, the sensor was successfully applied in the detection of glucose in fizzy and orange juice [97].

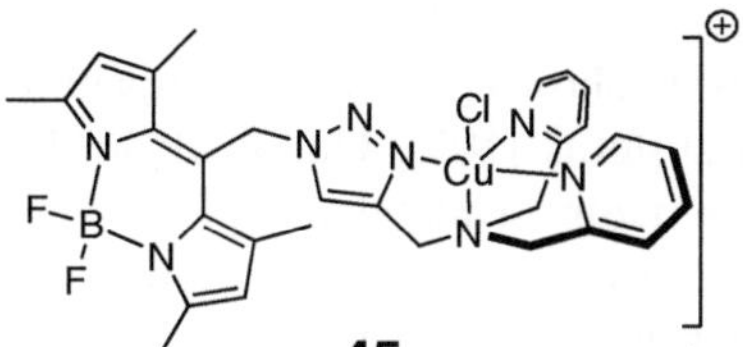

45

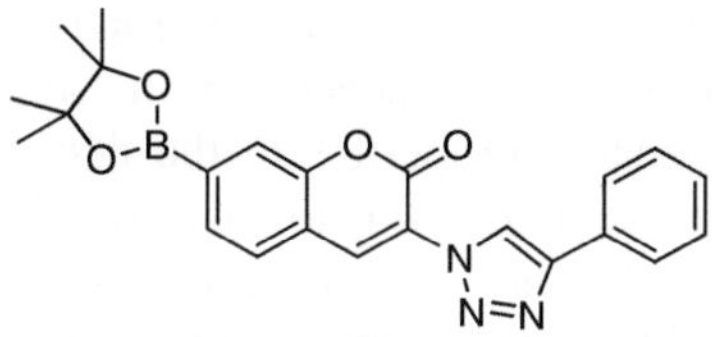

46

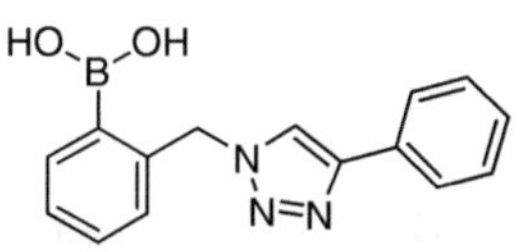

47

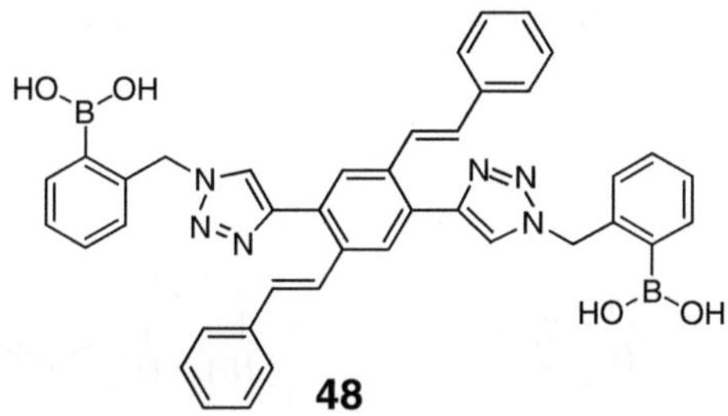

48

A number of other elegant sensor systems have been reported in which the triazole does not appear to be directly involved in the detection event. Fujimoto, Yamada and Inouye have reported that complementary oligonucleotide sequences with terminal pyrenes can be functionalised with cyclodextrins. Upon addition of a range of fatty acid guests, an increase in pyrene excimer emission occurs, allowing their detection; although there was no selectivity for the β-cyclodextrin systems, some selectivity for oleic acid was reported for the α-cyclodextrin analogues [98]. Similarly, click triazoles have been used to produce ferrocene–β-cyclodextrin conjugates which were shown to bind bile salts electrochemically as well as by isothermal calorimetry and NMR spectroscopy, although the triazole appears to only play a structural role [99]. Tetraphenylethene containing polytriazoles have been shown to be capable of detecting picric acid as an explosive surrogate in both solution and in model test-strip sensors. The sensor system operates through enhanced photoluminescence of the polymers due to aggregation-induced emission in the absence of the explosive surrogate. Addition of picric acid at concentrations as low as 0.1 ppm resulted in observable changes, but it is unclear whether the triazole is intimately involved in the sensing mechanism [100].

6 Biomolecule Sensing

As is the case for small molecule sensors, the detection of biological analytes presents a wealth for of sensing opportunities, and it is no surprise that click triazoles are employed to this end. Imperali and co-workers have previously developed a number of fluorophores which upon binding of Mg^{2+} show an enhanced fluorescence response which has been used as a means of detecting protein kinase activity [101]. Through the use of click chemistry, the excitation and emission wavelength of the original systems could be shifted bathochromically by 15 and 40 nm, respectively in **49**, and it was shown to have comparable chemosensing properties to the first-generation systems [102]. Our work in this area has centred on a novel form of allosteric sensing in metal complexes. We prepared the Cu^{2+} complex **50** of a biotinylated click cyclam ligand and showed that the weakly coordinated axial triazole could be perturbed by the addition of avidin which was detected by EPR and ENDOR spectroscopy. A model analogue system which did not contain the triazole showed no change in its EPR spectrum [103].

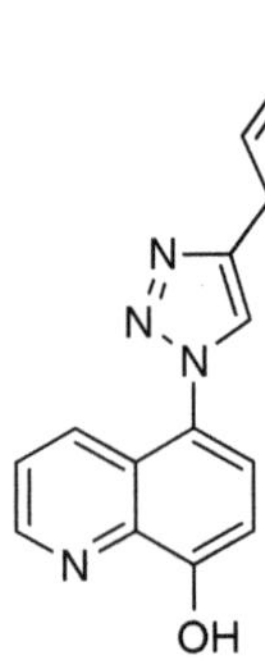

49

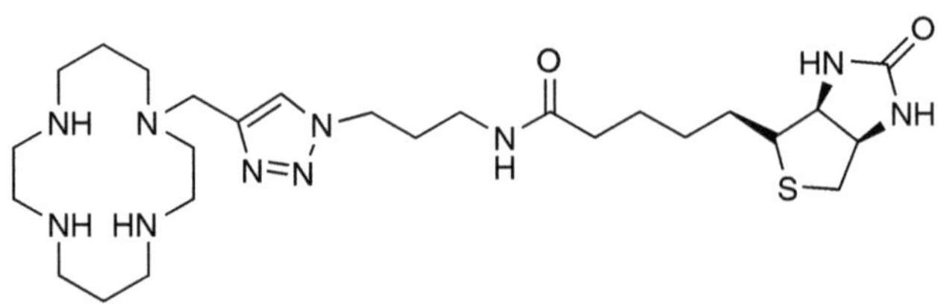

50

Methodology to functionalise two separate oligodeoxyribosenucleotides with a terminal azide and an alkyne-terminated naphthalimide which form a fluorescent triazole-containing dye in the presence of Cu^+ has been shown to be effective in the detection of the correct complementary DNA template. Whilst the click triazole formation was slow in the absence of the template, a significant enhancement was observed in the presence of the correct complementary template, resulting in a rapid increase in fluorescence, whereas a template with only a single nucleotide mismatch did not [104].

As is the case for small molecule sensors, a number of impressive systems have been developed, incorporating click triazoles, but it is unclear that the triazole plays a role in sensing other than providing a convenient means of sensor synthesis; nonetheless, these systems merit some mention. A Xe binding crytand has been functionalised with a matrix metalloprotein 7 (MMP7) selective peptide sequence. MMP7 activity was detected by both reduced tryptophan fluorescence upon peptide cleavage as well as a shift in the ^{129}Xe NMR chemical shift [105]. In keeping with other electrochemical systems already discussed, ferrocene–mannose conjugates have been shown to be effective electrochemical sensors for concanavalin A lectin, although the triazole role appears to be structural as a number of other non-triazole-containing conjugates were also effective sensors [106]. Microemulsion polymerisation has been used to prepare azide- and alkyne-terminated nanoparticles with sensitivity to pH changes and also stubilisin. Although the triazole only appears to act as a spacer in these systems, there is no reason why this approach could not be used to incorporate it as an integral part of the sensor system in other applications of this technology [107]. Similarly, gold nanoparticles have been functionalised with bioluminescent *Renilla* luciferase and a short peptide sequence susceptible to cleavage by MMP2. Quenching of luciferase fluorescence by the nanoparticles could be recovered through site-specific proteolytic cleavage of the peptide by MMP2 with the triazole apparently only acting as a convenient means of attaching the bioconjugate to the nanoparticle [108]. Finally, Lai and co-workers have utilised click triazoles to prepare electrochemical DNA (E-DNA) sensors. In the first example, the click triazole is employed to append a 25-base stem-loop DNA probe to a gold electrode which is labelled with methylene blue. The 15-base loop region targets the *K-ras* gene, and when DNA hybridisation occurs, the distance of the redox label from the electrode is increased, reducing electron transfer and consequent analyte detection [109]. The same approach was then adopted to produce a 3-pixel E-DNA sensor array through the sequential electrochemical generation of Cu^+ at each pixel to facilitate the specific click reaction of the alkyne and azide [110].

7 Concluding Remarks

Since the advent of the concept of "click" chemistry in 2001, it is without question that the click triazole has received most attention, and it has been shown to be an extremely versatile moiety with applications in a wide range of sensing applications, and there is little doubt that a number of other recent developments will soon be applied in sensing applications [6]. In the vast majority of sensors presented in this chapter, the triazole is integral to the sensing application; however, the utility of click chemistry is seen in its widespread application as a ligation motif in sensors, only some of which are presented herein. It therefore seems likely that in the coming years, practical applications of click triazole sensors in functional sensor materials for an array of applications will soon be realised.

Acknowledgements The author wishes to acknowledge Dr M.H. Todd of The School of Chemistry, The University of Sydney for the numerous useful discussions over several years.

References

1. Desvergne JP, Czarnik AW (eds) (1997) Chemosensors of ion and molecule recognition. Springer, Berlin
2. Anslyn EV (2007) Supramolecular analytical chemistry. J Org Chem 72:687–699
3. Fabbrizzi L, Poggi A (1995) Sensors and switches from supramolecular chemistry. Chem Soc Rev 24:197–202
4. Supramolecular chemistry of anionic species, themed issue (2010) Chem Soc Rev 39(10)
5. Fabbrizzi L (ed) Special volume on Luminescent Sensors (2000) Coord Chem Rev 205

6. Lau YH, Rutledge PJ, Watkinson M, Todd MH (2011) Chemical sensors that incorporate click-derived triazoles. Chem Soc Rev 40:2848–2866
7. de Silva AP, Gunaratne HQN, Gunnlaugsson T, Huxley AJM, McCoy CP, Rademacher JT, Rice TE (1997) Signaling recognition events with fluorescent sensors and switches. Chem Rev 97:1515–1566
8. Tornøe CW, Christensen C, Meldal M (2002) Peptidotriazoles on solid phase: [1,2,3]-triazoles by regiospecific copper(I)-catalyzed 1,3-dipolar cycloadditions of terminal alkynes to azides. J Org Chem 67:3057–3064
9. Rostovtsev VV, Green LG, Fokin VV, Sharpless KB (2002) A stepwise Huisgen cycloaddition process: copper(I)-catalyzed regioselective "ligation" of azides and terminal alkynes. Angew Chem Int Ed 41:2596–2599
10. Meldal M, Tornøe CW (2008) Cu-catalyzed azide – alkyne cycloaddition. Chem Rev 108:2952–3015
11. Kolb HC, Finn MG, Sharpless KB (2001) Click chemistry: diverse chemical function from a few good reactions. Angew Chem Int Ed 40:2004–2021
12. Moses JE, Moorhouse AD (2007) The growing applications of click chemistry. Chem Soc Rev 36:1249–1262
13. Applications of click chemistry, themed issue (2010) Chem Soc Rev 39(4)
14. Kim SH, Choi HS, Kim J, Lee SJ, Quang T, Kim JS (2010) Novel optical/electrochemical selective 1,2,3-triazole ring-appended chemosensor for the Al^{3+} ion. Org Lett 12:560–563
15. Maity D, Govindaraju T (2010) Pyrrolidine constrained bipyridyl-dansyl click fluoroionophore as selective Al^{3+} sensor. Chem Commun 46:4499–4501
16. Maity D, Govindaraju (2010) Conformationally constrained (Coumarin-Triazoyl-Bipyridyl) click fluoroionophore as a selective Al^{3+} sensor. Inorg Chem 49:7229–7231
17. Hsieh Y-C, Chir J-L, Wu H-H, Chang P-S, Wu A-T (2009) A sugar-aza-crown ether-based fluorescent sensor for Hg^{2+} and Cu^{2+}. Carbohydr Res 344:2236–2239
18. Hsieh Y-C, Chir J-L, Wu H-H, Guo C-Q, Wu A-T (2010) Synthesis of a sugar-aza-crown ether-based cavitand as a selective fluorescent chemosensor for Cu^{2+} ion. Tetrahedron Lett 51:109–111
19. Hrishikesan E, Saravanan C, Kannan P (2011) Bis-triazole-appended azobenzene chromophore for selective sensing of copper(II) ion. Ind Eng Chem Res 50:8225–8229
20. Sanji T, Nakamura M, Tanaka M (2011) Fluorescence 'turn-on' detection of Cu^{2+} ions with aggregation-induced emission-active tetraphenylethene based on click chemistry. Tetrahedron Lett 52:3283–3286
21. Tomat E, Lippard SJ (2010) Imaging mobile zinc in biology. Curr Opin Chem Biol 14:225–230
22. Xu Z, Yoon J, Spring DR (2010) Fluorescent chemosensors for Zn^{2+}. Chem Soc Rev 39:1996–2006
23. Zhu L-N, Gong S-L, Gong S-L, Yang C-L, Qin J-G (2008) Novel pyrene-armed Calix[4] arenes through triazole connection: ratiometric fluorescent chemosensor for Zn^{2+} and promising structure for integrated logic gates. Chin J Chem 26:1424–1430
24. Park SY, Yoon JH, Hong CS, Souane R, Kim JS, Matthews SE, Vicens J (2008) A pyrenyl-appended triazole-based Calix[4]arene as a fluorescent sensor for Cd^{2+} and Zn^{2+}. J Org Chem 73:8212–8218
25. Souchon V, Maisonneuve S, David O, Leray I, Xie J, Valeur B (2008) Photophysics of cyclic multichromophoric systems based on β-cyclodextrin and Calix[4]arene with appended pyridin-2'-yl-1,2,3-triazole groups. Photochem Photobiol Sci 7:1323–1331
26. Pathak RK, Ibrahim SM, Rao CP (2009) Selective Recognition of Zn^{2+} by Salicylaldimine appended triazole-linked Di-derivatives of Calix[4]arene by Enhanced fluorescence emission in aqueous-organic solutions: role of terminal $–CH_2OH$ moieties in conjunction with the imine in recognition. Tetrahedron Lett 50:2730–2734
27. Pathak RK, Dikundwar AG, Row TNG, Rao CP (2010) A lower rim triazole linked Calix[4] arene conjugate as a fluorescence switch on sensor for Zn^{2+} in blood serum milieu. Chem Commun 46:4345–4347

28. Ni X-L, Wang S, Zeng X, Tao Z, Yamato T (2011) Pyrene-linked triazole-modified Homooxacalix[3]arene: a unique C_3 symmetry ratiometric fluorescent chemosensor for Pb^{2+}. Org Lett 13:552–555
29. Ni X-L, Zeng X, Redshaw C, Yamato T (2011) Ratiometric fluorescent receptors for both Zn^{2+} and $H_2PO_4^-$ ions based on a pyrenyl-linked triazole-modified Homooxacalix[3]arene: a potential molecular traffic signal with an R-S latch logic circuit. J Org Chem 76:5696–5702
30. Shahgaldian P, Pieles U (2006) Cyclodextrin derivatives as chiral supramolecular receptors for enantioselective sensing. Sensors 6:593–615
31. Zhang YM, Chen Y, Li Z-Q, Li N, Liu Y (2010) Quinolinotriazole-β-cyclodextrin and its adamantane carboxylic acid complex as efficient water-soluble fluorescent Cd^{2+} sensors. Bioorg Med Chem 18:1415–1420
32. David O, Maisonneuve S, Xie J (2007) Generation of new fluorophore by click chemistry: synthesis and properties of β-cyclodextrin substituted by 2-pyridyl triazole. Tetrahedron Lett 48:6527–6530
33. Ballesteros-Garrido R, Abarca B, Ballesteros R, de Arellano CR, Leroux FR, Colobert F, García-España E (2009) [1,2,3]Triazolo[1,5-a]pyridine derivatives as molecular chemosensors for Zinc(II), nitrite and cyanide anions. New J Chem 33:2102–2106
34. Huang S, Clark RJ, Zhu L (2007) Highly sensitive fluorescent probes for zinc ion based on triazolyl-containing tetradentate coordination motifs. Org Lett 9:4999–5002
35. Tamanini E, Katewa A, Sedger LM, Todd MH, Watkinson M (2009) A synthetically simple, click-generated Cyclam-based zinc(II) sensor. Inorg Chem 48:319–324
36. Tamanini E, Flavin K, Motevalli M, Piperno S, Gheber LA, Todd MH, Watkinson M (2010) Cyclam-based "clickates": homogeneous and heterogeneous fluorescent sensors for Zn(II). Inorg Chem 49:3789–3800
37. Jobe K, Brennan CH, Motevalli M, Goldup SM, Watkinson M (2011) Modular 'click' sensors for zinc and their application in vivo. Chem Commun 47:6036–6038
38. Nguyen DM, Frazer A, Rodriguez L, Belfield KD (2010) Selective fluorescence sensing of zinc and mercury ions with hydrophilic 1,2,3-triazoyl fluorine probes. Chem Mater 22:3472–3481
39. Nguyen DM, Wang X, Ahn H-Y, Rodriguez L, Bondar MV, Belfield KD (2010) Novel hydrophilic Bis(1,2,3-triazolyl)fluorenyl probe for in vitro zinc ion sensing. Appl Mater Int 2:2978–2981
40. Jiao LJ, Meng T, Chen YM, Zhang M, Wang XL, Hao EH (2010) Triazolyl-linked 8-hydroxyquinoline dimer as a selective turn-on fluorosensor for Cd^{2+}. Chem Lett 39: 803–805
41. Hung HC, Cheng C-W, Ho IT, Chung W-S (2009) Dual-mode recognition of transition metal ions by bis-triazoles chained pyrenes. Tetrahedron Lett 50:302–305
42. Hung H-C, Cheng C-W, Wang Y-Y, Chen Y-J, Chung W-S (2009) Highly selective fluorescent sensors for Hg^{2+} and Ag^+ based on bis-triazole-coupled polyoxyethylenes in MeOH solution. Eur J Org Chem 6360–6366
43. Zhu LL, Lynch VM, Anslyn EV (2004) FRET Induced by an 'allosteric' cycloaddition reaction regulated with exogenous inhibitor and effectors. Tetrahedron 60:7267–7275
44. Varazo K, Le Droumaguet C, Fullard K, Wang Q (2009) Metal ion detection using a fluorogenic 'click' reaction. Tetrahedron Lett 50:7032–7034
45. Jin J, Zhang X-B, Xie D-X, Gong Y-J, Zhang J, Chen X, Shen G-Li, Yu R-X (2010) Clicking fluoroionophores onto mesoporous silicas: a universal strategy towards efficient fluorescent surface sensors for metal ions. Anal Chem 82:6343–6346
46. Chang K-C, Su I-H, Senthilvelan A, Chung W-S (2007) Triazole-modified Calix[4]crown as a novel fluorescent on-off switchable chemosensor. Org Lett 9:3363–3366
47. Kim H, Lee S, Lee J, Tae J (2010) Rhodamine triazole-based fluorescent probe for the detection of Pt^{2+}. Org Lett 12:5342–5345
48. Schweinfurth D, Hardcastle KI, Bunz UHF (2008) 1,3-Dipolar cycloaddition of alkynes to azides. Construction of operationally functional metal responsive fluorophores. Chem Commun 2203–2205

49. Shi J, Liu L, He J, Meng X, Guo Q (2007) Facile derivatization of pyridyloxazole-type fluorophore via click chemistry. Chem Lett 36:1142–1143
50. Zhang J, Luo J, Zhu XX, Junk MJN, Hinderberger D (2009) Molecular pockets derived from cholic acid as chemosensors for metal ions. Langmuir 26:2958–2962
51. Zhang J, Junk MJN, Luo J, Hinderberger D, Zhu XX (2010) 1,2,3-Triazole-containing molecular pockets derived from cholic acid: the influence of structure on host-guest coordination properties. Langmuir 26:13415–13421
52. Varazo K, Xie F, Gulledge D, Wang Q (2008) Synthesis of triazolyl anthracene as a selective fluorescent chemosensor for the Cu(II) ion. Tetrahedron Lett 49:5293–5296
53. Zhang Y-J, He X-P, Hu M, Li Z, Shi X-X, Chen G-R (2011) Highly optically selective and electrochemically active chemosensor for copper(II) based on triazole-linked glucosyl anthraquinone. Dyes Pigments 88:391–395
54. Maisonneuve S, Fang Q, Xie J (2008) Benzothiadiazoyl-triazoyl cyclodextrin: a selective fluoroionophore for Ni(II). Tetrahedron 64:8716–8720
55. Song Z, He X-P, Jin X-P, Gao L-X, Sheng L, Zhou Y-B, Li J, Chen G-R (2011) 'Click' to bidentate bis-triazoyl sugar derivatives with promising biological and optical features. Tetrahderon Lett 52:894–898
56. Brombosz SM, Appleton AL, Zappas AJ II, Bunz UHF (2010) Water-soluble Benzo- and Naphtho-thiadiazole-based bistriazoles and their metal sensing properties. Chem Commun 46:1419–1421
57. Huang X, Meng J, Dong Y, Cheng Y, Zhu C (2010) Polymer-based fluorescence sensor incorporating triazole moieties for Hg^{2+} detection via click reaction. Polymer 51:3064–3067
58. Zheng L, Huang X, Shen Y, Cheng Y (2010) Click chemistry approach to fluorescence-based polybinaphthyls incorporating a triazole moiety for Hg^{2+} recognition. Synlett 453–456
59. Liu X, Yang X, Peng H, Zhu C, Cheng Y (2011) A fluorescent sensor for Hg^{2+} and Ag^{+} functions as a modular switch based on click-generated triazole moiety. Tetrahedron Lett 52:2295–2298
60. Liu X, Yang X, Fu Y, Zhu C, Cheng Y (2011) Novel fluorescent sensor for Ag^{+} and Hg^{2+} based on the BINOL-pyrene derivative. Tetrahderon 67:3181–3186
61. He X-P, Song Z, Wang Z-Z, Shi X-X, Chen K (2011) Creation of 3,4-bis-triazolocoumarin-sugar conjugates via fluorogenic dual click chemistry and their quenching specificity with silver(I) in aqueous media. Tetrahderon 67:3343–3347
62. Tian D, Yan H, Li H (2010) A selective fluorescent probe of Hg^{2+} based on triazole-linked 8-Oxyquinoline Calix[4]arene by click chemistry. Supramol Chem 22:249–255
63. Ruan Y-B, Maisonneuve S, Xie J (2011) Highly selective fluorescent and colorimetric sensor for Hg^{2+} based on triazole-linked NBD. Dyes Pigments 90:239–244
64. Yao Y, Tian D, Li H (2010) Cooperative binding of bifunctionalized and click-synthesized silver nanoparticles for colorimetric Co^{2+} sensing. Appl Mater Int 2:684–690
65. Zhou Y, Wang S, Jiang X (2008) Visual detection of copper(II) by azide- and alkyne-functionalised gold nanoparticles using click chemistry. Angew Chem Int Ed 47:7454–7456
66. Li H, Yao Y, Han H, Zhan J (2009) Triazole-ester modified silver nanoparticles: click synthesis and Cd^{2+} colorimetric sensing. Chem Commun 4812–4814
67. Li H, Zheng Q, Han C (2010) Click synthesis of Podand triazole-linked gold nanoparticles as highly selective and sensitive colorimetric probes for lead(II) ions. Analyst 135:1360–1364
68. Chang K-C, Su I-H, Lee G-H, Chung W-S (2007) Triazole- and Azo-coupled Calix[4]arene as a highly sensitive chromogenic sensor for Ca^{2+} and Pb^{2+} ions. Tetrahedron Lett 48:7274–7278
69. Chang K-C, Su I-H, Wang Y-Y, Chung W-S (2010) A bifunctional chromogenic Calix[4] arene chemosensor for both cations and anions: a potential Ca^{2+} and F^{-} switched INHIBIT logic gate with a YES logic function. Eur J Org Chem 4700–4704
70. Hu J, Zhang M, Yu LB, Ju Y (2010) Synthesis and binding ability of 1,2,3-triazole-based triterpenoid receptors for recognition of Hg^{2+} ion. Biorg Med Chem Lett 20:4342–4345

71. Kumar A, Pandey PS (2009) Steroidal 1,2,3-triazole-based sensors for Hg^{2+} ion and their logic gate behaviour. Tetrahedron Lett 50:5842–5845
72. Zhan J, Tian D, Li H (2009) Synthesis of Calix[4]crowns containing soft and hard ion binding sites via click chemistry. New J Chem 33:725–728
73. Li H, Zhan J, Chen M, Tian D, Zou Z (2010) Metal ions recognition by 1,2,3-Triazolium Calix [4]arene esters synthesised via click chemistry. J Incl Phenom Macrocyc Chem 66:43–47
74. Viguier RFH, Hulme A (2006) A sensitized Europium complex generated by micromolar concentrations of copper(I): toward the detection of copper(I) in biology. J Am Chem Soc 128:11370–11371
75. Devaraj NK, Dinolfo PH, Chidsey CED, Collman JP (2006) Selective functionalization of independently addressable microelectrodes by electrochemical activation and deactivation of a coupling catalysts. J Am Chem Soc 128:1794–1795
76. Qiu S, Gao S, Zhu X, Lin Z, Qiu B, Chen G (2011) Development of ultra-high sensitive and selective electrochemiluminescent sensor for copper(II) ions: a novel strategy for modification of gold electrodes using click chemistry. Analyst 136:1580–1585
77. Zahran EM, Hua Y, Li Y, Flood AH, Bachas LG (2010) Triazolophanes: a new class of halide-selective ionophores for potentiometric sensors. Anal Chem 82:368–375
78. Hua Y, Flood AH (2010) Flipping the switch on chloride concentrations with a light-active foldamer. J Am Chem Soc 132:12838–12840
79. Romero T, Caballero A, Tárraga A, Molina P (2009) A click-generated triazole tethered ferrocene-pyrene dyad for dual-mode recognition of the pyrophosphate anion. Org Lett 11:3466–3469
80. Gassensmith JJ, Barr L, Baumes JM, Paek A, Nguyen A, Smith BD (2008) Synthesis and photophysical investigation of squaraine rotaxanes by "clicked capping". Org Lett 10:3343–3346
81. Gassensmith JJ, Matthys S, Lee J-J, Wojcik A, Kamat PV, Smith BD (2010) Squaraine rotaxane as a reversible optical chloride sensor. Chem Eur J 16:2916–2921
82. Kumar A, Chhatra RK, Pandey PS (2010) Synthesis of click bile acid polymers and their application in stabilization of silver nanoparticles showing iodide sensing property. Org Lett 12:24–27
83. Kim JS, Park SY, Kim SH, Thuery P, Souane R (2010) A pyrenyl-appended triazole-based Calix[4]arene as a fluorescent sensor for iodide ion. Bull Kor Chem Soc 31:624–629
84. Morales-Sanfrutos J, Ortega-Muñoz M, Lopez-Jaramillo J, Hernandez-Mateo F, Santoyo-Gonzalez F (2008) Synthesis of Calixerene-based cavitands and nanotubes by click chemistry. J Org Chem 73:7768–7771
85. Meudtner RM, Hecht S (2008) Helicity inversion in responsive foldamers induced by achiral halide ion guests. Angew Chem Int Ed 47:4926–4930
86. Wang Y, Li F, Han Y, Wang F, Jiang H (2009) Folding and aggregation of cationic Oligo (aryl-triazole)s in aqueous solution. Chem Eur J 15:9424–9433
87. Wang L, Dai C, Chen W, Liu Wang S, Wang B (2011) Facile derivatization of azide ions using click chemistry for their sensitive detection with LC-MS. Chem Commun 47:10377–10379
88. Lau YH, Price JR, Todd MH, Rutledge PJ (2011) A click fluorophore sensor that can distinguish Cu^{II} and Hg^{II} via selective anion-induced demetallation. Chem Eur J 17:2850–2858
89. Ornelas C, Ruiz Aranzaes J, Cloutet E, Alves S, Astruc D (2007) Click assembly of 1,2,3-triazole-linked dendrimers, including ferrocenyl dendrimers, which sense both oxo and metal cations. Angew Chem Int Ed 46:872–877
90. Astruc D, Ornelas C, Aranzaes JR (2008) Ferrocenyl-terminated dendrimers: design for applications in molecular electronics, molecular recognition and catalysis. J Inorg Organomet Polym Mater 18:4–17
91. Camponovo J, Ruiz J, Cloutet E, Astruc D (2009) New polyalkynyl dendrons and dendrimers: "click" chemistry with azidomethylferrocene and specific anion and cation electrochemical sensing properties of the 1,2,3-triazole-containing dendrimers. Chem Eur J 15:2990–3002

92. Rosenthal J, Lippard SJ (2010) Direct detection of nitroxyl in aqueous solution using a tripodal copper(II) BODIPY complex. J Am Chem Soc 132:5536–5537
93. Lupei D, Nanting N, Minyong L, Wang B (2010) A fluorescent hydrogen peroxide probe based on a 'click' modified coumarin fluorophore. Tetrahedron Lett 51:1152–1154
94. Zhang Y, Li B, Xu C (2010) Visual detection of ascorbic acid via alkyne-azide click reaction using gold nanoparticles as a colorimetric probe. Analyst 135:1579–1584
95. Scrafton DK, Taylor JE, Mahon MF, Fossey JS, James TD (2008) "Click-fluors": modular fluorescent saccharide sensors based on a 1,2,3-triazole ring. J Org Chem 73:2871–2874
96. Mulla K, Dongare P, Zhou N, Chen G, Thompson DW, Zhao Y (2011) Highly sensitive detection of saccharides under physiological conditions with click synthesized boronic acid-oligomer fluorophores. Org Biomol Chem 9:1332–1336
97. Odaci D, Gacal BN, Gacal B, Timur S, Yagci Y (2009) Fluorescence sensing of glucose using glucose oxidase modified by PVA-pyrene prepared via "click" chemistry. Biomacromolecules 10:2928–2934
98. Fujimoto K, Yamada S, Inouye M (2009) Synthesis of versatile fluorescent sensors based on click chemistry: detection of unsaturated fatty acids by their pyrene-emission switching. Chem Commun 7164–7166
99. Casas-Solva JM, Ortiz-Salmerón E, Fernández I, García-Fuentes L, Santoyo-González F, Vargas-Berenguel A (2009) Ferrocene-β-cyclodextrin conjugates: synthesis, supramolecular behavior, and use as electrochemical sensors. Chem Eur J 15:8146–8162
100. Qin A, Lam JWY, Tang L, Jim CKW, Zhao H, Sun J, Tang BZ (2009) Polytriazoles with aggregation-induced emission characteristics: synthesis by click polymerization and application as explosive chemosensors. Macromolecules 42:1421–1424
101. Lukovíc E, González-Vera JA, Imperiali B (2008) Recognition-domain focused chemosensors: versatile and efficient reporters of protein kinase activity. J Am Chem Soc 130:12821–12827
102. González-Vera JA, Lukovíc E, Imperiali B (2009) Synthesis of red-shifted 8-hydroxyquinoline derivatives using click chemistry and their incorporation into phosphorylation chemosensors. J Org Chem 74:7309–7314
103. Tamanini E, Rigby SEJ, Motevalli M, Todd M, Watkisnon M (2009) Responsive metal complexes: a click-based "allosteric scorpionate" complex permits the detection of a biological recognition event by EPR/ENDOR spectroscopy. Chem Eur J 15:3720–3728
104. Jentzsch E, Mokhir A (2009) A fluorogenic, nucleic acid directed "click" reaction. Inorg Chem 48:9593–9595
105. Wei Q, Seward GK, Hill PA, Patton B, Dimitrov IE, Kuzma NN, Dmochowski IJ (2006) Designing ^{129}Xe NMR biosensors for matrix metalloproteinase detection. J Am Chem Soc 128:13274–13283
106. Casas-Solvas JM, Ortiz-Salmerón E, García-Fuentes L, Vargas-Berenguel A (2008) Ferrocene-mannose conjugates as electrochemical molecular sensors for concanavalin A lectin. Org Biomol Chem 6:4230–4235
107. Welser K, Ayal Perera MD, Aylott JW, Chan WC (2009) A facile method to clickable sensing polymeric nanoparticles. Chem Commun 6601–6603
108. Kim Y-P, Daniel WL, Xia Z, Xie H, Mirkin CA, Rao J (2010) Bioluminescent nanosensors for protease detection based upon gold nanoparticle-luciferase conjugates. Chem Commun 46:76–78
109. Cañete SJP, Yang W, Lai RY (2009) Folding-based electrochemical DNA sensor fabricated by "click" chemistry. Chem Commun 4835–4837
110. Cañete SJP, Lai RY (2010) Fabrication of an electrochemical DNS sensor array via potential assisted "click" chemistry. Chem Commun 46:3941–3943

Top Heterocycl Chem (2012) 28: 137–162
DOI: 10.1007/7081_2011_66

Published online: 25 January 2012

Triazole-Based Polymer Gels

Hak-Fun Chow, Chui-Man Lo, and Yuan Chen

Abstract This chapter summarizes the recent progress of the chemistry of triazole-based polymer gels from 2000 to present. Based on the gelation mechanism, the gels are classified into reversible physical and irreversible chemical gels. The design principles of these gel systems were examined, and the various driving forces for gelation, such as hydrogen bonding, π–π stacking interaction, hydrophobic effect, covalent linkage, and metal coordination, were described. Structural factors that affect the swelling ability of such gels and their applications in controlled drug release were revealed. The advantages and drawbacks of using CuAAC chemistry to create such gelating systems, as compared to other conventional synthetic methodologies, were also discussed.

Keywords 1,2,3-Triazole · Click chemistry · Controlled release · Polymer gels

Contents

H.-F. Chow (✉) and Y. Chen
Department of Chemistry, Center of Novel Functional Molecules, and Institute of Molecular Functional Materials, AoE Scheme, The Chinese University of Hong Kong, Shatin, NT, Hong Kong, SAR
e-mail: hfchow@cuhk.edu.hk

C.-M. Lo
School of Science and Technology, The Open University of Hong Kong, Homantin, Kowloon, Hong Kong, SAR

1 Introduction

The heterocyclic 1,2,3-triazole ring is one of the key structural elements in many drug molecules and polymers [1]. This functionality becomes increasingly popular due to the discovery that it could be readily accessible via the copper(I)-catalyzed cycloaddition between an acetylene and an azide (CuAAC) (Scheme 1) [2, 3]. Several excellent review articles have already been published on the use of CuAAC chemistry in the preparation of polymers [4–18] and biomaterials [19–27]. It is now becoming clear that the 1,2,3-triazole ring not only functions as a linker between two molecular fragments but also contributes to the overall physical and chemical properties of the resulting molecules. There are abundant examples showing that oligo- and polytriazole compounds possess many interesting conformational and supramolecular properties [28]. This article aims to focus on one supramolecular aspect of such triazole-based polymers, namely, their gelation properties. The survey covers the literature on this subject from 2000 to present. A search on the Web of Science database using the keywords triazole and gel revealed no reports on this particular subject prior to 2000.

The immobilization of solvent molecules by the addition of a small amount of another molecular species (gelator) is called gelation. This phenomenon had long been noted in the history of human activities, such as food preparation and preservation. However, the scientific rationale and the molecular mechanism behind this phenomenon have just been realized. Broadly speaking, gelators can be classified into two categories based on their molecular weight – low molecular weight gelators (LMWG) and polymer gelators (PG). The molecular basis of gelation by LMWG has been extensively reviewed [29–36]. It is believed that the gelators self-assemble into a three-dimensional (3D) network structure through noncovalent interactions such as hydrogen bonding, π–π stacking, hydrophobic, and/or van der Waals interactions. The solvent molecules are then entrapped and immobilized within the matrix via capillary force. The resulting network structure can therefore be considered as a supramolecular polymer and can be broken down to the individual LMWG reversibly upon external stimuli such as change of temperature and solvent conditions. This type of gel is also known as physical gel, and the sol–gel process is generally reversible. On the other hand, the 3D network created by most polymer gelators is covalent in nature. Most often, the network is formed during polymerization using polyvalent monomer units. In this case, the gel is called a chemical gel, and the sol–gel process is mostly irreversible unless the 3D network can be easily broken down by thermal degradation. As it turns out, polymer physical gels are rare as compared to LMWG physical gels [37].

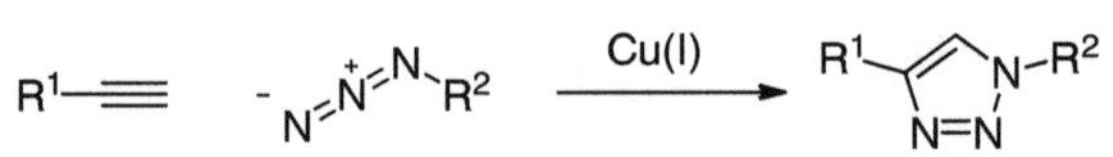

Scheme 1 Preparation of 1,4-disubstituted 1,2,3-triazole via CuAAC reaction

In this chapter, we will discuss the chemistry and properties of triazole-based polymer gels according to whether they belong to physical or chemical gels.

2 Triazole-Based Physical Gels

Back in 2006, Finn first reported the gelating properties of three *trans*-1,2-diamidocyclohexanes LMWG **1–3** (Scheme 2) [38]. The gelating properties of the diacetylene **1** and diazide **2** were slightly inferior to that of **3**. Hence, compounds **1** and **2** did not form gels as a 3% solution in CH_3CN, but compound **3** did at the same concentration. On the other hand, all three compounds failed to form gels in a mixed solvent of 2,6-lutidine–CH_3CN (1:20). In order to improve the gelation property of **1** and **2**, they were subjected to CuAAC polymerization with various reaction partners to produce poly(amide–triazole)s **4** and **5** with much improved gelation power in the same mixed solvent system. This work provided the first examples that such poly (amide–triazole) compounds were better organogelators than their constituent monomers. On the other hand, the in situ–prepared poly(amide–triazole) **6** from the

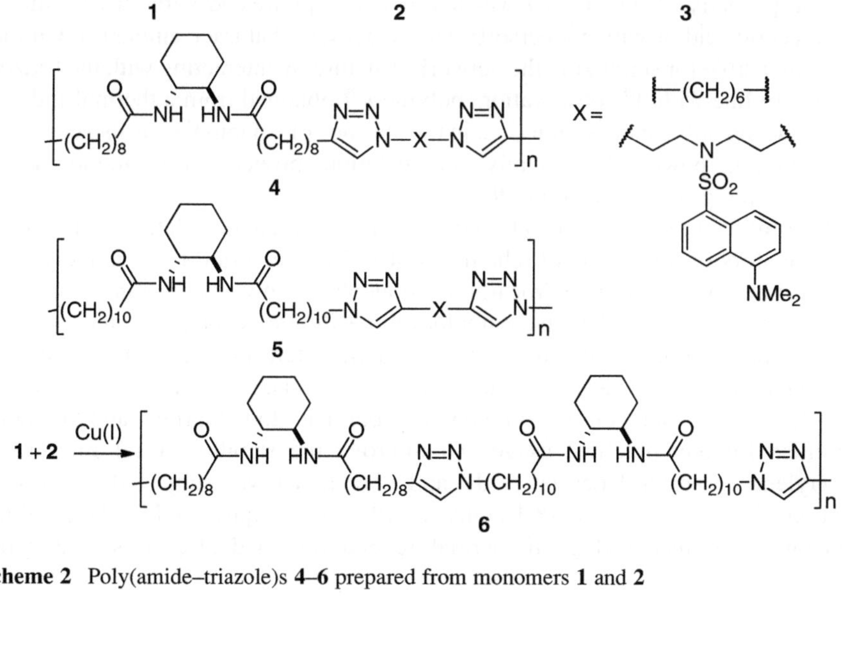

Scheme 2 Poly(amide–triazole)s **4–6** prepared from monomers **1** and **2**

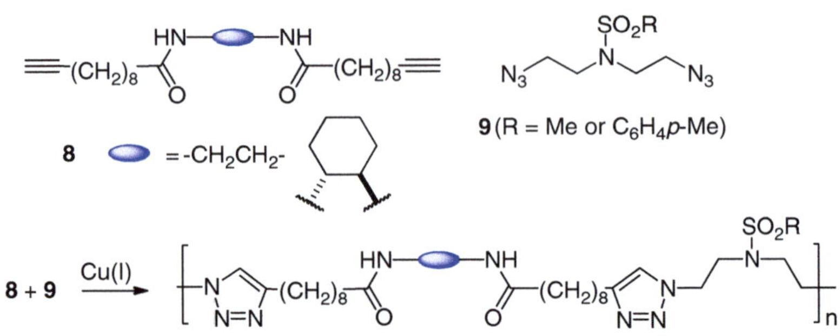

7

Scheme 3 Poly(amide–triazole)s **7** formed copolymerization of monomers **8** and **9**

CuAAC polymerization between **1** and **2** was found to be a relatively weak organogelator in the mixed solvent system with a T_{gel} of 47°C. However, the thermal stability of the organogel could be improved ($T_{gel} = 69$°C) if it was prepared under the same conditions in the presence of 9 equiv. of **3**. This suggested that a synergistic gelation effect was in operation between LWMG **3** and polymer **6**. It should be noted that in both cases, the metal catalyst Cu(I) ion was incorporated into the gel and made it more thermally stable. Hence, the gel mixture of **3**:**6** (10:1) without Cu(I) ions had a lower T_{gel} of 63°C.

As an extension of the above work, Díaz reported another series of poly (amide–triazole)s **7** from the AABB type CuAAC copolymerization of diacetylenes **8** and diazides **9** (Scheme 3) [39]. Polymers **7** were selective organogelators for dimethylsulfoxide (DMSO) but formed brittle gels in a mixture of DMSO and a variety of cosolvents such as toluene, ethyl acetate, and dimethylformamide. The gelation property of polymers **7** was lost if it was pretreated with ethylenediaminetetraacetic acid or with coppersorb. This suggested that trace amount of Cu ions was responsible for stabilizing the network structure by interacting with the triazole ligands. In line with this observation, polymers **7** obtained from a thermal-induced cycloaddition polycondensation (i.e., in the absence of Cu ions) were also devoid of any gelating power. Incidentally, no gelation property was found if the sulphonamide group in monomer **9** was removed.

Poly(amide–triazole)s could also be used to strengthen the gelation properties of weak gelators or nongelators (Scheme 4) [40]. The cholesterol derivative **10** is a powerful organogelator in hydrocarbon and alcoholic solvents but cannot immobilize aromatic solvents. The zinc–phthalocyanine(Zn–Pc)–cholesterol derivative **11** is also a nongelator in the same solvent system. Interestingly, a 20:1 mixture of compounds **10** and **11** was able to gel aromatic solvents. Due to the presence of the zinc–Pc moiety, the gel was photoactive but had low thermal and temporal stabilities for any practical usage. To improve the stability, a 1:1 mixture of diacetylene cholesterol derivative **12** and diazide **13** was subjected to CuAAC reactions in the presence of 18.1 equiv. of **10** and 1.8 equiv. of **11**. The resulting organogels **14** displayed good thermal reversibility and also possessed better

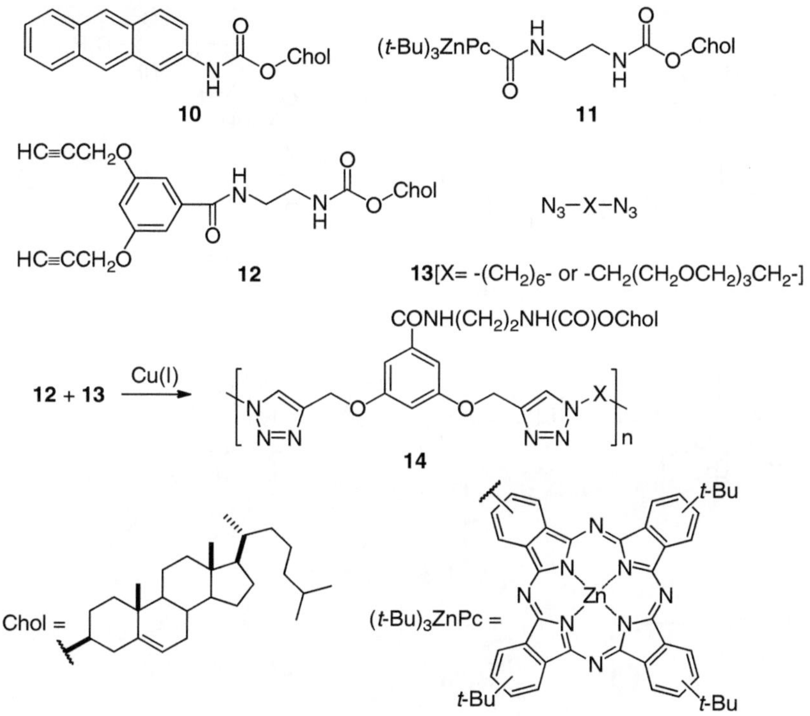

Scheme 4 Oligo(amide–triazole) gel **14** formed from the CuAAC copolymerization of monomers **12** and **13** in the presence of a LMWG **10** and a photoactive compound **11**

stability than the gels obtained from the 20:1 mixture of **10** and **11**. Careful analysis of the CuAAC product **14** by MALDI-TOF revealed that it consisted mainly of short oligomers. This example hence showed that such oligotriazole compounds could be used to reinforce the strength of multicomponent gels.

In 2007, Binder reported a supramolecular network **15** formed from the hydrogen bonding interactions between two types of triazole-containing polymers **16** ($M_n = 5{,}200$) and **17** ($M_n = 700$ or 2,000) (Scheme 5) [41]. Polymer **16** was a trifunctional motif possessing three bis-(2,6-diamidopyridine) units, while polymer **17** was a bifunctional compound containing two 5,5-disubstituted barbiturate functionalities. The barbiturate–bis-(2,6-diamidopyridine) partner has been established to have a strong binding constant ($K_b = 10^5\ M^{-1}$) in nonpolar organic solvent. Upon mixing the two polymers **16** and **17** in a 2:3 molar ratio in solvents such as CH_2Cl_2, hexane, or toluene, a gel **15** was formed. Gelation was reversible, and the gel state could be transformed back to solution upon warming. As expected, gelation was due to supramolecular hydrogen bonding interactions as confirmed by attenuated total internal reflection spectroscopic study. However, there was little evidence to suggest that the triazole units contributed any role to gel formation. Their sole function appeared to be just a linker to connect the hydrogen bonding

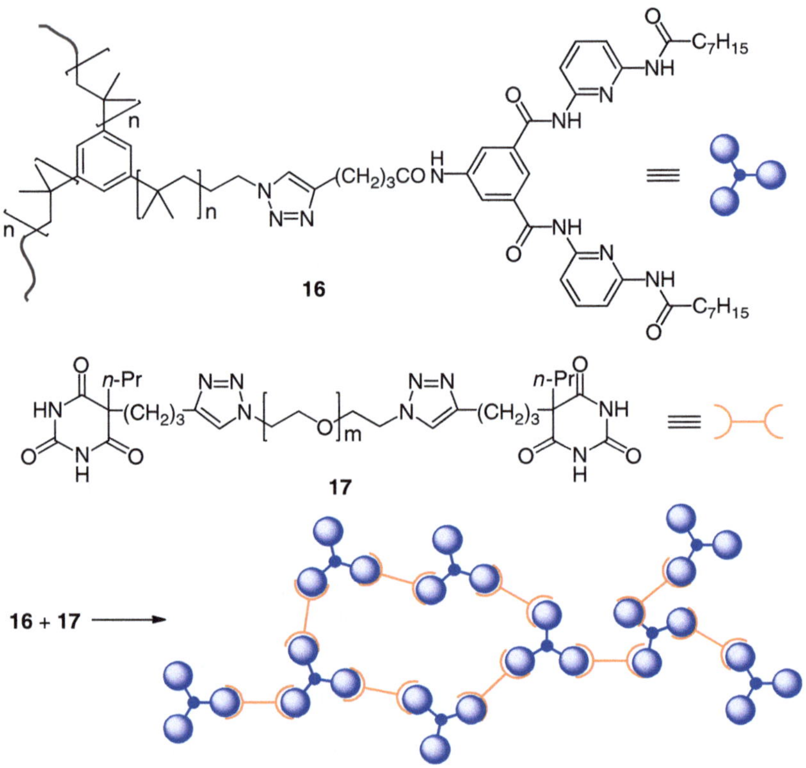

Scheme 5 Physical polymer gel **15** formed from hydrogen bonding interactions between polymers **16** and **17**

motifs to the polymer backbone. Entrapment of octylamine-coated iron oxide nanoparticles inside the gel was demonstrated provided that the amount of iron oxide was less than 5% wt. Diffusion studies using different dye molecules suggested that the gel could function as a controlled release agent.

The first example of a triazole-based dendronized polymer physical gel was reported by Chow [42]. A G1–G3 series of dendronized poly(amide–triazole)s **18** was obtained from the CuAAC promoted polymerization of AB type macromonomers **19** (Scheme 6). As expected, due to steric inhibition, the polymerization efficiency was inversely affected by the size of the dendron appendages. These three dendronized polymers exhibited strikingly different physical and supramolecular properties. The **18**-G1 polymer was an intractable solid, while the **18**-G3 polymer possessed extremely good solubility in many nonpolar organic solvents. On the other hand, only the **18**-G2 possessed gelating property among the three dendronized polymers. It had very broad-spectrum gelation power and could immobilize ethyl acetate, acetone, ethanol, THF, and, in particular, aromatic solvents at 0.5% w/v concentration level. Accordingly to FT-IR analysis, it was found that all monomers **19** were devoid of

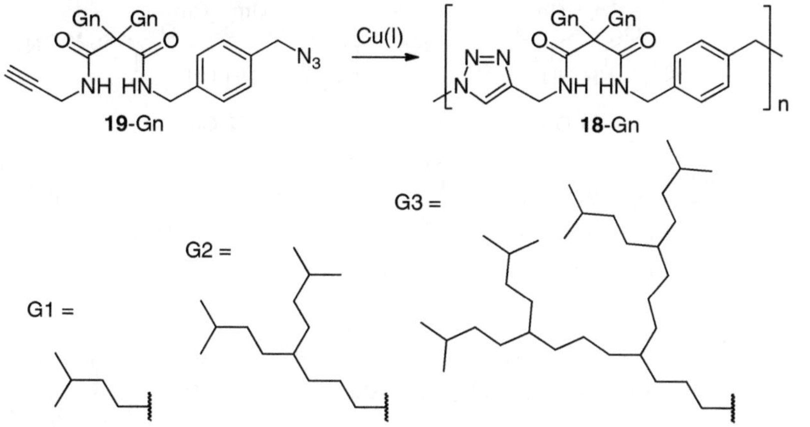

Scheme 6 Dendronized poly(amide–triazole)s **18** from the CuAAC polymerization of AB monomers **19**

intermolecular hydrogen bonding interactions and gelating property. After CuAAC polymerization, the resulting poly(amide–triazole)s **18** all exhibited strong interchain hydrogen bonding interactions. This kind of synergistic binding upon polymerization was explained by a zip templating effect – in which placement of many amide units along a polymer chain in regular intervals facilitated multiple hydrogen bonding interactions and enhanced interchain association. It was also noted that the strength of interchain association was offset by the size of the dendron. For **18**-G1, the dendrons were too small to dampen the strong interchain hydrogen bonding, and this rendered the polymer poorly soluble. For **18**-G3, the dendrons were too big, and hence, interchain association was the weakest, leading to its dissolution in organic solvents. It appeared that only **18**-G2 possessed the optimal dendron size to sustain a 3D network structure without dissolution by solvent molecules. At the same time, the internal voids could entrap guest solvent molecules and resulted in gel formation. The formation of strong, transparent gels in aromatic solvents suggested that the phenyl and triazole rings might be involved in π–π interactions.

Chow later extended the above work by preparing another series of dendronized polymers **20** via the AABB CuAAC copolymerization of diacetylenes **21** and diazides **22** (Scheme 7) [43]. In contrast to dendronized polymers **18** where the triazole dipoles were aligned in a head-to-tail fashion along the polymer chain, the triazole dipoles were now arranged in either a head-to-head or a tail-to-tail manner in dendronized polymers **20**. As it turned out, this difference in polymer backbone symmetry played a subtle role in determining their self-assembling and gelation properties. Surprisingly, the G2G2-derived dendronized polymer **20**-G2G2 was nongelating, although it possessed the closest structural similarity to the previous gelating dendronized polymer **18**-G2. On the other hand, polymers **20**-G1G2 and **20**-G2G1 were the only organogelators in this new series of compounds. Their gels were found to be weaker as compared to those of **18**-G2. It was suggested that the overall polymer dipole moment, in addition to the dendron size and density of

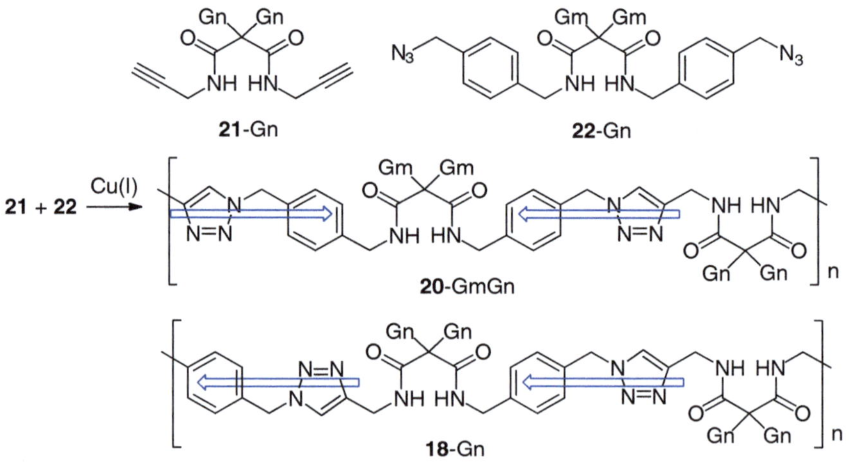

Scheme 7 Dendronized poly(amide–triazole)s **20** from the CuAAC copolymerization of AA **21** and BB monomers **22** (the *arrows* show the relative arrangement of the triazole dipoles along the polymer chain of **18** and **20**)

amide hydrogen bonding motif, was an additional factor in determining the strength of interchain association. It should be noted that such dendronized polymers should adopt a relative rigid rod structure due to interdendron repulsion [44]. The asymmetric dendronized polymers **18** would have a large cumulative dipole moment due to the head-to-tail alignment of individual triazole dipoles and resulted in stronger interchain dipole–dipole interactions. On the other hand, in the symmetrical dendronized polymers **20,** the cumulative polymer dipole moment should be minimal as the individual triazole dipoles canceled out due to their symmetrical arrangement along the polymer backbone. Hence, interchain association between polymers **18** should be inherently weaker than that of polymers **20.** In order to compensate for the diminishing dipole–dipole interaction, the strength of the network could only be reinforced if a smaller dendron was in place. As a result, only the **20**-G1G2 and **20**-G2G1 had the optimal dendron size to generate a stable 3D gel network.

There are also scattered examples of triazole-linked polymers that functioned as hydrogelators. It should be noted that in most cases, the single triazole unit functioned as a linker between different polymer blocks and most likely did not participate in the gelation mechanism. A triazole-linked diblock copolymer **23** consisting of a poly(ethylene oxide) (PEO) block and a poly(L-glutamic acid) (PLG) block was prepared (Fig. 1) [45]. The polymer itself was devoid of any hydrogelation property. It was found that normal micelles consisting of a PLG core and PEO corona were formed in water at pH 7. However, when the micellar solution was treated with α-cyclodextrin (α-CD), complexation of α-CD with the PEO block resulted in a change of solubility property of the PEO block and led to the formation of a thermally reversible hydrogel. Interestingly, in the presence of α-CD at pH 8, reverse micelles in which the core was made up of the α-CD-bound

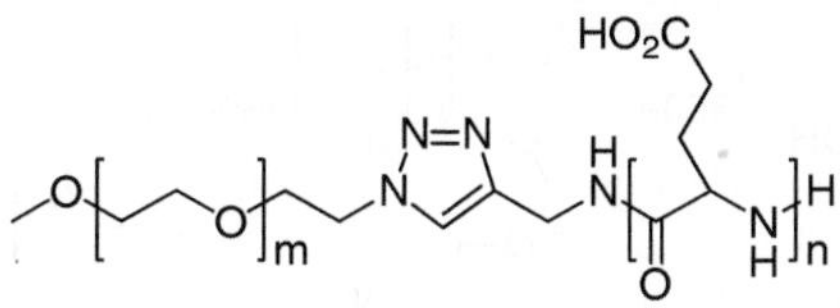

23

Fig. 1 Diblock copolymer **23** capable of forming normal and reverse micellar hydrogels

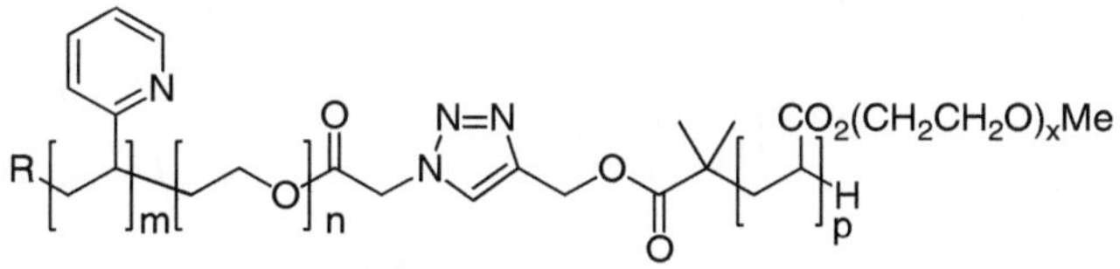

24

Fig. 2 Triblock copolymer hydrogels **24**

PEO block and an anionic L-glutamate corona were formed. When the pH of the solution was adjusted to 5, protonation of the L-glutamate back to L-glutamic acid occurred. This then led to extensive interchain hydrogen bonding and the formation of reverse micellar hydrogels.

Another example was a triblock terpolymer **24** which could function as a pH-dependent hydrogel (Fig. 2) [46]. Here, the triazole ring simply functioned as a linker to connect the two separate polymer blocks together.

3 Triazole-Based Chemical Gels

3.1 Hydrogels

Most polymer hydrogels are based on a water-miscible PEO skeleton. The advantages of using PEOs lie on their excellent biocompatibility, low immunogenicity, and good water-swelling properties. PEOs are also available in a wide MW range (200 to 35000 kDa) to cater for different applications. In addition, functionalization of the two hydroxyl terminal groups for further manipulations is straightforward. Hence, they are the most popular choice of base materials for the construction of hydrogels for use in controlled drug delivery and tissue engineering.

One of the first PEO chemical gels constructed through triazole cross-linking was disclosed by Hilborn (Scheme 8) [47]. Poly(vinyl alcohol) derivatives with acetylene **25** and azide side chains **26** were prepared and cross-linked to produce a hydrogelating network **27**. For comparative study, a similar type of hydrogelator **29**

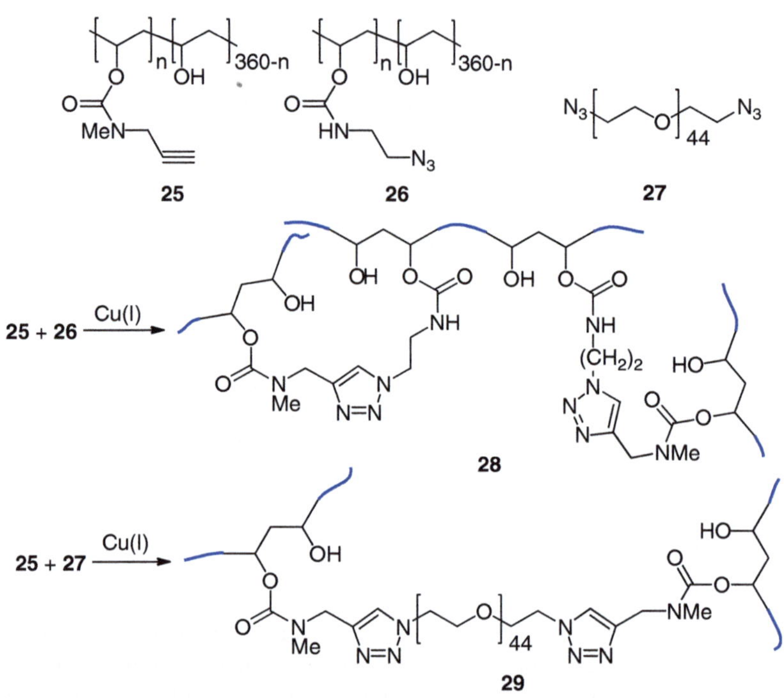

Scheme 8 Poly(vinyl alcohol)-based chemical hydrogels **28** and **29** prepared by CuAAC cross-linking

was also prepared from the cross-linking of diacetylene **25** and a PEO-derived diazide **27**. It was found that gelation by **28** was instantaneous while that by **29** was much slower (1 h). In addition, the gels formed from **28** possessed higher gelation capacity than that from **29**. Gel fractions for the gels prepared from the former were also higher (70–89%) than those prepared from the latter (37–64%). It was suggested that the bifunctional diazide **27** was more prone to form single-end linkages which reduced the extent of interchain cross-linking and the use of multivalent azides such as **26** could provide better cross-linking efficiency.

Hawker employed a similar strategy to prepare well-defined PEO-based hydrogel networks **32** (Scheme 9) [48]. Hence, a PEO-based diacetylene **30** and tetraazide **31** were cross-linked to afford the target hydrogel network **32**. The cross-linking was extremely efficient and as the gel contained less than 0.2% unreacted azide residues as determined by derivatization study. It was also found that gels prepared from this click strategy possessed better mechanical strength as compared to those obtained from photochemical cross-linking. One advantage of this preparation procedure was that the chemical and physical nature of the resulting gels could be fine-tuned by variations of the azide/acetylene ratio and the length of the PEO blocks, allowing greater control on the quality of the gel for various applications. Furthermore, various additives such as carbon black,

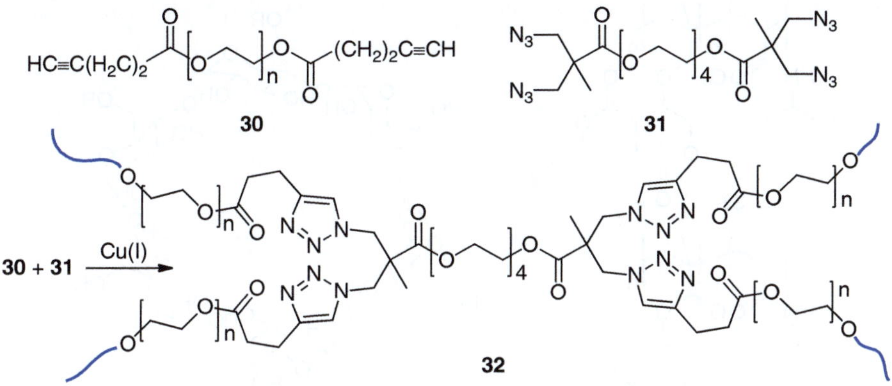

Scheme 9 PEO-based CuAAC cross-linked hydrogels **32**

4-phosphonooxy-2,2,6,6-tetramethylpiperidyloxy nitroxide, and titanium oxide nanoparticles could be incorporated within the gel network during CuAAC polymerization to further enrich their functional property.

Zhang utilized in situ CuAAC cross-linking between acetylene-modified **33** and azide-modified **34** poly(*N*-isopropylacrylamide-*co*-hydroxyethyl methacrylate) to form thermoresponsive hydrogels **36** (Scheme 10) [49]. Based on scanning electron microscopic study, all hydrogels had a similar porous network structure. The pore size increased gradually with decreasing density of cross-linking acetylene unit in polymer **33** or **34** (i.e., smaller *n*/*m* ratio). The temperature-dependent swelling profile of these gels was similar to that of poly(*N*-isopropylacrylamide) (PNIPAAm). Hence, below the lower critical solution temperature, all hydrogels exhibited a swollen state with a high swelling ratio. As the temperature increased, the swelling ratio decreased, and the gel turned into a shrunk state. Interestingly, the shrinking and reswelling kinetics of the hydrogels were faster than those of the homopolymer PNIPAAm.

The azide-modified poly(*N*-isopropylacrylamide-*co*-hydroxylethyl methacrylate) **34** was also used to cross-link with a hexaacetylene derivative of α-CD **35** to generate structurally similar porous hydrogels [50].

Due to the high cytotoxicity of Cu(I) ion and the difficulty in completely removing it after the CuAAC reaction, it becomes highly desirable to promote the [2+3] cycloaddition between the acetylene and the azide under Cu ion–free conditions. In this connection, Kiser reported the in situ preparation of cross-linked hydrogels by using a Cu(I)-free Huisgen cycloaddition between an azide-functionalized methacrylate polymer **37** and an activated diacetylene **38** (Scheme 11) [51]. However, even with **38** as the activated dipolarophile, the rate of cross-linking was still too slow (10% completion in 49 h at 24°C) to be useful in most in situ applications.

Hydrogels constructed from the cross-linking of polysaccharide derivatives were disclosed by Crescenzi (Scheme 12) [52]. Water-soluble hyaluronic acid derivatives with acetylene **40** and azide side chains **41** were cross-linked to afford

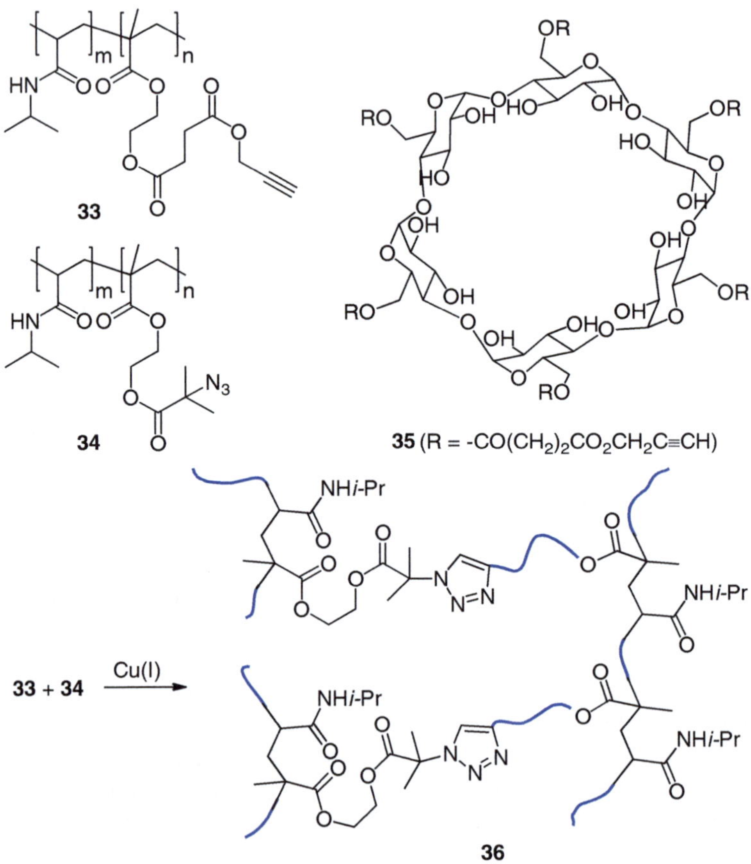

Scheme 10 Poly(*N*-isopropylacrylamide-*co*-hydroxylethyl methacrylate)-based CuAAC cross-linked chemical hydrogels **36**

the hydrogel **42** in situ. The resulting gels could entrap small molecules such as benzidamine and doxorubicin. Subsequent controlled release of these molecules under physiological conditions was realized. Furthermore, *Saccharomyces cerevisiae* yeast cells could also be homogeneously entrapped inside the gel with good adhesion property. Most importantly, the cells showed proliferating activity inside the gel matrix. This type of click gels was therefore very useful for tissue engineering applications. Other structurally similar hydrogels based on CuAAC cross-linking between azide derivative **40** and 1,4-diethynylbenzene, 1,6-heptadiyne, or 1,8-nonadiyne were also prepared, and their controlled release properties were studied [53].

Peptide derivatives were used by Anseth in the construction of hydrogel networks. For example, the peptidyl enediyne **43** was used to cross-link with a PEO-derived tetraazide **44** to produce peptide hydrogels **45** (Scheme 13) [54]. The CuAAC-inert alkene functionalities were subsequently subjected to the second functionalization

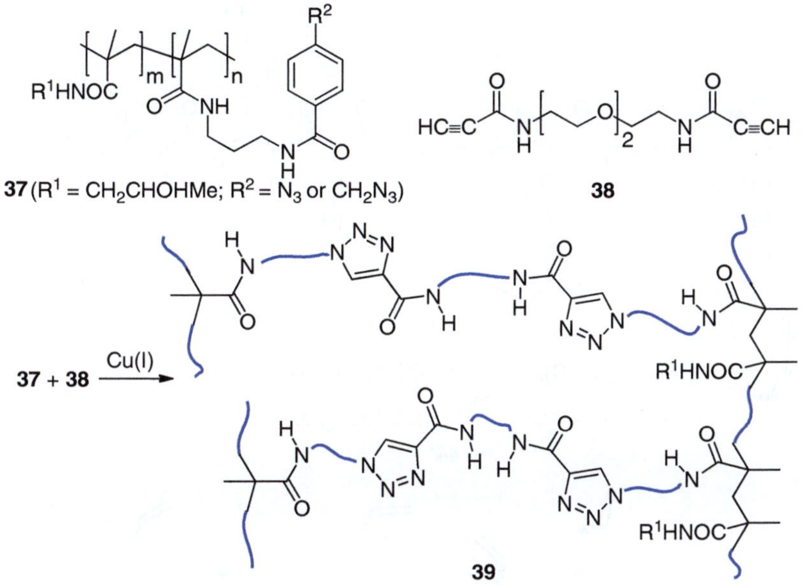

Scheme 11 Copper(I)-free preparation of cross-linking hydrogels **39**

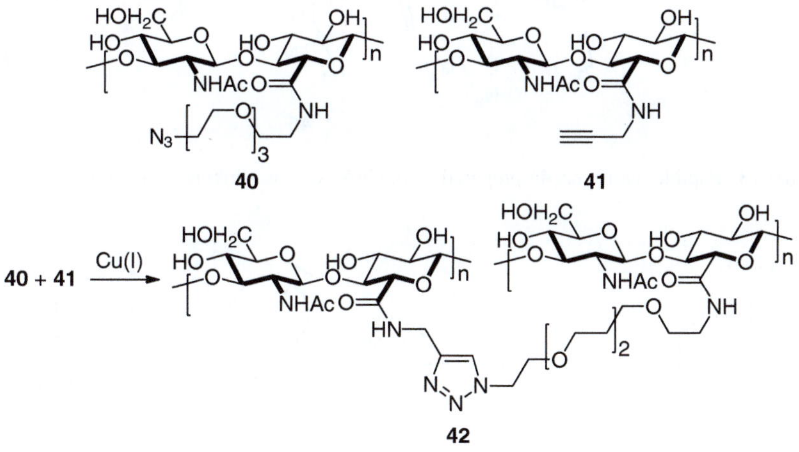

Scheme 12 Hyaluronan-derived CuAAC cross-linked hydrogels **42**

reaction – thiol-ene click chemistry [55] with a cysteine containing compound to create gels with a variety of 2D and 3D biochemical gradient patterns.

Anseth later reported a copper-free strain-promoted coupling [56] towards the preparation of another peptide hydrogels. In this example, the terminal alkyne moieties at the peptide ends were replaced by the more reactive electron deficient

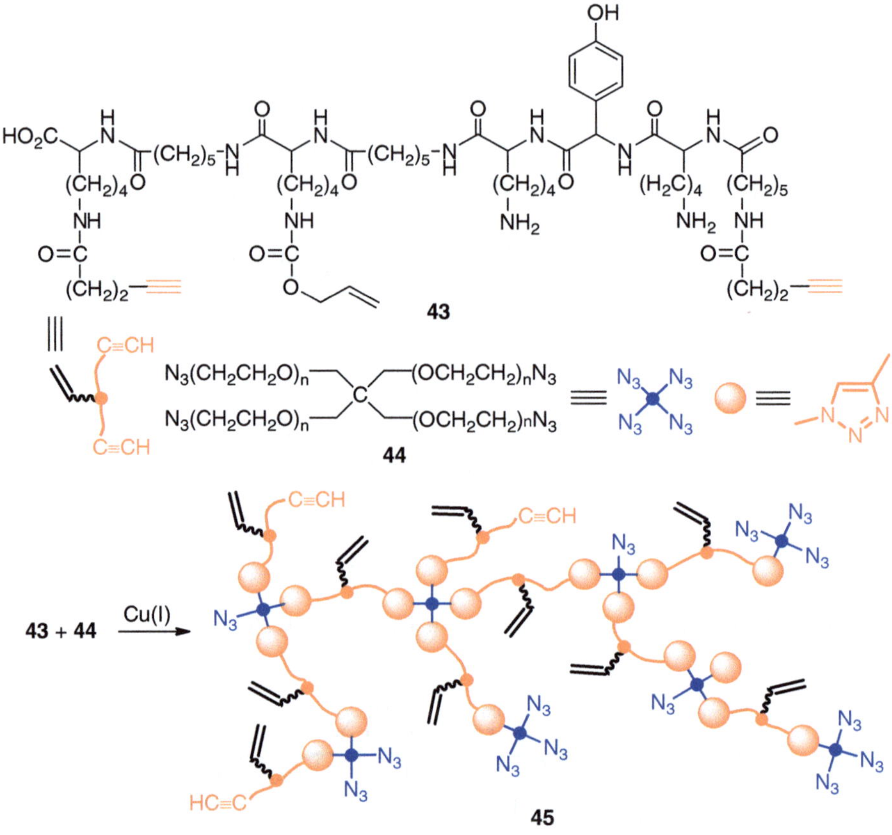

Scheme 13 Peptide hydrogels **45** prepared from CuAAC cross-linking of **43** and **44**

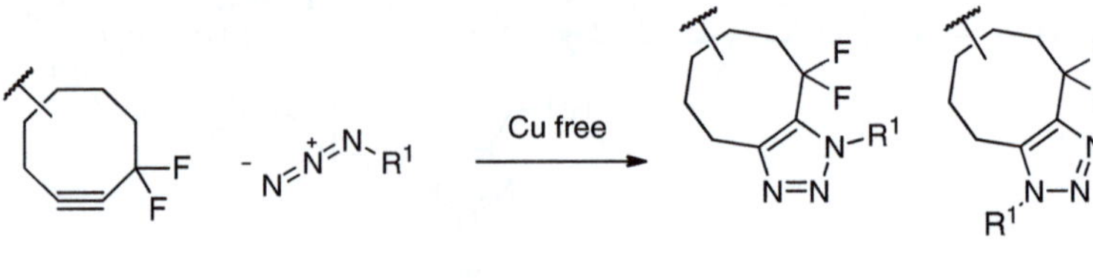

Scheme 14 Copper-free cycloaddition reaction between a 3,3-difluorocyclooctyne derivative and an azide

3,3-difluorocyclooctyne residues (Scheme 14) [57, 58]. The efficiency of the copper-free coupling reaction was comparable to that of the CuAAC reaction, but the intrinsic toxicity problem of Cu ion was eliminated. This new protocol should be of practical use in living systems.

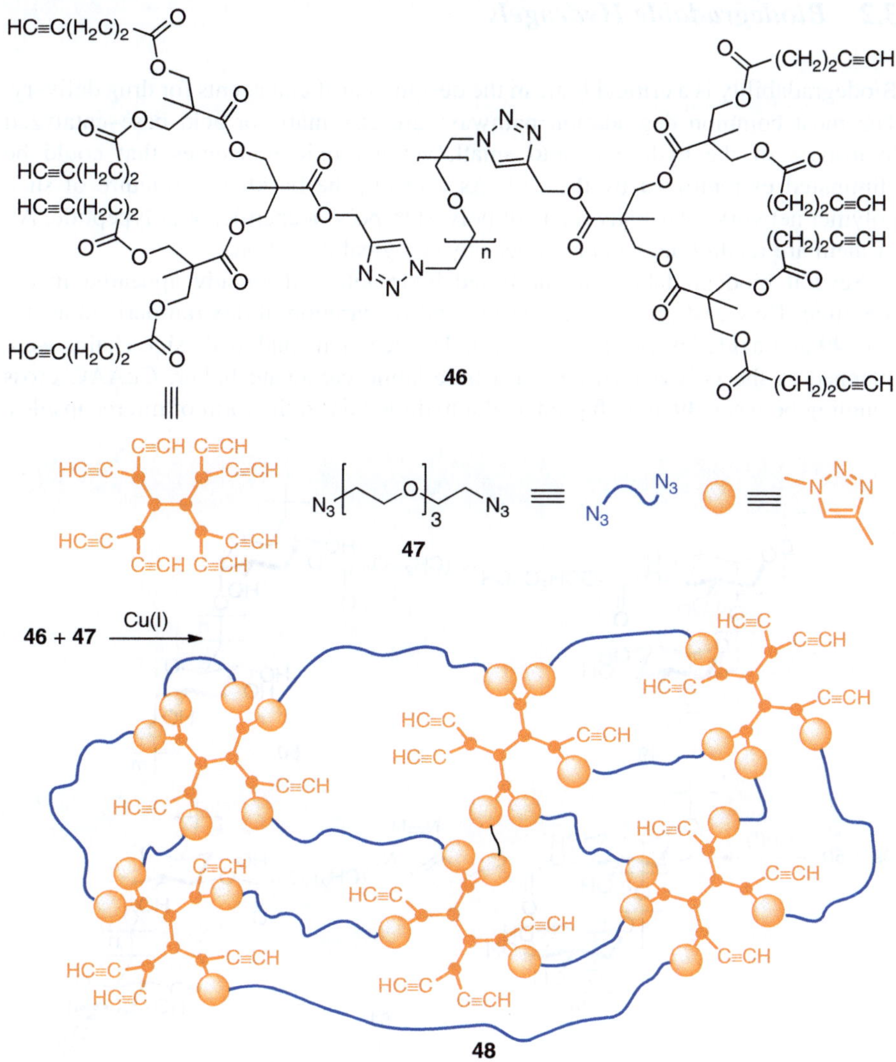

Scheme 15 Dendritic hydrogel **48** formed from CuAAC cross-linking of a multifunctional dendrimer **46** and a PEO-derived diazide **47**

A chemically cross-linked hydrogel constructed from a dendritic skeleton was recently reported by Sanyal (Scheme 15) [59]. Polyester dendrimers (e.g., **46**) functionalized with acetylene surface groups were cross-linked with a PEO-derived diazide **47** to furnish hydrogel networks **48**. Their swelling properties were strongly dependent on the size of the dendrimer and of the diazide linker **47**. The excess leftover acetylene moieties, the amount of which could be fine-tuned by changing the ratio of **46**:**47**, could be further functionalized with biomolecules such as streptavidin.

3.2 Biodegradable Hydrogels

Biodegradability is a critical issue in the design of artificial agents for drug delivery. The most common degradation pathways are enzymatic or acid base–catalyzed hydrolysis of the hydrogels into small, yet nontoxic molecules that could be eliminated or reutilized by the cell. As a result, the backbone structure of such polymer networks is usually made of polyester, polysaccharide, or polypeptide. All of them are readily degradable under physiological conditions.

Several biodegradable triazole-based hydrogels had already appeared in the literature. De Geest reported the design and preparation of dextran-derived acetylene **49** and azide **50** (Scheme 16) [60]. The acetylene and azide side chains were attached to the polysaccharide via a base-labile carbonate linker. CuAAC cross coupling between **49** and **50** produced a hydrogel **51** in the form of microcapsules.

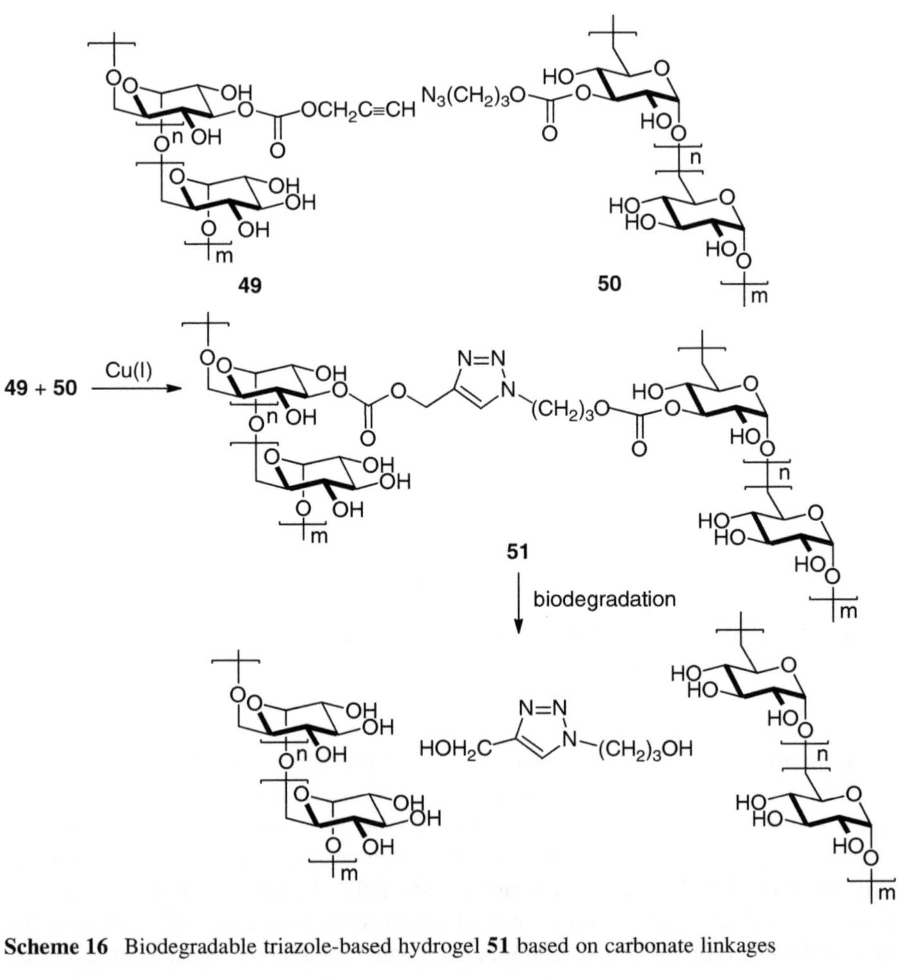

Scheme 16 Biodegradable triazole-based hydrogel **51** based on carbonate linkages

As expected, the microcapsules gradually swelled and dissolved in the aqueous environment upon addition of one drop of 1 M NaOH. Microcapsules pre-encapsulated with a model drug – fluorescein isothiocyanate dextran – were also found to degrade gradually under physiological conditions at pH = 7.4, accompanied by a sustained release of the drug molecule.

pH-Responsive hydrogels based on the acid-labile poly(α-azido-ε-caprolactone) skeleton **52** were reported by Jérôme (Scheme 17) [61]. In the first part of the synthesis, part of the azido units was converted into amino residues to provide a pH-sensitive amphiphilic polymer **53**. In the second step, the remaining azide units were then cross-linked with a PEO-based diacetylene **54** to create the hydrogel network **55**. The swelling behavior of the network **55** displayed different profiles according to the pH of the medium. Hence, water uptake was higher at pH 5.5 than at pH 9 due to protonation of the amino side chains, which conferred better hydrophilicity and water encapsulation capability to the network. The controlled release property of the network was also demonstrated by studying the release profile of a dye molecule at pH 1. The biodegradable property of the gel was due to cleavage of the polyester backbone under such strongly acidic conditions.

Yang recently disclosed a class of biodegradable PEO-peptide hydrogels with well-defined structure and their application in cell delivery (Scheme 18) [62].

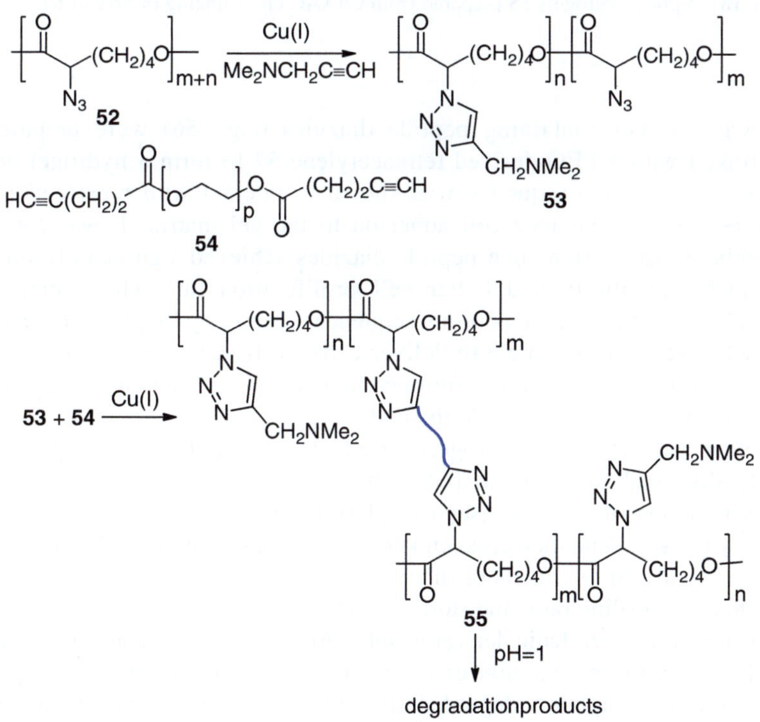

Scheme 17 pH-Responsive biodegradable gel **39** based on a poly(ε-caprolactone) skeleton

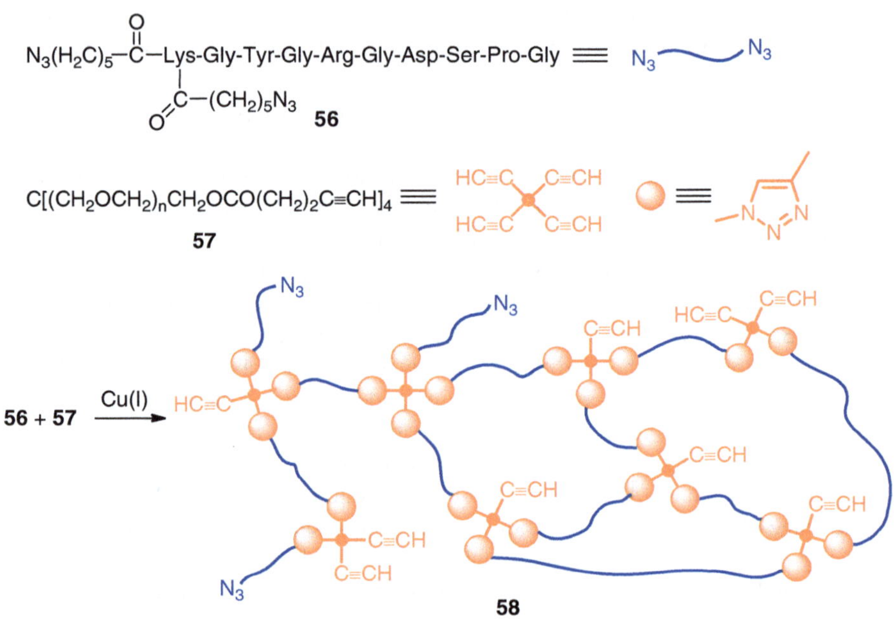

Scheme 18 Peptide hydrogels **58** prepared from CuAAC cross-linking of **56** and **57**

Two Arg-Gly-Asp–containing peptide diazides (e.g., **56**) were prepared and cross-linked with a PEO-derived tetraacetylene **57** to form a hydrogel network **58**. The Arg-Gly-Asp sequence was known to interact with most cell integrin receptors [63] and enhance cell adhesion to the gel matrix. It was found that hydrogels prepared from such peptide diazides achieved significantly improved cell attachment ability and better cell proliferation rate when compared to hydrogels without such a peptide sequence. These hydrogels may have the potential to serve as a vehicle to deliver cells for tissue regeneration. However, their biodegradability was not mentioned in this study; presumably, they could be broken down by proteases inside the cell.

Another peptide-based hydrogels based on a D-Ala-Phe-Lys sequence were recently disclosed by Hennink [64]. This triad is particularly sensitive to the hydrolytic action of proteases such as plasmin and trypsin. A diazide tripeptide **59** was synthesized and then cross-linked with various multivalent PEO derivatives **60** and **61** to afford hydrogels of different degree of cross-linking (Fig. 3). It was found that the swelling ratio and storage modulus of the hydrogels were affected by the valency of the acetylenic derivative **60** or **61** and the initial concentrations of **59** and **60** used in the preparation of the hydrogels. In the presence of trypsin, the hydrogels were completely degraded after 40–80 h. However, no gel degradation was observed when the much bigger size plasmin was used. This could be due to steric inaccessibility of this larger enzyme towards the tripeptide substrate.

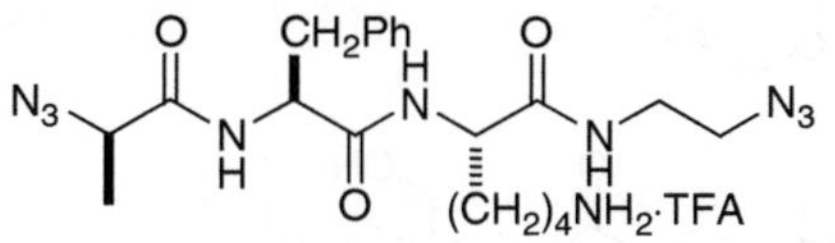

59

60

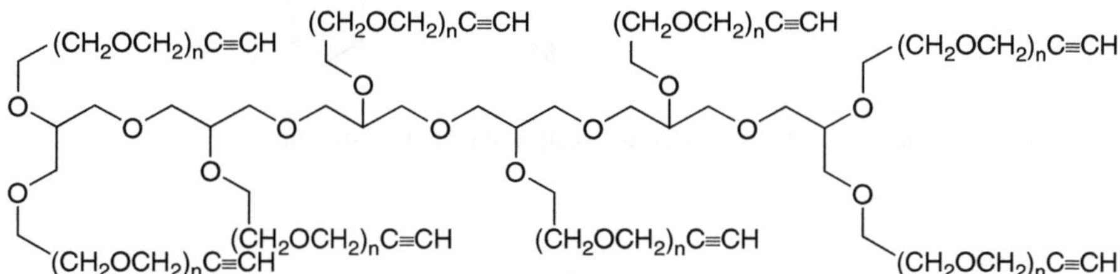

61

Fig. 3 Peptide diazide **59**, PEO tetraacetylene **60**, and PEO octaacetylene **61**

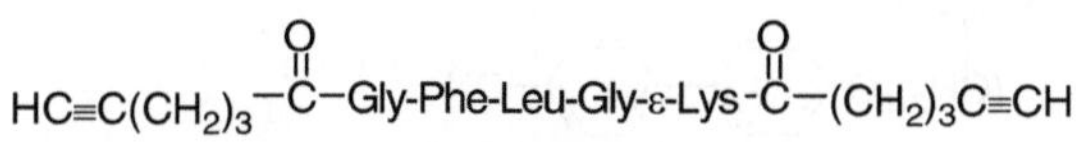

62

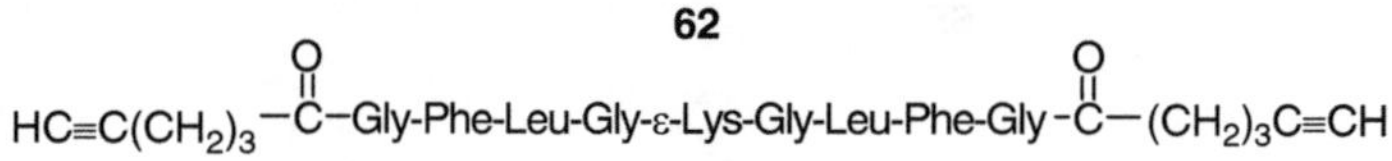

63

Fig. 4 Peptide diacetylenes **62** and **63**

An alternative strategy for making peptide hydrogels is to anchor the acetylene residues onto the peptide and have the azide moieties switched to the PEO skeleton. Kopeček disclosed the preparation of a Gly-Phe-Leu-Gly–derived diacetylene **62** and a Gly-Phe-Leu-Gly-ε-Lys-Gly-Leu-Phe-Gly–derived diacetylene **63** (Fig. 4) [65]. These peptides were then coupled to the PEO-modified tetraazide **44** to produce the target polytriazole-based hydrogels for degradation study in the presence of papain at pH 6.0 and at 37°C. It was found that the length of the PEO chain (i.e., n in structure **44**) had a large effect on the rate of degradation. Hydrogels prepared from a higher n value (~50) were degraded much faster than those prepared using a smaller *n* value (~12).

Malkoch recently reported the accelerated synthesis of a new class of degradable polyester dendrimers that were decorated with internal and surface acetylenic moieties (Fig. 5) [66]. The G2 dendrimer **64** was then cross-linked with 1 equiv. of N_3–PEO_{8000}–N_3 to give a hydrogel. Due to the acid- and base-labile ester linkages, the hydrogel could be degraded within 1 h at pH 11 or 4 days at pH 4 in aqueous solutions.

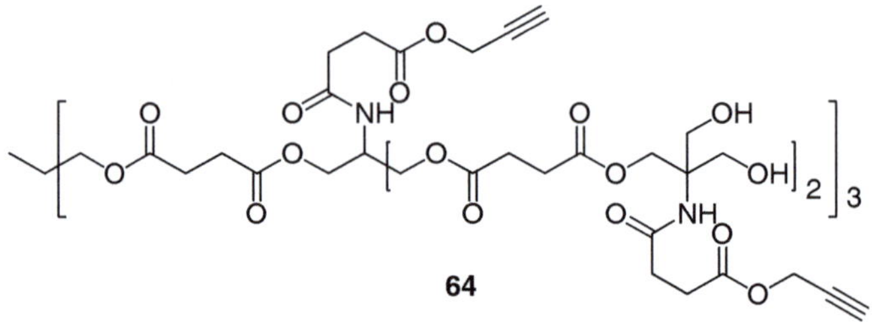

Fig. 5 Polyester dendrimer **64** with multiple acetylene units for cross-linking

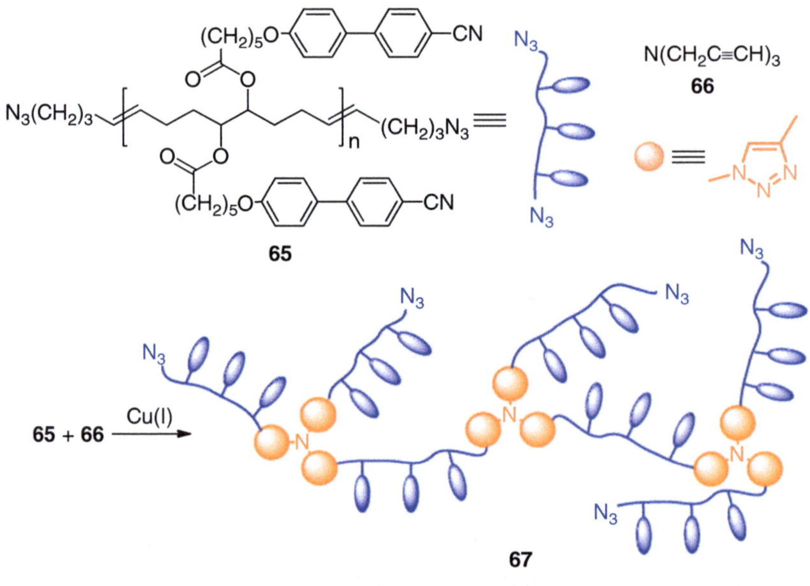

Scheme 19 Liquid crystalline organogel **67** prepared from CuAAC cross-linking between diazide **65** and triacetylene **66**

3.3 Organogels

There are only a handful of polytriazole-based organogels that exist in the form of a chemically cross-linked network. One of the earliest examples was a class of liquid crystal (LC) gels reported by Grubbs [67]. A series of telechelic diazide polymers (e.g., **65**) with liquid crystalline side chains were prepared by ring-opening metathesis polymerization following by end-group functionalization (Scheme 19). By controlling the number of mesogen pendant in each repeating unit and using a chain transfer agent, telechelic polymers **65** with different mesogen density and strand lengths were prepared. These diazides were then cross-linked with tripropargylamine **66** to give a network structure which could immobilize the liquid crystal 4-*n*-pentyl-4′-cyanobiphenyl to form LC gels **67**. The unconstrained sample gels showed fast, reversible, and low-threshold electro-optical switching properties. It was also shown that LC gels prepared from longer-strand polymers (larger *n* values) exhibited a higher degree of switching than those made from shorter ones.

An interesting class of metal-induced cross-linking gel was reported by Hecht [68]. A poly(triazole-pyridine) **68** was prepared by AABB CuAAC copolymerization of a pyridyl diacetylene **69** and a chiral aryl diazide **70** (Scheme 20). Due to the repulsive interactions between the nitrogen lone-pair electrons on the triazole and pyridine units [69], the polymer tended to adopt a helical conformation in the absence of metal ions. On the other hand, metallogels **71** were formed when Zn (II), Fe(II), or Eu(III) ion was added to a solution of the polymer in acetonitrile. It was believed that the 2,6-bis(triazolyl)pyridine ligands chelated to the metal ion to form a hexacoordinate metal complex, which then led to a metal-promoted cross-linked 3D network that entrapped acetonitrile solvents.

Several chemically degradable triazole-based polymer networks (e.g., **72**) were reported by Finn and coworkers [70]. The network was constructed by the click cross-linking of a linear α,ω-bifunctional diazide polymer **73** containing a double bond in the polymer main chain with either a triacetylene **74** or a tetraacetylene **75**

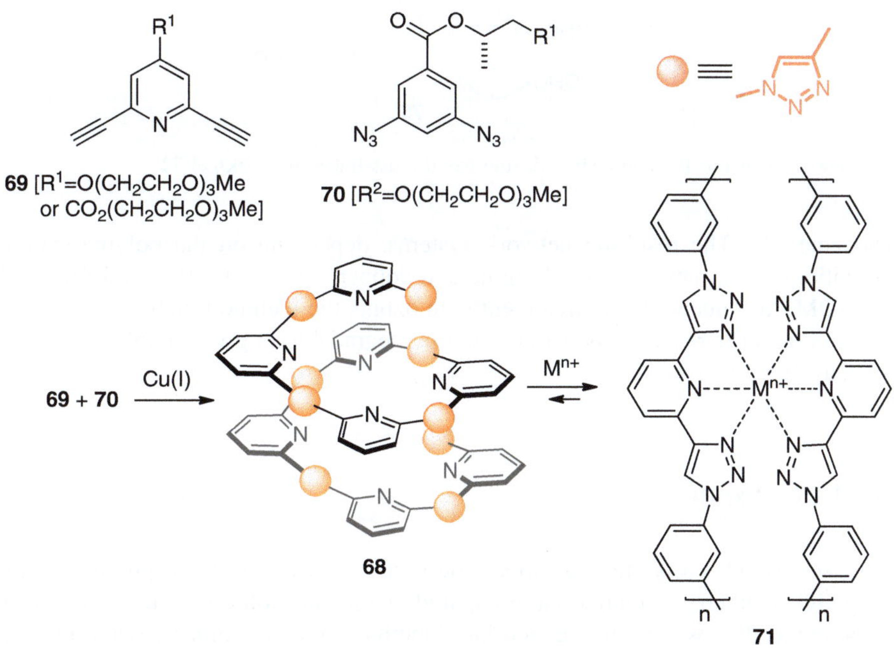

Scheme 20 Metalloorganogel **71** prepared from metal-induced cross-linking of click polymer **68** (the substituents R^1 and R^2 in some of the structures are omitted for clarity)

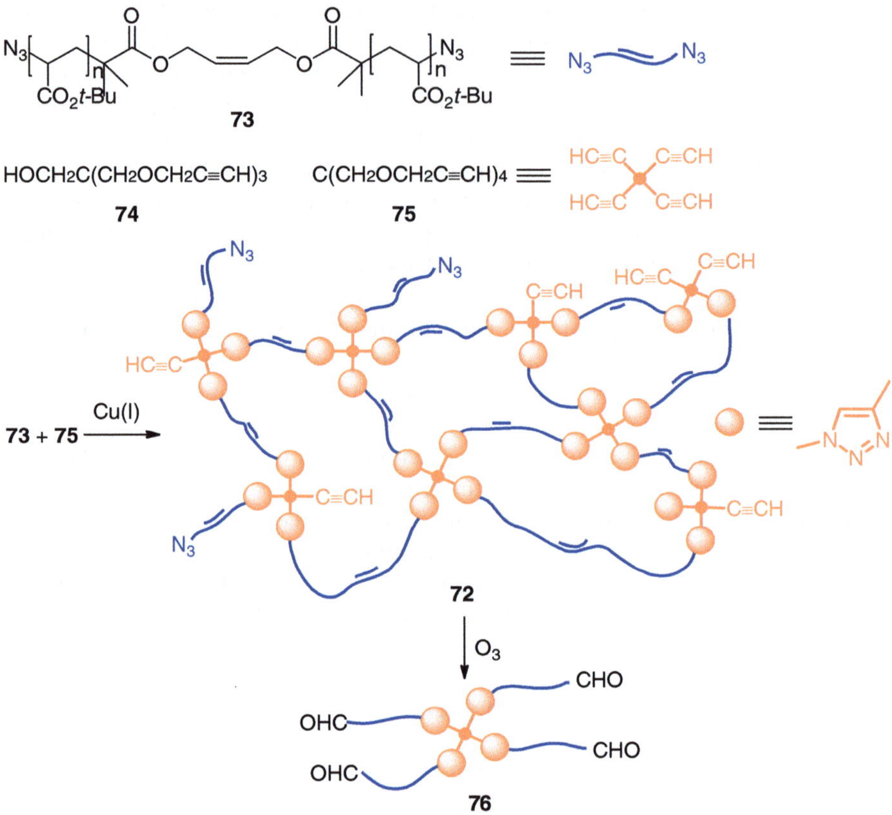

Scheme 21 Chemically degradable triazole-based cross-linked network/gel **72**

(Scheme 21). The resulting network systems, depending on the polymerization conditions (e.g., nature of alkyl amine and copper catalyst used), could form gel with DMF or toluene. Upon treatment with ozone, the olefinic functionalities were cleaved, and the resulting network could be degradable to give soluble oligomers such as **76**.

4 Conclusion

CuAAC provides a facile entry to a wide variety of triazole-based polymers with very high synthetic efficiencies as compared to other cross-linking methods such as photoinduced cross-linking of acrylate-functionalized monomers, photoinduced thiol-ene conjugation, or Michael addition type conjugation between thiol and acrylate or vinyl sulfone precursors. Due to the presence of the acidic triazole C–H and lone pair electrons on the nitrogen atoms, such triazole-based polymer materials could also promote intra- and interpolymer chain interactions and enhance gel formation. Although the biocompatibility and toxicity profiles of such triazole-based materials await further evaluations, they have emerged as an important class of compounds in many applications such as drug delivery and tissue engineering. New triazole-based polymer gels with added functions such as pH sensitivity, metal complexation ability, sensing capability, and/or photochemical property will undoubtedly enrich their practical usefulness. Nonetheless, the discovery of new coupling conditions without employing toxic Cu ion remains a key issue to be addressed before such novel polymers can have genuine practical biomedical applications.

Acknowledgment We thank the Research Grants Council, HKSAR, for the financial support (Project Number: 400810).

References

1. Finley KT (1980) Triazoles: 1,2,3. In: Montgomery JA (ed) The chemistry of heterocyclic compounds, vol 39. Wiley, New York
2. Tornøe CW, Christensen C, Meldal M (2002) Peptidotriazoles on solid phase: [1,2,3]-Triazoles by regiospecific copper(I)-catalyzed 1,3-dipolar cycloadditions of terminal alkenes to azides. J Org Chem 67:3057–3064
3. Rostovtsev VV, Green LG, Fokin VV, Sharpless KB (2002) A stepwise Huisgen cycloaddition process: Copper(I)-catalyzed regioselective "ligation" of azides and terminal alkynes. Angew Chem Int Ed 41:2596–2599
4. Binder WH, Kluger C (2006) Azide/alkyne-"click" reactions: applications in material science and organic synthesis. Curr Org Chem 10:1791–1815
5. Lutz JF (2007) 1,3-Dipolar cycloadditions of azides and alkynes: a universal ligation tool in polymer and materials science. Angew Chem Int Ed 46:1018–1025
6. Voit B (2007) The potential of cycloaddition reactions in the synthesis of dendritic polymers. New J Chem 31:1139–1151
7. Nandivada H, Jiang X, Lahann J (2007) Click chemistry: versatility and control in the hands of materials scientists. Adv Mater 19:2197–2208
8. Golas PL, Matyjaszewski K (2007) Click chemistry and ATRP: a beneficial union for the preparation of functional materials. QSAR Comb Sci 26:1116–1134
9. Fournier D, Hoogenboom R, Schubert US (2007) Clicking polymers: a straightforward approach to novel macromolecular architectures. Chem Soc Rev 36:1369–1380
10. Binder WH, Sachsenhofer R (2007) 'Click' chemistry in polymer and materials science. Macromol Rapid Commun 28:15–54
11. Binder WH, Sachsenhofer R (2008) 'Click' chemistry in polymer and material science: an update. Macromol Rapid Commun 29:952–981
12. Lundberg P, Hawker CJ, Hult A, Malkoch M (2008) Click assisted one-pot multi-step reactions in polymer science: accelerated synthetic protocols. Macromol Rapid Commun 29:998–1015
13. Meldal M (2008) Polymer "clicking" by CuAAC reactions. Macromol Rapid Commun 29:1016–1051
14. Johnson JA, Finn MG, Koberstein JT, Turro NJ (2008) Construction of linear polymers, dendrimers, networks, and other polymeric architectures by copper-catalyzed azide-alkyne cycloaddition "click" chemistry. Macromol Rapid Commun 29:1052–1072

15. Carlmark A, Hawker C, Hult A, Malkoch M (2009) New methodologies in the construction of dendritic materials. Chem Soc Rev 38:352–362
16. Iha RK, Wooley KL, Nyström AM, Burke DJ, Kade MJ, Hawker CJ (2009) Applications of orthogonal "click" chemistries in the synthesis of functional soft materials. Chem Rev 109:5620–5686
17. Billiet L, Fournier D, Du Prez F (2009) Step-growth polymerization and 'click' chemistry: the oldest polymers rejuvenated. Polymer 50:3877–3886
18. Golas PL, Matyjaszewski K (2010) Marrying click chemistry with polymerization: expanding the scope of polymeric materials. Chem Soc Rev 39:1338–1354
19. Dondoni A (2007) Triazoles: the keystone in glycosylated molecular architectures constructed by a click reaction. Chem Asian J 2:700–708
20. Angell YL, Burgess K (2007) Peptidomimetics via copper-catalyzed azide-alkyne cycloadditions. Chem Soc Rev 36:1674–1689
21. Pieters RJ, Rijkers DTS, Liskamp RMJ (2007) Application of the 1,3-dipolar cycloaddition reaction in chemical biology: approaches toward multivalent carbohydrates and peptides and peptide-based polymers. QSAR Comb Sci 26:1181–1190
22. Dirks AJ, Cornelissen JJLM, van Delft FL, van Hest JCM, Nolte RMJ, Rowan AE, Rutjes FPJT (2007) From (bio)molecules to biohybrid materials with the click chemistry approach. QSAR Comb Sci 26:1200–1210
23. Le Droumaguet B, Velonia K (2008) Click chemistry: a powerful tool to create polymer-based macromolecular chimeras. Macromol Rapid Commun 29:1073–1089
24. Gramlich PME, Wirges CT, Manetto A, Carell T (2008) Postsynthetic DNA modification through the copper-catalyzed azide-alkyne cycloaddition reaction. Angew Chem Int Ed 47:8350–8358
25. Santoyo-Gonzalez F, Hernandez-Mateo F (2009) Silica-based clicked hybrid glyco materials. Chem Soc Rev 38:3449–3462
26. Amblard F, Cho JH, Schinazi RF (2009) Cu(I)-catalyzed Huisgen azide-alkyne 1,3-dipolar cycloaddition reaction in nucleoside, nucleotide and oligonucleotide chemistry. Chem Rev 109:4207–4220
27. van Dijk M, Rijkers DTS, Liskamp RMJ, van Nostrum CF, Hennink WE (2009) Synthesis and applications of biomedical and pharmaceutical polymers via click chemistry methodologies. Bioconjugate Chem 20:2001–2016
28. Chow HF, Lau KN, Ke Z, Liang Y, Lo CM (2010) Conformational and supramolecular properties of main chain and cyclic click oligotriazoles and polytriazoles. Chem Commun 46:3437–3453
29. Terech P, Weiss RG (1997) Low molecular mass gelators of organic liquids and the properties of their gels. Chem Rev 97:3133–3160
30. van Esch JH, Feringa BL (2000) New functional materials based on self-assembling organogels: from serendipity towards design. Angew Chem Int Ed 39:2263–2266
31. Estroff LA, Hamilton AD (2004) Water gelation by small organic molecules. Chem Rev 104:1201–1217
32. Fages F (ed) (2005) Low molecular mass gelators: design, self-assembly, function. In: Topics in current chemistry, vol 256. Springer, Berlin
33. Sangeetha NM, Maitra U (2005) Supramolecular gels: Functions and uses. Chem Soc Rev 34:821–836
34. de Loos M, Feringa BL, van Esch JH (2005) Design and application of self-assembled low molecular weight hydrogels. Eur J Org Chem 3615–3631
35. Weiss RG, Terech P (eds) (2006) Molecular gels. Materials with self-assembled fibrillar networks. Springer, Dordrecht
36. Dastidar P (2008) Suparmolecular gelling agents: can they be designed? Chem Soc Rev 37:2699–2715
37. Suzuki M, Hanabusa K (2010) Polymer organogelators that make supramolecular organogels through physical cross-linking and self-assembly. Chem Soc Rev 39:455–463

38. Díaz DD, Rajagopal K, Strable E, Schneider J, Finn MG (2006) "Click" chemistry in a supramolecular environment: stabilization of organogels by copper(I)-catalyzed azide-alkyne [3+2] cycloaddition. J Am Chem Soc 128:6056–6057
39. Díaz DD, Tellado JJM, Velázquez DG, Ravelo AG (2008) Polymer thermoreversible gels from organogelators enabled by 'click' chemistry. Tetrahedron Lett 49:1340–1343
40. Díaz DD, Cid JJ, Vázquez P, Torres T (2008) Strength enhancement of nanostructured organogels through inclusion of phthalocyanine-containing complementary organogelator structures and in situ cross-linking by click chemistry. Chem Eur J 14:9261–9273
41. Binder WH, Petraru L, Roth T, Groh PW, Pálfi V, Keki S, Ivan B (2007) Magnetic and temperature-sensitive release gels from supramolecular polymers. Adv Funct Mater 17:1317–1326
42. Lau KN, Chow HF, Chan MC, Wong KW (2008) Dendronized polymer organogels from click chemistry: a remarkable gelation property owing to synergistic function-group binding and dendritic size effects. Angew Chem Int Ed 47:6912–6916
43. Chow HF, Lau KN, Chan MC (2011) Click dendronized poly(amide–triazole)s: effect of dendron size and polymer backbone symmetry on self assembling and gelation properties. Chem Eur J. 17:8395–8403
44. Schlüter AD (2005) A covalent chemistry approach to giant macromolecules with cylindrical shape and an engineerable interior and surface. Top Curr Chem 245:151–191
45. Chen Y, Pang XH, Dong CM (2010) Dual stimuli-responsive supramolecular polypeptide-based hydrogel and reverse micellar hydrogel mediated by host-guest chemistry. Adv Funct Mater 20:579–586
46. Reinicke S, Schmalz H (2011) Combination of living anionic polymerization and ATRP via "click" chemistry as a versatile route to multiple responsive triblock terpolymers and corresponding hydrogels. Colloid Polym Sci 289:497–512
47. Ossipov DA, Hilborn J (2006) Poly(vinyl alcohol)-based hydrogels formed by "click chemistry". Macromolecules 39:1709–1718
48. Malkoch M, Vestberg R, Gupta N, Mespouille L, Dubois P, Mason AF, Hedrick JL, Liao Q, Frank CW, Kingsbury K, Hawker CJ (2006) Synthesis of well-defined hydrogel networks using click chemistry. Chem Commun 2774–2776
49. Xu XD, Chen CS, Wang ZC, Wang GR, Cheng SX, Zhang XZ, Zhuo RX (2008) "Click" chemistry for in situ formation of thermoresponsive p(NIPAAm-co-HEMA)-based hydrogels. J Polym Sci Polym Chem 46:5263–5277
50. Xu XD, Chen CS, Lu B, Wang ZC, Cheng SX, Zhang XZ, Zhuo RX (2009) Modular synthesis of thermosensitive p(NIPAAm-co-HEMA)-β-CD based hydrogels via click chemistry. Macromol Rapid Commun 30:157–164
51. Clark M, Kiser P (2009) In situ crosslinked hydrogels formed using Cu(I)-free Huisgen cycloaddition reaction. Polym Int 58:1190–1195
52. Crescenzi V, Cornelio L, Di Meo C, Nardecchia S, Lamanna R (2007) Novel hydrogels via click chemistry: synthesis and potential biomedical applications. Biomacromolecules 8:1844–1850
53. Testa G, Di Meo C, Nardecchia S, Capitani D, Mannina L, Lamanna R, Barbetta A, Dentini M (2009) Influence of dialkyne structure on the properties of new click-gels based on hyaluronic acid. Int J Pharm 378:86–92
54. Polizzotti BD, Fairbanks BD, Anseth KS (2008) Three-dimensional biochemical patterning of click-based composite hydrogels via thiolene photopolymerization. Biomacromolecules 9:1084–1087
55. Hoyle CE, Bowman CN (2010) Thiol-ene click chemistry. Angew Chem Int Ed 49:1540–1573
56. DeForest CA, Sims EA, Anseth KS (2010) Peptide-functionalized click hydrogels with independently tunable mechanics and chemical functionality for 3D cell culture. Chem Mater 22:4783–4790
57. Baskin JM, Prescher JA, Laughlin ST, Agard NJ, Chang PV, Miller IA, Lo A, Codelli JA, Bertozzi CR (2007) Copper-free click chemistry for dynamic in vivo imaging. Proc Natl Acad Sci USA 104:16793–16797

58. Codelli JA, Baskin JM, Agard NJ, Bertozzi CR (2008) Second-generation difluorinated cyclooctynes for copper-free click chemistry. J Am Chem Soc 130:11486–11493
59. Altin H, Kosif I, Sanyal R (2010) Fabrication of "clickable" hydrogels via dendron–polymer conjugates. Macromolecules 43:3801–3808
60. De Geest BG, Van Camp W, Du Prez FE, De Smedt SC, Demeester J, Hennink WE (2008) Biodegradable microcapsules designed via 'click' chemistry. Chem Commun 190–192
61. Zednik J, Riva R, Lussis P, Jérôme C, Jérôme R, Lecomte P (2008) pH-Responsive biodegradable amphiphilic networks. Polymer 49:697–702
62. Liu SQ, Ee PLR, Ke CY, Hedrick JL, Yang YY (2009) Biodegradable poly(ethylene glycol)-peptide hydrogels with well-defined structure and properties for cell delivery. Biomaterials 30:1453–1461
63. Ruoslahti E, Pierschbacher MD (1987) New perspectives in cell adhesion: RGD and integrins. Science 238:491–497
64. van Dijk M, van Nostrum CF, Hennink WE, Rijkers DTS, Liskamp RMJ (2010) Synthesis and characterization of enzymatically biodegradable PEG and peptide-based hydrogels prepared by click chemistry. Biomacromolecules 11:1608–1614
65. Yang J, Jacobsen MT, Pan H, Kopeček J (2010) Synthesis and characterization of enzymatically degradable PEG-based peptide-containing hydrogels. Macromol Biosci 10:445–454
66. Antoni P, Hed Y, Nordberg A, Nyström D, von Holst H, Hult A, Malkoch M (2009) Bifunctional dendrimers: from robust synthesis and accelerated one-pot postfunctionalization strategy to potential applications. Angew Chem Int Ed 48:2126–2130
67. Xia Y, Verduzco R, Grubbs RH, Kornfield JA (2008) Well-defined liquid crystal gels from telechelic polymers. J Am Chem Soc 130:1735–1740
68. Meudtner RM, Hecht S (2008) Responsive backbones based on alternating triazole-pyridine/benzene copolymers: from helically folding polymers to metallosupramolecularly crosslinked gels. Macromol Rapid Commun 29:347–351
69. Meudtner RM, Ostermeier M, Goddard R, Limberg C, Hecht S (2007) Multifunctional "clickates" as versatile extended heteroaromatic building block: efficient synthesis via click chemistry, conformational preferences, and metal coordination. Chem Eur J 13:9834–9840
70. Johnson JA, Lewis DR, Díaz DD, Finn MG, Koberstein JT, Turro NJ (2006) Synthesis of degradable model networks via ATRP and click chemistry. J Am Chem Soc 128:6564–6565

Top Heterocycl Chem (2012) 28: 163–184
DOI: 10.1007/7081_2011_72

Published online: 17 February 2012

Click Triazoles for Bioconjugation

Tianqing Zheng, Sara H. Rouhanifard, Abubakar S. Jalloh, and Peng Wu

Abstract Click chemistry is a set of rapid, selective and robust reactions that give near-quantitative yield of the desired product in aqueous solutions. The Cu(I)-catalyzed azide-alkyne cycloaddition (CuAAC) that forms 1,4-disubstituted triazoles is a prototypical example of click chemistry that features exquisite selectivity and bioorthogonality—that is, non-interacting with biological components while proceeding under physiological conditions. Over the past ten years, CuAAC has found extensive applications in the field of chemical biology. In this chapter, we describe the discovery of Cu(I) catalysts for this transformation and the recent development of the strain-promoted azide-alkyne cycloaddition that eliminate the use of copper. We also highlight several recent applications toward conjugating biomolecules, including proteins, nucleic acids, lipids and glycans, with biophysical probes for both in vitro and in vivo studies.

Keywords Bioconjugation · Bioorthogonal · Click chemistry

Contents

T. Zheng, S.H. Rouhanifard, A.S. Jalloh, and P. Wu (✉)
Department of Biochemistry, Albert Einstein College of Medicine, Yeshiva University, 1300 Morris Park Ave, Bronx, NY 10461, USA
e-mail: peng.wu@einstein.yu.edu

1 Introduction

Over the past forty years, a growing synergy has developed between the field of organic chemistry and the modern biosciences as a direct result of a recent push for translational research. This area of research is intended to transform the findings from basic scientific research into direct, medicinal applications, and in many cases requires a creative use of organic chemistry to address major biological problems. For example, the discovery of chemoselective transformations has enabled the incorporation of probes and tags onto biomolecules to create new therapeutic agents or research tools. Covalent attachment of polyethylene glycol (PEG) chains, a process termed PEGylation, is routinely practiced in pharmaceutical companies to improve pharmacokinetics of therapeutic proteins. This modification is often realized by exploiting the electrophilic reactivity of succinimidyl carbonate-functionalized PEGs for ε-amino groups of lysine residues or the N-terminal amino acid of target proteins. However, since proteins often possess multiple copies of lysine residues, site-specific labeling can be difficult to achieve. PEG molecules randomly introduced to a target protein may interfere with its desired biological function, hampering its therapeutic application. To address this challenge, researchers have developed genetically encoded unnatural amino acids [1] and peptide tags [2, 3] for introducing novel functional groups beyond those found in the 20 canonical amino acids. These new tagging methods, combined with tools from the emerging field of bioorthogonal click chemistry [4], have conferred site specificity to protein modification manipulations. Indeed, orthogonality and selectivity, two central factors that govern the success of these approaches, have been increasingly adopted by chemical biologists in recent years. Today, bioorthogonal click chemistry has found applications beyond protein science and offers reliable methods for bioconjugation in the covalent labeling of other biological macromolecules, including nucleic acids, glycans, and lipids [5]. Featuring benign reaction conditions and unprecedented selectivity, bioorthogonal click chemical transformations are the first choice for in situ labeling of biomolecules in living systems.

2 CuAAC and Copper-Free click chemistry

Discovered by Sharpless-Fokin and Meldal in 2002, the Cu(I)-catalyzed azide-alkyne cycloaddition (CuAAC) that forms 1,4-disubstituted triazoles is a prototypical example of bioorthogonal click chemistry (Fig. 1a) [6, 7]. This

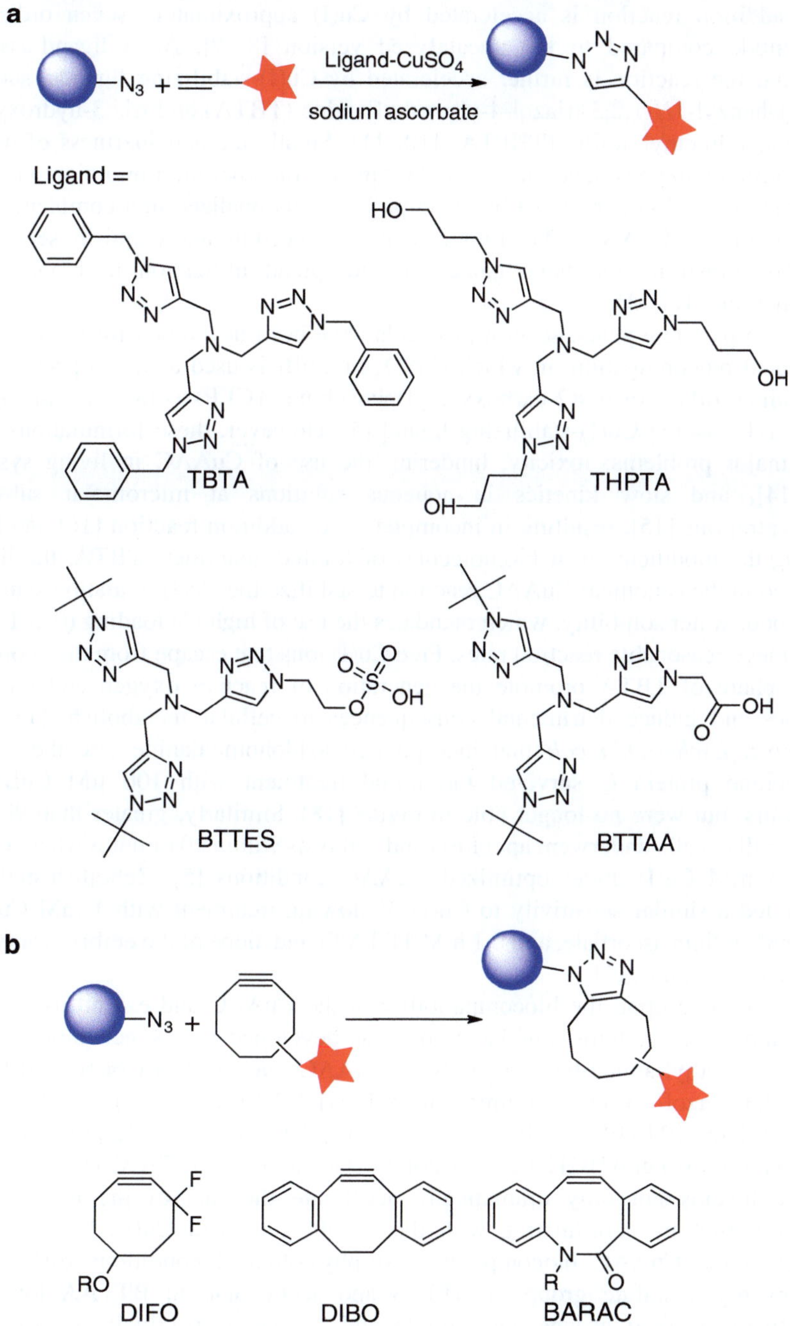

Fig. 1 (**a**) Cu(I)-catalyzed azide-alkyne cycloaddition (CuAAC)- and (**b**) copper-free click chemistry

cycloaddition reaction is accelerated by Cu(I) approximately seven orders of magnitude compared to the uncatalyzed version [8, 9]. As a ligand-assisted process, the reaction is further accelerated by Cu(I)-stabilizing ligands, such as tris[(1-benzyl-1*H*-1,2,3-triazol-4-yl)methyl]amine (TBTA) and tris(3-hydroxypropyltriazolylmethyl)amine (THPTA) [10, 11]. Small size and inertness of azides and terminal alkynes make them popular tags for incorporation into biomolecules, and their detection is realized by using probes functionalized in a complementary fashion via the CuAAC. The simplicity of the procedure and exquisite selectivity of this transformation have gained it widespread utilization in bioorthgonal conjugation [12, 13].

In the past few years, several protocols have been developed for the CuAAC-mediated bioconjugation, in which $CuSO_4$ or CuBr is used as the copper source, sodium ascorbate or tris(2-carboxyethyl)phosphine (TCEP) as the reducing agent, and TBTA as the Cu(I)-stabilizing ligand [5]. However, these formulations have two major problems: toxicity, hindering the use of CuAAC in living systems [5, 14], and slow kinetics in aqueous solutions at micromolar substrate concentrations [15], resulting in incomplete cycloaddition reaction [16] and hampering the modification of biomolecules of limited quantities. TBTA, the ligand utilized in the canonical CuAAC reaction to stabilize the Cu(I) oxidation state, has very poor water solubility, which mandates the use of high Cu loading (0.2–1 mM) to achieve reasonable reaction rates. Free Cu(I) ions that escape from the coordination sphere of TBTA promote the generation of reactive oxygen and nitrogen species and induce detrimental consequences to cellular metabolism [17]. For example, *Escherichia coli* that incorporated azidohomoalanine into their outer membrane protein C survived the initial treatment with 100 μM CuBr for 16 hours, but were no longer able to divide [18]. Similarly, greater than 90% of mammalian cells underwent apoptosis and cell lysis within 20 minutes when treated with 1 mM Cu(I) under optimized CuAAC conditions [5]. Zebrafish embryos exhibited a similar sensitivity to Cu(I). Following treatment with 1 mM $CuSO_4$, 1.5 mM sodium ascorbate, and 0.1 mM TBTA ligand, none of the embryos survived beyond 15 minutes [5].

To improve upon the biocompatibility of the CuACC and extend its application into living systems, our laboratory has developed a new generation of tris (triazolylmethyl)amine-based ligands for CuAAC, such as 2-(4-((bis((1-tert-butyl-1H-1,2,3-triazol-4-yl)methyl)amino)methyl)-1H-1,2,3-triazol-1-yl) ethyl hydrogen sulfate (BTTES) [15] and 2-[4-({bis[(1-tert-butyl-1H-1,2,3-triazol-4-yl) methyl]amino}methyl)-1H-1,2,3-triazol-1-yl]acetic acid (BTTAA) (Fig. 1a) [19]. These ligands not only dramatically accelerate the rate of the azide-alkyne cycloaddition by coordinating with the in situ generated Cu(I), but they also rendered the CuAAC biocompatible. At physiological conditions (pH = 7.4), the hydrogen sulfate group in BTTES and acetic acid in BTTAA ionize to negatively charged groups, i.e., sulfate and acetate, respectively, securing the solubility of the ligand–Cu(I) complex and preventing cellular internalization of the coordinated copper ions.

Alternatively, the Bertozzi laboratory pioneered the development of a copper-free [3+2] cycloaddition reaction by employing ring strain as a driving force for alkyne activation (Fig. 1b) [20, 21]. Among the cyclooctyne derivatives developed by the Bertozzi and the Boons groups, 3,3-difluorocyclooct-1-yne (DIFO) [22], 4-dibenzocyclooctynol (DIBO) [23], and biarylazacyclooctynone (BARAC) [24] showed rapid kinetics in biomolecular labeling experiments. Particularly, fluorescently labeled DIFO reagents have been elegantly used for imaging azide-tagged glycans within complex biological systems, including live cells [22], *C. elegans* [25], and zebrafish embryos [26, 27], with very low background fluorescence. However, recent in vivo studies revealed that DIFO-based probes have nonspecific reactivity toward mouse serum albumin, presumably via covalent-bond formation between the cyclooctyne and cysteine residues [28].

Today, CuAAC and copper-free click chemistry are serving as complementary approaches to modify biomolecules for both in vitro and in vivo applications. In this chapter, we will highlight the recent applications of these triazole-forming reactions to the imaging and profiling of biomolecules. Comprehensive reviews of click chemistry-based approaches for bioconjugation have been published recently [12, 29].

3 Triazole Formation-Based Bioconjugation for Imaging Cell-Surface Proteins and Newly Synthesized Proteins

The homeostasis of cellular proteins is tightly regulated by synthesis and degradation machineries in response to their ever-changing microenvironment. Within live cells, proteins are continuously synthesized to replenish those that have been naturally metabolized or degraded in response to stimuli. Techniques that enable the visualization of proteins in their cellular environment are powerful tools for analyzing their cellular dynamics and functions. Popular methods for visualizing proteins in their native environments rely on genetically encoded fluorescent proteins (e.g., GFP). By fusing a target protein to GFP, a protein of interest can be visualized in live cells or organisms. However, the large size of GFP may interfere with the structure and function of the fused protein.

Alternatively, a consensus peptide sequence that harbors a specific site for bioorthogonal conjugation can be exploited to modify a target protein for in vivo imaging. By treating the tagged protein on the cell surface with the enzyme that adds a synthetic prosthetic group analog bearing the desired orthogonal functionality, the protein can then be labeled with an imaging probe via bioorthogonal click reactions. Biotin ligase [30], formylglycine-generating enzyme [31, 32], and transglutaminase [33] have all been employed for this purpose. For example, Ting and coworkers redirected a microbial lipoic acid ligase (LplA) [34] to specifically attach an alkyl azide onto an engineered LplA acceptor peptide (LAP) fused to cell-surface proteins in live mammalian cells. The alkyl azide was then selectively labeled

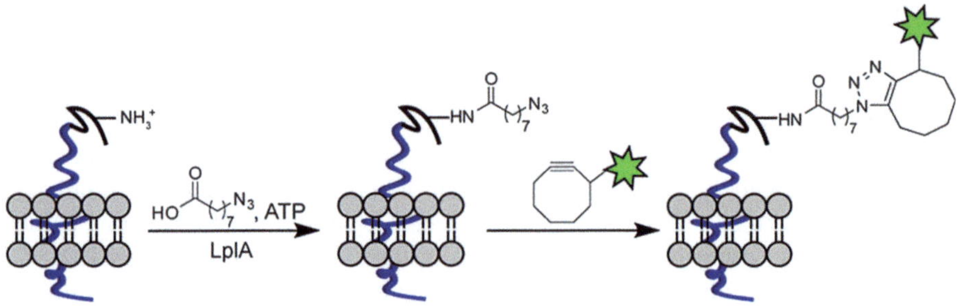

Fig. 2 Site-specific labeling of cell-surface proteins using lipoic acid ligase

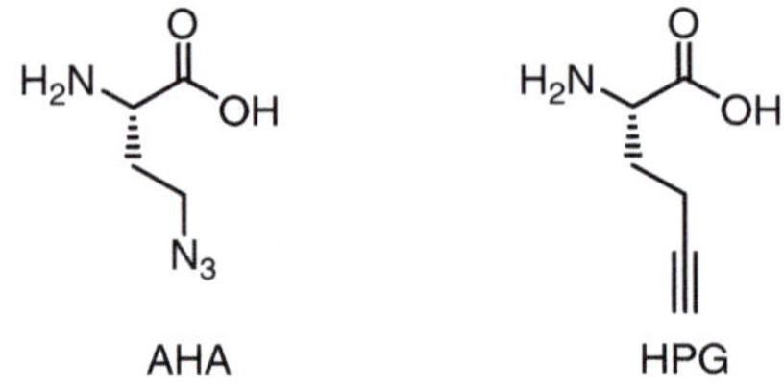

Fig. 3 Methionine analogues used for metabolically labeling of newly synthesized proteins

with fluorescently conjugated cyclooctyne probes to image the dynamics of the labeled proteins (Fig. 2) [35]. Such methods are ideal for studying the dynamics and distribution of one or a few preselected proteins. Since genetic manipulation is required for adding a peptide tag to the protein of interest, it is difficult to apply these methods to monitor the newly synthesized proteins at a global level.

In order to analyze the global protein synthesis, researchers in the Tirrell group have developed a method called BONCAT (bioorthogonal noncanonical amino acid tagging), a two-step labeling approach. In the first step, the methionine analogue azidohomoalanine (AHA) or homopropargylglycine (HPG) (Fig. 3) is metabolically incorporated into the newly synthesized proteins. In the second step, an alkyne- or azide-bearing fluorophore is used to detect these proteins in fixed cells via TBTA-mediated CuAAC [36].

They successfully applied this method for visualizing newly synthesized proteins in rat hippocampal neurons. After a 1-h incubation with either AHA or HPG, newly synthesized proteins can be detected in both the somata and dendrites. Newly synthesized proteins in spine-like protrusions from dendrites can be detected after a 2-hour exposure to AHA or HPG.

In order to monitor the newly synthesized proteins inside live cells, the Tirrell lab has developed a set of cell-permeable coumarin-functionalized cyclooctynes. These coumarin cyclooctynes would ligate to the cellular azide-bearing proteins, allowing visualization of the dynamic turnover of the newly synthesized proteins [37].

4 Triazole Formation-Based Bioconjugation for Imaging and Profiling Protein Posttranslational Modifications

We have come a long way in the journey toward understanding the molecular basis of life. Half a century ago, Francis Crick proposed the central dogma as a working model that credits the three-dimensional structure and function of proteins to their one-dimensional progenitors: DNAs and RNAs [38]. However, the discovery of posttranslational modification (PTM) makes it clear that the structure and function of proteins and polypeptides can be dictated by factors other than their encoding nucleic acids and amino acid building blocks [39].

A major task in the postgenomic era is to assign specific functions of individual proteins [40]. Genetic tools such as gene microarrays and RNA interference have played major roles in revealing the expression patterns and functions of target genes [41]. However, these techniques are insensitive to most PTMs that are under spatiotemporal control—processes that modulate protein folding [42], functional states [43], and their subcellular localizations [41].

In the past decade, bioconjugation techniques based on CuAAC have emerged as new tools for profiling protein PTMs, including poly(ADP-ribosyl)ation, methylation, acylation, and glycosylation. In these methods, an azide or an alkyne moiety is first introduced to a natural substrate (e.g., cofactor monosaccharide, etc.) of a PTM enzyme. The modified substrate is then processed by its respective endogenous PTM enzymes to modify the substrate proteins in their native environment or by a recombinant PTM enzyme in crude cell lysates. In a subsequent step, an imaging probe can be introduced, allowing the visualization of the labeled biomolecules in living systems or in sodium dodecyl sulfate-polyacrylamide gel. Alternatively, the tagged proteins can be conjugated with an affinity probe via CuAAC and isolated for proteomic analysis and functional studies.

4.1 Profiling Protein Poly(ADP-ribosyl)ation

In 2010, the Lin group created a clickable analogue of nicotinamide adenine dinucleotide (NAD), an ADP-ribose donor, to identify novel substrates of poly (ADP-ribose) polymerases (PARPs). PARPs mediate protein poly(ADP-ribosyl) ation by adding one or more ADP-ribose moieties posttranslationally to target proteins. Seventeen different variants of PARP have been identified in humans, and they are generally known to play an important role in DNA repair, transcription, and mitosis. In order to understand the functions of individual PARPs, it is necessary to know which substrate proteins each PARP is modifying. Using the alkyne-tagged NAD analogue specific to PARP-1 in a PARP-catalyzed reaction, Lin and coworkers enabled the conjugation of affinity tags to the substrates of PARP-1 and performed affinity purification under denaturing conditions in order to minimize nonspecific interactions within the cell lysate and give fewer

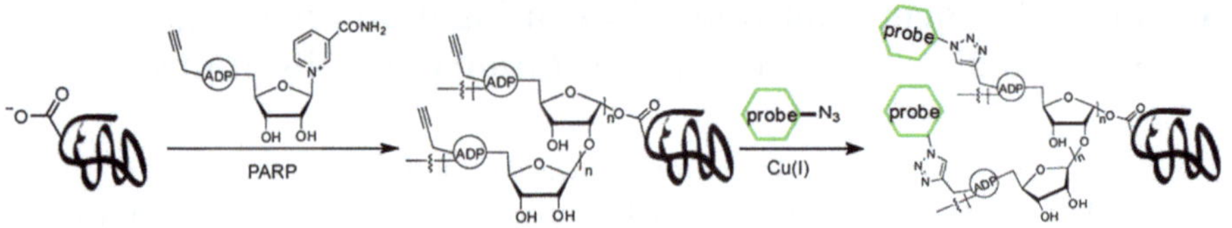

Fig. 4 Identifying substrates of poly(ADP-ribose) polymerases using a CuAAC-based strategy

false-positive results than the alternative methods (Fig. 4). Following affinity purification, they were able to identify over 70 candidate substrates for PARP-1 by tandem mass spectrometry, 40% of which had been previously identified by other methods [44].

4.2 Profiling Protein Fatty Acylation

Protein fatty acylation plays an important role in normal eukaryotic cell function and signaling as well as in human disease. Conventional approaches for detecting protein fatty acylation involve metabolic labeling with radiolabeled fatty acids and subsequent analysis using autoradiography. Such methods are hazardous and not compatible with downstream proteomic studies. To circumvent these limitations, the Hang group developed an alternative approach to metabolically treat mammalian cells with alkyne-modified fatty acids and labeled the alkyne-incorporated proteins with a biotin-azide probe by CuAAC for affinity capture and MS identification (Fig. 5). Using this approach, Hang and coworkers were able to identify from Jurkat cells 57 known fatty-aceylated proteins and 109 new candidate proteins with high confidence, including histone H3 variants [45]. Recently, the same approach was applied to profile palmitoylated proteins in dendritic cells, a type of phagocytic cells found in the mammalian immune system. Protein S-palmitoylation is a PTM that occurs on cysteine residues that facilitates targeting and function of membrane-associated proteins [46]. Due to the diversity of S-palmitoylation in eukaryotic cells, it was suggested that multiple cellular pathways could be regulated by this PTM. Hang et al. performed large-scale profiling of S-palmitoylated proteins in dendritic cell line DC2.4 and found that interferon-induced transmembrane protein 3 (IFITM3), a protein associated with the host immune response to microbial infection, was palmitoylated. Further study revealed that S-palmitoylation of IFITM3 is responsible for the clustering of this protein in the ER membrane, and this PTM is also critical for IFITM3's antiviral activity against influenza viruses [47].

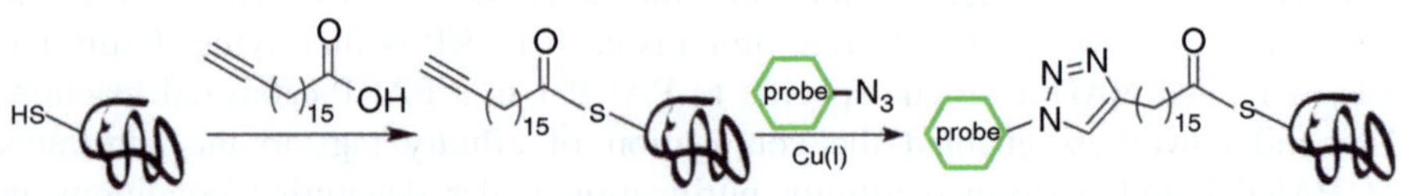

Fig. 5 Two-step labeling strategy for detecting S-palmitoylated proteins

4.3 Imaging Cellular Glycans

In nature, nine monosaccharides are used as building blocks to produce a variety of polysaccharides that decorate the surface of eukaryotic cells. These glycans are critical mediators of many biological processes, including angiogenesis, fertilization, cell–cell communication, and neuronal development [48]. Inside the cell, glycans regulate fundamental cellular processes including transcription, translation, and protein trafficking [49]. Glycosylation is a dynamic process that reports on a cell's physiological state. When a cell undergoes a transformation of state, the glycans on its cell surface will change. In many cases, altered glycosylation, i.e., over- and down-expression of naturally occurring glycans and the appearance of glycans normally restricted to embryonic expression, is a hallmark of tumor phenotype [50]. Therefore, the ability to visualize the dynamic changes of glycans in vivo will augment our understanding of glycans' physiological roles in developmental processes and in malignant transformation.

Glycosylation is a posttranslational modification, not under direct genetic control. Therefore, genetic reporters (e.g., GFP) used for in vivo protein imaging cannot be exploited for visualizing glycans. Instead, imaging of glycans is rendered possible by integrating bioorthogonal click chemistry with metabolic oligosaccharide engineering. By hijacking a cell's biosynthetic machinery, monosaccharide building blocks functionalized with bioorthogonal chemical tags can be incorporated into target glycans. In a subsequent step, a complementary imaging probe is conjugated via a bioorthogonal click reaction, enabling the visualization of the tagged glycoconjugates (Fig. 6) [51].

Since the breakthrough discovery by the Bertozzi group in 1997 that bioorthogonal ligation can take place on the surface of live cells [52], several different classes of glycans have been successfully detected using this method by

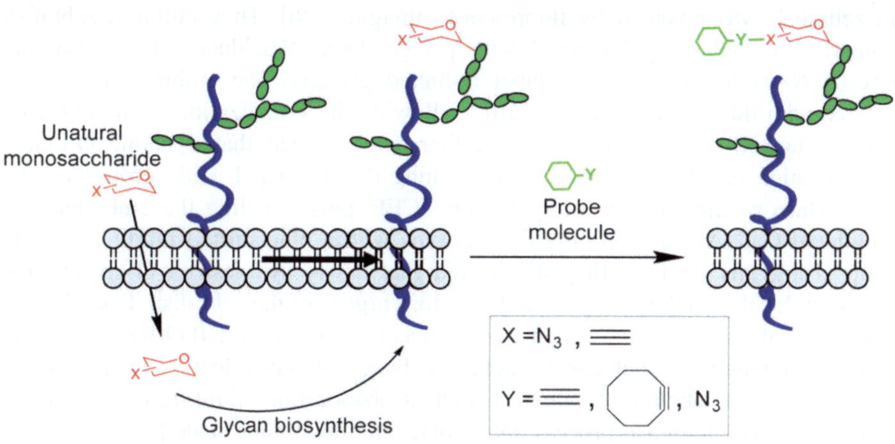

Fig. 6 Detection of cell-surface glycans using metabolic labeling and triazole-forming click chemistry

Table 1 Azide- and alkyne-bearing monosaccharides used for metabolic labeling of glycans

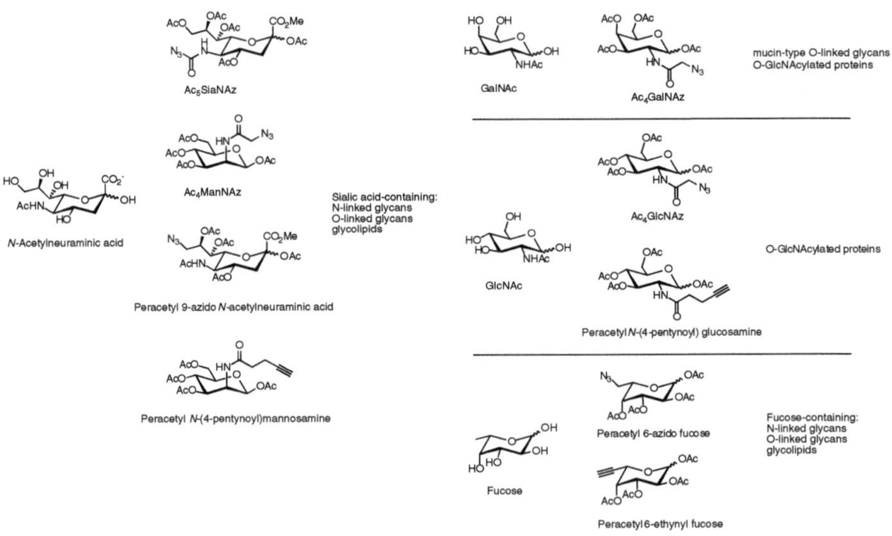

Natural monosaccharide target	Unnatural monosaccharide	Glycan types labeled
N-Acetylneuraminic acid	$Ac_5SiaNAz$ $Ac_4ManNAz$ Peracetyl 9-azido *N*-acetylneuraminic acid Peracetyl *N*-(4-pentynoyl)mannosamine	Sialic acid-containing: N-linked glycans O-linked glycans glycolipids
GalNAc	$Ac_4GalNAz$	mucin-type O-linked glycans O-GlcNAcylated proteins
GlcNAc	$Ac_4GlcNAz$ Peracetyl *N*-(4-pentynoyl) glucosamine	O-GlcNAcylated proteins
Fucose	Peracetyl 6-azido fucose Peracetyl 6-ethynyl fucose	Fucose-containing: N-linked glycans O-linked glycans glycolipids

employing the azide or the alkyne as the chemical reporter (tag) (Table 1) [51]. Detection of the azide-bearing glycans in live cells or living organism is realized using the Bertozzi–Staudinger ligation [53] or the triazole-forming reactions [51]. Likewise, detection of the alkyne-bearing glycans in live cells can be accomplished using the biocompatible CuAAC (Table 1) [15].

The first example of triazole-formation enabled glycan imaging in a living organism was reported by the Bertozzi group in 2007, in which the dynamic glycosylation in zebrafish was revealed by fluorescence imaging [26]. They cultured zebrafish embryos in media supplemented with peracetylated *N*-azidoacetylgalactosamine ($Ac_4GalNAz$) to metabolically label O-linked glycans. The embryos were then reacted with fluorescently labeled DIFO, allowing the visualization of the O-linked glycans tagged by the azide. They further demonstrated that dynamic O-linked glycosylation can be visualized by quenching the unreacted azide groups on the cell surface of the labeled embryos with TCEP, pulse-labeling the embryos with additional GalNAz, and probing the embryos with a blue-shifted DIFO conjugate [26]. Using this method, they discovered that newly synthesized O-glycans are primarily located in the fins, jaw, and olfactory organs of the zebrafish. One obvious advantage of this technique is better tissue accessibility of the DIFO-based imaging probes compared to what can be achieved by using bulky lectins or antibodies. However, robust labeling requires a 1-hour incubation time, limiting its application to follow a fast biological process taking place in minute time scale [26].

The discovery of the BTTES- and BTTAA-mediated CuAAC overcomes this obstacle and paves the way for rapid imaging of glycans in vivo. This method

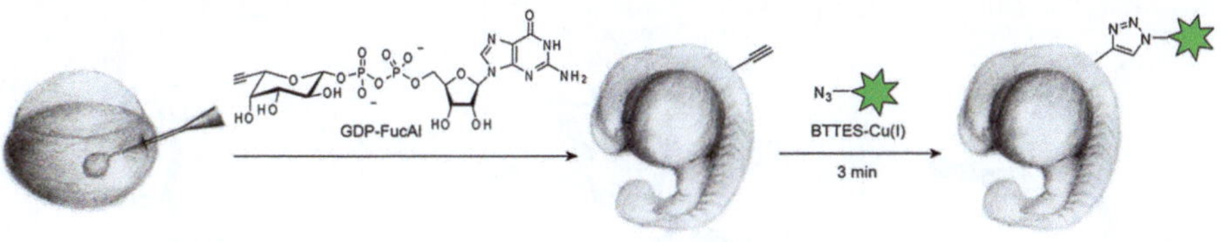

Fig. 7 Imaging fucosylated glycans in developing zebrafish embryos

allowed the visualization of fucosylated glycans in developing zebrafish for the first time [15]. Our lab showed that by using guanidine 5′-diphosphate-β-6-ethynyl fucose (GDP-FucAl) directly as a metabolic precursor to bypass the guanidine 5′-diphosphate-β-L-fucose (GDP-fucose) salvage pathway, fucosylated glycans in zebrafish embryos can be labeled with an alkyne tag. Upon a 3-minute click reaction with a fluorophore-conjugated azide, the fucosylated glycans in enveloping layer, the embryos' outermost layer of cells, were lightened up for fluorescence visualization, and the labeled glycans were detectable as early as 2.5 hour postfertilization (hpf) (Fig. 7) [15]. Notably, a single dose of GDP-FucAl could yield a detectable fluorescent signal up to 96 hpf.

However, the metabolic method is only applicable for monosaccharide labeling. Since each monosaccharide is usually found in a variety of cell-surface polysaccharide glycans, higher-order glycans, such as disaccharides or trisaccharides, of specific composition cannot be uniquely labeled by hijacking their biosynthetic pathways with unnatural monosaccharides. As the first step in overcoming this challenge, our lab developed a chemoenzymatic approach that has been successfully applied to label cell-surface polysaccharides bearing *N*-acetyllactosamine (LacNAc; Galβ1,4GlcNAc) building blocks.

This method employs a recombinant *Helicobacter pylori* α(1,3)fucosyltransferase [54] to transfer a C-6 azide- or alkyne-tagged fucose residue to the 3-OH of *N*-acetylglucosamine unit of the LacNAc disaccharide. The tagged LacNAc can then be selectively derivatized with probes via triazole-forming click reactions. Using this method, we discovered that murine lymphocytes with an activated memory phenotype ($CD44^{high}CD62L^{low}$, $CD25^{+}$) exhibited higher levels of LacNAc compared to their naïve counterparts ($CD44^{low}CD62L^{high}$, $CD25^{-}$) [55]. Thus, this chemoenzymatic approach may serve as a valuable tool to discriminate differentially activated cell subsets ex vivo (Fig. 8).

4.4 Profiling Sialylated Glycoproteins

The metabolic labeling and chemical reporter strategy also found applications in targeted glycoproteomics. Based on the pioneering work of Bertozzi and coworkers, the Wong laboratory designed a glycoproteomic method for identifying sialylated glycoproteins from crude cell lysates [56]. In this method, mammalian cells are metabolically treated with Ac_4ManNAl to label sialylated glycoproteins

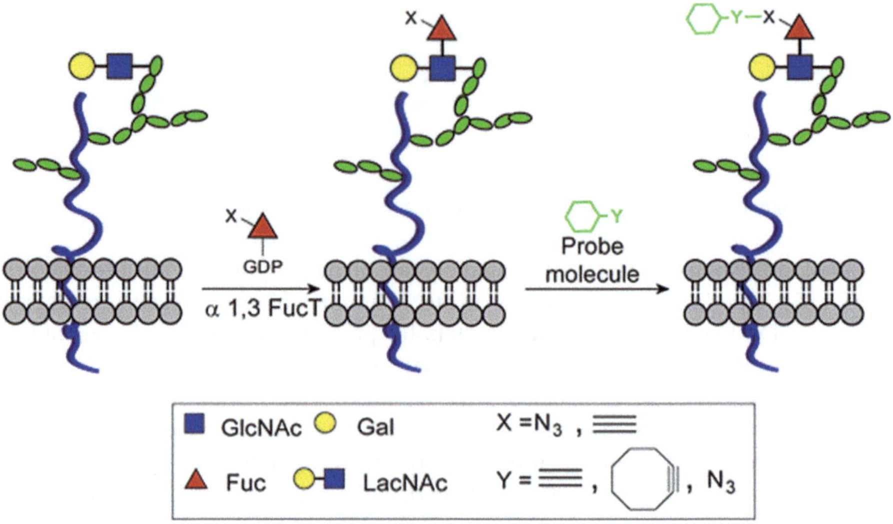

Fig. 8 A chemoenzymatic method for the detection of LacNAc disaccharides

with alkyne tags. These proteins are then biotinylated via CuAAC and enriched by affinity capture on streptavidin beads. Following on-bead trypsin digestion and PNGase F treatment, N-linked glycoproteins that are sialylated are selectively released and identified by LC-MS^2. Adopting this method to profile sialylated glycoproteins in syngeneic prostate cancer cell lines PC3-N2 (nonmetastatic) and PC3-ML2 (highly metastatic), Semmes and coworkers identified and annotated 64 (26% of total nonredundant proteins) and 72 (29%) extracellular or membrane-bound glycoproteins from N2 and ML2 cells, respectively. Using ingenuity pathway analysis, they discovered that the majority of glycoproteins unique to the nonmetastatic PC3-N2 cell line were mediators of basic cell functions, such as vesicular transport, lipid metabolism, and mRNA processing and splicing. By contrast, the majority of glycoproteins overexpressed in the metastatic ML2 cell line were involved in cell motility, migration, and invasion [57].

4.5 Profiling O-Linked-β-N-Acetylglucosamine (O-GlcNAc) Glycosylation

O-GlcNAcylation is a dynamic glycosylation that is found on cytosolic and nuclear proteins. Today, more than 500 proteins are found to be *O*-GlcNAcylated [58]. These proteins participate in numerous cellular processes, including transcription, apoptosis, signal transduction, nutrient sensing, and proteasomal degradation. To identify *O*-GlcNAcylated proteins at the proteome level, Hsieh-Wilson and coworkers developed a chemoenzymatic approach, which involves transferring *N*-azidoacetyl galactosamine (GalNAz) to *O*-GlcNAc residues of *O*-GlcNAcylated

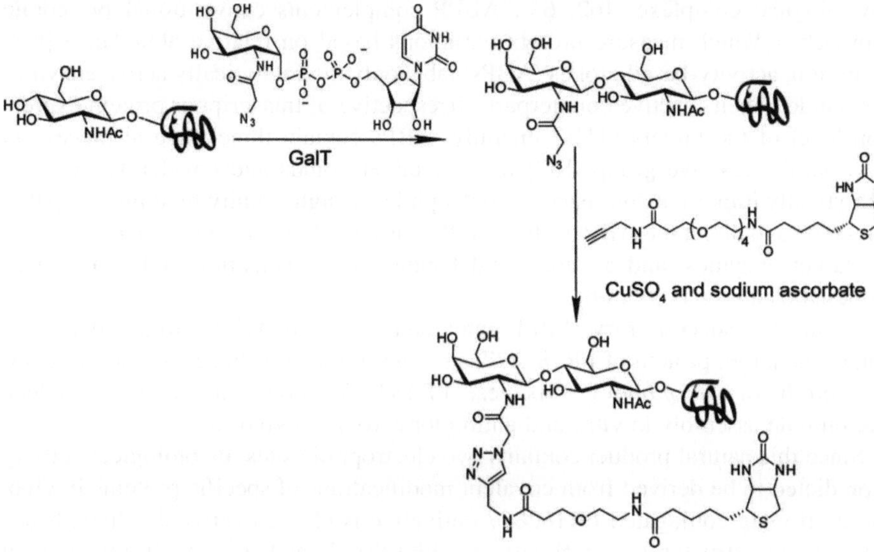

Fig. 9 A chemoenzymatic method for the detection of protein *O*-GlcNAcylation

proteins in crude cell lysates. The azido-tagged proteins can then be modified with biotin tags via CuAAC and be purified for mass spectrometry identification (Fig. 9) [59].

Using this method, the Hart lab discovered that histones, nuclear proteins in eukaryotic cells that assemble DNAs into nucleosomes, are *O*-GlcNAcylated [60]. Histones isolated from Hela cells were first subjected to tryptic digestion, followed by the chemoenzymatic labeling and mass spectrometry identification. Among the four core histones, three were found to be modified by *O*-GlcNAc, i.e., H2A, H2B, and H4. Threonine101 of H2A, serine36 of H2B, and serine47 of H4 were identified as the sites of *O*-GlcNAcylation. Interestingly, *O*-GlcNAcylation of serine36 of H2B and serine47 of H4 regulate histone-tail dynamics and may induce changes in the nucleosomal structure. It was discovered that histone *O*-GlcNAcylation level increased with heat shock and this increase accompanies chromatin condensation.

5 Triazole Formation-Based Bioconjugation for Activity-Based Protein Profiling

Another area in chemical biology that the triazole-forming click reactions have revolutionized is activity-based protein profiling (ABPP). ABPP is a chemical biology platform that uses active site-directed probes [61] to study the dynamic expression and function of mechanistically similar enzymes in their various physiological complexes [62, 63]. ABPP complements conventional proteomic approaches, which measure protein variations based on relative abundance [64]. By design, activity-based probes (ABPs) label only the catalytically active enzymes to exclude – their inactive counterparts, irrespective of transcript or protein expression level of the targets [41]. Generally, ABPs contain three integral pieces: an electrophilic reactive group (RG) that selectively binds and covalently modifies catalytically important and reactive nucleophiles, a high-affinity binding group that can guide and direct the ABP to structurally conserved features in the active sites of the target enzymes, and an analytical handle for visualization and isolation of probe-labeled targets [65, 66].

As an early success story, ABPP, integrated with CuAAC, led to the identification of the target protein of FR182877, a natural product isolated from a fermentation broth of *Streptomyces* species. FR182877 shows activities to induce microtubule assembly in vitro and antitumor activity in vivo [67].

Since this natural product contains two electrophilic sites, its biological activity is predicted to be derived from covalent modifications of specific proteins in vivo. An affinity tag-conjugated FR182877 derivative is ideal to verify this hypothesis. Due to the structural complexity of FR182877 and its sensitivity toward nucleophiles, conventional methods for its functionalization were complicated by side reactions. Using the Cu(I)-catalyzed triazole formation, a short peptide containing both a rhodamine tag and a biotin tag was introduced to (−)-FR182877 to generate a trifunctional probe (Fig. 10) [68]. Incubating the probe with a mouse heart proteome in vitro, followed by isolation of the biotin-labeled proteins by avidin chromatography, carboxylesterase-1 (CE-1) was identified as a specific target of this natural product. Competition experiments with the serine hydrolase-directed probe fluorophosphonate rhodamine determined that (−)-FR182877 is a highly potent (IC_{50} = 34 nM) and selective inhibitor of CE-1.

Despite the successful development of ABPP probes for several enzyme classes, the large reporter tags, e.g., fluorophore or biotin, prevents the cellular uptake and distribution of many ABPP probes in living organisms. To solve this problem,

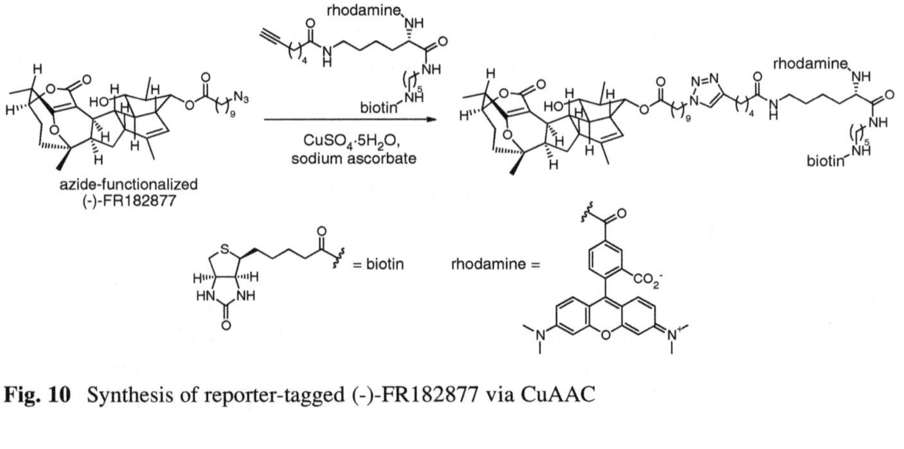

Fig. 10 Synthesis of reporter-tagged (-)-FR182877 via CuAAC

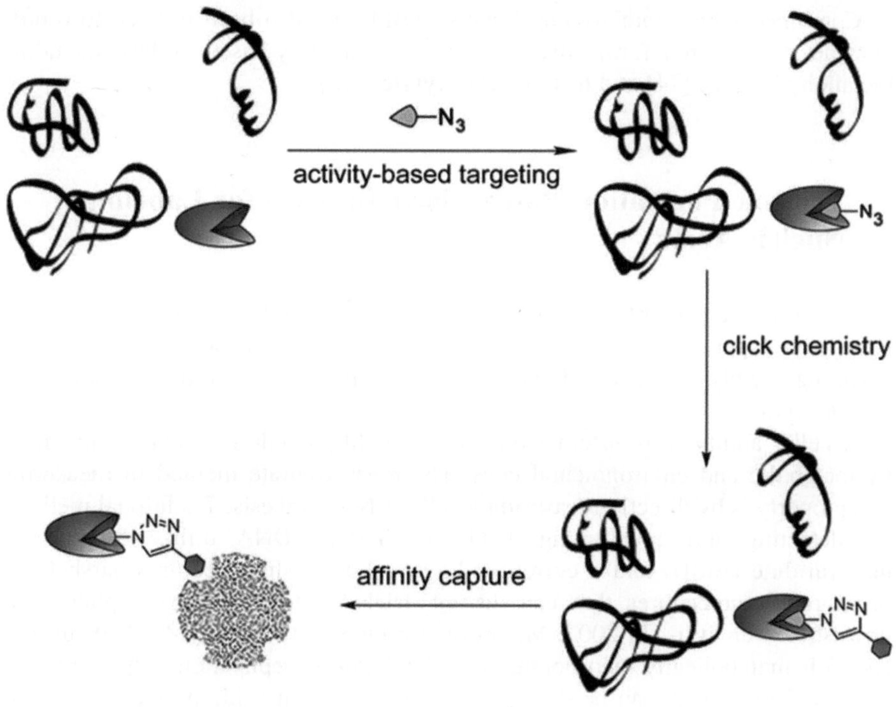

Fig. 11 The workflow of activity-based protein profiling

Cravatt [69] and Overkleeft [70] independently designed a second generation probes, in which the large reporter tag is replaced with a small, bioorthogonal chemical handle (i.e., an alkyne or an azide). After covalent modification of target proteins in crude cell lysates, the alkyne or azide tag can be further conjugated to affinity probes via CuAAC or Bertozzi–Staudinger ligation [53], allowing the enrichment of tagged proteins for further analysis (Fig. 11).

Using an azide-bearing active-site probe, azidohexyl benzenesulfonate and a rhodamine-alkyne detection probe and this two-step labeling strategy, Cravatt and coworkers were able to identify a number of serine hydrolases, including ECH-1, in the human breast cancer cell line MDA-MB-435 [69]. Azidohexyl benzenesulfonate was also active in live mice, enabling the capture of ECH-1 and its subsequent detection via rhodamine-alkyne and mass spectrometry analyses ex vivo in mice tissue homogenates [69]. Further studies revealed that background noise in proteome labeling can be significantly reduced by replacing the azidohexyl benzenesulfonate/rhodamine-alkyne cycloaddition pair by the hex-5-ynyl benzenesulfonate/rhodamine-azide pair [66]. The higher sensitivity of the latter enabled the detection of low-abundance enzyme activity that had eluded detection in azidohexyl benzenesulfonate/rhodamine-alkyne reaction. Since 2004, the optimized ABPP conditions have been successfully applied to profile glycosidases [71], cytochrome P450 [72], and other enzyme classes [73].

Combined with photocrossing linkers, ABPP has also been utilized to profile enzymes that do not form covalent bond with activity-based probes, including metallohydrolases [74] and histone deacetylases [75].

6 Triazole Formation-Based Bioconjugation for Labeling Nucleic Acids

Likewise, the remarkable selectivity of triazole-forming bioconjugation reactions have been expoited to modify nucleic acids for a number of applications such as detecting newly synthesized DNA and determining genomic locations of 5-hydroxymethylcytosine.

A cell's ability to proliferate reflects its healthy conditions and is influenced by metabolic and environmental cues. The most accurate method of measuring this property is by directly measuring a cell's DNA synthesis. Traditional methods for detecting and quantitating newly synthesized DNA utilize 5-bromo-2′-deoxyuridine (BrdU) and a corresponding antibody, which requires harsh DNA denaturation procedures that can disrupt dsDNA integrity and complicate the downstream analysis. In 2007, Mitchison reported that 5-ethynyl-2′-deoxyuridine (EdU) is metabolically incorporated into DNA during replication. After cell fixation, EdU in dsDNA can be detected with a fluorescent azide probe via CuAAC [76]. By using this procedure, newly synthesized DNA can be visualized with excellent sensitivity under milder conditions than the BrdU-based method.

5-Hydroxymethylcytosine (5-hmc), an oxidized form of 5-methylcytosine (5-mc), is an epigenetic mark discovered recently in certain mammalian tissues [77, 78]. The function of 5-hmC in epigenetic regulation is thought to be different

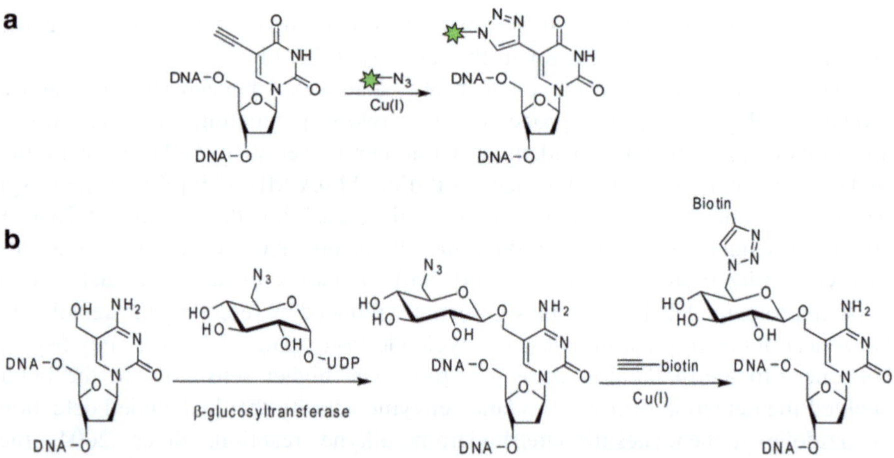

Fig. 12 Detecting newly synthesized DNA (**a**) and determining genomic locations of 5-hydroxymethylcytosine (**b**) via CuAAC

from 5-mC and currently remains a mystery. Determination of genomic locations of 5-hmc serves as the first step to dissecting its functional roles. He and coworkers developed a chemoenzymatic method for site-specific labeling of 5-hmc [79]. This method utilizes β-glucosyltransferase to transfer an azido-containing glucose from the corresponding uridine 5′-diphospho-α-D-glucopyranoside to the 5-hydroxymethyl group of 5-hmc. The resulting azide-modified 5-hmc can then be conjugated to biotin probes via the triazole-formation reactions and quantified using avidin-based assays. Using this method, they discovered that the expression of 5-hmc in mouse cerebellum is developmentally regulated and is increased accompanying neuronal maturation (Fig. 12).

7 Summary and Outlook

In the past ten years, the discovery of the triazole-forming click reactions has paved the way for a number of breakthroughs in the field of chemical biology. However, several technique challenges are waiting to be overcome. First, CuAAC, as presently formulated, cannot be applied to label biomolecules in the cytoplasm of live cells. To mitigate the toxicity associated with CuAAC, we incorporated functional groups that bear negative charges (at physiological pH) into Cu(I)-coordinating ligands to minimize the membrane permeability of the coordinated Cu(I). Accordingly, only cell-surface biomolecules are efficiently labeled. In a few studies, copper-free click chemistry has been adopted to modify azide-containing cytosolic proteins. However, high background noise is associated with this methodology [37]. Second, the invention of cyclooctyne probes eliminates the use of Cu(I) catalysts. Nevertheless, elevated ground-state energy of the cyclooctynes by ring strain inevitably increases their nonspecific reactivity. Through the joint efforts of chemists and biologists, we believe that these hurdles will be overcome in the near future by the discovery of new accelerating ligands for Cu(I) or new bioorthogonal reactions with better kinetics and selectivity than the triazole-forming reactions.

Acknowledgement The author's work that was highlighted in this book chapter was supported by the National Institutes of Health to P.W. (GM080585 and GM093282), the Mizutani Foundation for Glycoscience, and DuPont (DuPont Young Professor Award).

References

1. Wang L, Schultz PG (2004) Expanding the genetic code. Angew Chem Int Ed 44:34–66
2. O'Hare HM, Johnsson K, Gautier A (2007) Chemical probes shed light on protein function. Curr Opin Struct Biol 17:488–494
3. Rabuka D (2010) Chemoenzymatic methods for site-specific protein modification. Curr Opin Chem Biol 14:790–796

4. Baskin JM, Bertozzi CR (2007) Bioorthogonal click chemistry: covalent labeling in living systems. Qsar Comb Sci 26:1211–1219
5. Sletten EM, Bertozzi CR (2009) Bioorthogonal chemistry: fishing for selectivity in a sea of functionality. Angew Chem Int Ed 48:6974–6998
6. Rostovtsev VV, Green LG, Fokin VV, Sharpless KB (2002) A stepwise Huisgen cycloaddition process: copper(I)-catalyzed regioselective "ligation" of azides and terminal alkynes. Angew Chem Int Ed 41:2596–2599
7. Tornoe CW, Christensen C, Meldal M (2002) Peptidotriazoles on solid phase: [1,2,3]-triazoles by regiospecific copper(I)-catalyzed 1,3-dipolar cycloadditions of terminal alkynes to azides. J Org Chem 67:3057–3064
8. Wu P, Fokin VV (2007) Catalytic azide-alkyne cycloaddition: reactivity and applications. Aldrichimica Acta 40:7–17
9. Meldal M, Tornoe CW (2008) Cu-catalyzed azide-alkyne cycloaddition. Chem Rev 108:2952–3015
10. Chan TR, Hilgraf R, Sharpless KB, Fokin VV (2004) Polytriazoles as copper(I)-stabilizing ligands in catalysis. Org Lett 6:2853–2855
11. Hong V, Presolski SI, Ma C, Finn MG (2009) Analysis and optimization of copper-catalyzed azide-alkyne cycloaddition for bioconjugation. Angew Chem Int Ed 48:9879–9883
12. Kolb HC, Sharpless KB (2003) The growing impact of click chemistry on drug discovery. Drug Discov Today 8:1128–1137
13. Lutz JF (2007) 1,3-dipolar cycloadditions of azides and alkynes: a universal ligation tool in polymer and materials science. Angew Chem Int Ed 46:1018–1025
14. Prescher JA, Bertozzi CR (2005) Chemistry in living systems. Nat Chem Biol 1:13–21
15. Soriano del Amo D, Wang W, Jiang H, Besanceney C, Yan A, Levy M, Liu Y, Marlow FL, Wu P (2010) Biocompatible copper(I) catalysts for in vivo imaging of glycans. J Am Chem Soc 132:16893–16899
16. Kaltgrad E, Sen Gupta S, Punna S, Huang CY, Chang A, Wong CH, Finn MG, Blixt O (2007) Anti-carbohydrate antibodies elicited by polyvalent display on a viral scaffold. Chembiochem 8:1455–1462
17. Gaetke LM, Chow CK (2003) Copper toxicity, oxidative stress, and antioxidant nutrients. Toxicology 189:147–163
18. Link AJ, Vink MK, Tirrell DA (2004) Presentation and detection of azide functionality in bacterial cell surface proteins. J Am Chem Soc 126:10598–10602
19. Besanceney-Webler C, Jiang H, Zheng T, Feng L, Soriano Del Amo D, Wang W, Klivansky LM, Marlow FL, Liu Y, Wu P (2011) Increasing the efficacy of bioorthogonal click reactions for bioconjugation: a comparative study. Angew Chem Int Ed 50:8051–8056
20. Agard NJ, Prescher JA, Bertozzi CR (2004) A strain-promoted [3 + 2] azide-alkyne cycloaddition for covalent modification of biomolecules in living systems. J Am Chem Soc 126:15046–15047
21. Jewett JC, Bertozzi CR (2010) Cu-free click cycloaddition reactions in chemical biology. Chem Soc Rev 39:1272–1279
22. Baskin JM, Prescher JA, Laughlin ST, Agard NJ, Chang PV, Miller IA, Lo A, Codelli JA, Bertozzi CR (2007) Copper-free click chemistry for dynamic in vivo imaging. Proc Natl Acad Sci USA 104:16793–16797
23. Ning X, Guo J, Wolfert MA, Boons GJ (2008) Visualizing metabolically labeled glycoconjugates of living cells by copper-free and fast Huisgen cycloadditions. Angew Chem Int Ed 47:2253–2255
24. Jewett JC, Sletten EM, Bertozzi CR (2010) Rapid Cu-free click chemistry with readily synthesized biarylazacyclooctynones. J Am Chem Soc 132:3688–3690
25. Laughlin ST, Bertozzi CR (2009) In vivo imaging of Caenorhabditis elegans glycans. ACS Chem Biol 4:1068–1072
26. Laughlin ST, Baskin JM, Amacher SL, Bertozzi CR (2008) In vivo imaging of membrane-associated glycans in developing zebrafish. Science 320:664–667

27. Baskin JM, Dehnert KW, Laughlin ST, Amacher SL, Bertozzi CR (2010) Visualizing enveloping layer glycans during zebrafish early embryogenesis. Proc Natl Acad Sci USA 107:10360–10365
28. Chang PV, Prescher JA, Sletten EM, Baskin JM, Miller IA, Agard NJ, Lo A, Bertozzi CR (2010) Copper-free click chemistry in living animals. Proc Natl Acad Sci USA 107:1821–1826
29. Lallana E, Riguera R, Fernandez-Megia E (2011) Reliable and efficient procedures for the conjugation of biomolecules through Huisgen azide-alkyne cycloadditions. Angew Chem Int Ed 50:8794–8804
30. Chen I, Howarth M, Lin W, Ting AY (2005) Site-specific labeling of cell surface proteins with biophysical probes using biotin ligase. Nat Methods 2:99–104
31. Carrico IS, Carlson BL, Bertozzi CR (2007) Introducing genetically encoded aldehydes into proteins. Nat Chem Biol 3:321–322
32. Wu P, Shui W, Carlson BL, Hu N, Rabuka D, Lee J, Bertozzi CR (2009) Site-specific chemical modification of recombinant proteins produced in mammalian cells by using the genetically encoded aldehyde tag. Proc Natl Acad Sci USA 106:3000–3005
33. Lin CW, Ting AY (2006) Transglutaminase-catalyzed site-specific conjugation of small-molecule probes to proteins in vitro and on the surface of living cells. J Am Chem Soc 128:4542–4543
34. Green DE, Morris TW, Green J, Cronan JE Jr, Guest JR (1995) Purification and properties of the lipoate protein ligase of Escherichia coli. Biochem J 309(Pt 3):853–862
35. Fernandez-Suarez M, Baruah H, Martinez-Hernandez L, Xie KT, Baskin JM, Bertozzi CR, Ting AY (2007) Redirecting lipoic acid ligase for cell surface protein labeling with small-molecule probes. Nat Biotechnol 25:1483–1487
36. Dieterich DC, Hodas JJ, Gouzer G, Shadrin IY, Ngo JT, Triller A, Tirrell DA, Schuman EM (2010) In situ visualization and dynamics of newly synthesized proteins in rat hippocampal neurons. Nat Neurosci 13:897–905
37. Beatty KE, Fisk JD, Smart BP, Lu YY, Szychowski J, Hangauer MJ, Baskin JM, Bertozzi CR, Tirrell DA (2010) Live-cell imaging of cellular proteins by a strain-promoted azide-alkyne cycloaddition. Chembiochem 11:2092–2095
38. Crick F (1970) Central dogma of molecular biology. Nature 227:561–563
39. Dove A (1999) Proteomics: translating genomics into products? Nat Biotech 17:233–236
40. Pandey A, Mann M (2000) Proteomics to study genes and genomes. Nature 405:837–846
41. Jessani N, Liu Y, Humphrey M, Cravatt BF (2002) Enzyme activity profiles of the secreted and membrane proteome that depict cancer cell invasiveness. Proc Natl Acad Sci 99:10335–10340
42. Seo J, Lee K-J (2004) Post-translational modifications and their biological functions: proteomic analysis and systematic approaches. J Biochem Mol Biol 37:35–44
43. Kobe B, Kemp BE (1999) Active site-directed protein regulation. Nature 402:373–376
44. Jiang H, Kim JH, Frizzell KM, Kraus WL, Lin H (2010) Clickable NAD analogues for labeling substrate proteins of poly(ADP-ribose) polymerases. J Am Chem Soc 132:9363–9372
45. Wilson JP, Raghavan AS, Yang YY, Charron G, Hang HC (2011) Proteomic analysis of fatty-acylated proteins in mammalian cells with chemical reporters reveals S-acylation of histone H3 variants. Mol Cell Proteomics 10:M110 001198
46. Linder ME, Deschenes RJ (2007) Palmitoylation: policing protein stability and traffic. Nat Rev Mol Cell Biol 8:74–84
47. Yount JS, Moltedo B, Yang YY, Charron G, Moran TM, Lopez CB, Hang HC (2010) Palmitoylome profiling reveals S-palmitoylation-dependent antiviral activity of IFITM3. Nat Chem Biol 6:610–614
48. Varki A, Cummings RD, Esko JD, Freeze HH, Stanley P, Bertozzi CR, Hart GW, Etzler ME (2008) Eseentials of glycobiology, 2nd edn. Cold Spring Harbor, New York
49. Zachara NE, Hart GW (2002) The emerging significance of O-GlcNAc in cellular regulation. Chem Rev 102:431–438
50. Taylor-Papadimitriou J, Epenetos AA (1994) Exploiting altered glycosylation patterns in cancer: progress and challenges in diagnosis and therapy. Trends Biotechnol 12:227–233

51. Laughlin ST, Bertozzi CR (2009) Imaging the glycome. Proc Natl Acad Sci USA 106:12–17
52. Mahal LK, Yarema KJ, Bertozzi CR (1997) Engineering chemical reactivity on cell surfaces through oligosaccharide biosynthesis. Science 276:1125–1128
53. Saxon E, Bertozzi CR (2000) Cell surface engineering by a modified Staudinger reaction. Science 287:2007–2010
54. Wang W, Hu T, Frantom PA, Zheng T, Gerwe B, Del Amo DS, Garret S, Seidel RD III, Wu P (2009) Chemoenzymatic synthesis of GDP-L-fucose and the Lewis X glycan derivatives. Proc Natl Acad Sci USA 106:16096–16101
55. Zheng T, Jiang H, Gros M, del Amo DS, Sundaram S, Lauvau G, Marlow F, Liu Y, Stanley P, Wu P (2011) Tracking N-acetyllactosamine on cell-surface glycans in vivo. Angew Chem Int Ed Engl 50:4113–4118
56. Hanson SR, Hsu TL, Weerapana E, Kishikawa K, Simon GM, Cravatt BF, Wong CH (2007) Tailored glycoproteomics and glycan site mapping using saccharide-selective bioorthogonal probes. J Am Chem Soc 129:7266–7267
57. Yang L, Nyalwidhe JO, Guo S, Drake RR, Semmes OJ (2011) Targeted identification of metastasis-associated cell-surface sialoglycoproteins in prostate cancer. Mol Cell Proteomics 10:M110 007294
58. Hart GW, Housley MP, Slawson C (2007) Cycling of O-linked beta-N-acetylglucosamine on nucleocytoplasmic proteins. Nature 446:1017–1022
59. Clark PM, Dweck JF, Mason DE, Hart CR, Buck SB, Peters EC, Agnew BJ, Hsieh-Wilson LC (2008) Direct in-gel fluorescence detection and cellular imaging of O-GlcNAc-modified proteins. J Am Chem Soc 130:11576–11577
60. Sakabe K, Wang Z, Hart GW (2010) Beta-N-acetylglucosamine (O-GlcNAc) is part of the histone code. Proc Natl Acad Sci USA 107:19915–19920
61. Cravatt BF, Sorensen EJ (2000) Chemical strategies for the global analysis of protein function. Curr Opin Chem Biol 4:663–668
62. Liu Y, Patricelli MP, Cravatt BF (1999) Activity-based protein profiling: the serine hydrolases. Proc Natl Acad Sci 96:14694–14699
63. Shields DJ, Niessen S, Murphy EA, Mielgo A, Desgrosellier JS, Lau SKM, Barnes LA, Lesperance J, Bouvet M, Tarin D, Cravatt BF, Cheresh DA (2010) RBBP9: a tumor-associated serine hydrolase activity required for pancreatic neoplasia. Proc Natl Acad Sci 107:2189–2194
64. Corthals GL, Wasinger VC, Hochstrasser DF, Sanchez J-C (2000) The dynamic range of protein expression: a challenge for proteomic research. Electrophoresis 21:1104–1115
65. Jessani N, Cravatt BF (2004) The development and application of methods for activity-based protein profiling. Curr Opin Chem Biol 8:54–59
66. Speers AE, Cravatt BF (2004) Profiling enzyme activities in vivo using click chemistry methods. Chem Biol 11:535–546
67. Sato B, Muramatsu H, Miyauchi M, Hori Y, Takase S, Hino M, Hashimoto S, Terano H (2000) A new antimitotic substance, FR182877. I. Taxonomy, fermentation, isolation, physico-chemical properties and biological activities. J Antibiot (Tokyo) 53:123–130
68. Adam GC, Vanderwal CD, Sorensen EJ, Cravatt BF (2003) (-)-FR182877 is a potent and selective inhibitor of carboxylesterase-1. Angew Chem Int Ed Engl 42:5480–5484
69. Speers AE, Adam GC, Cravatt BF (2003) Activity-based protein profiling in vivo using a copper(i)-catalyzed azide-alkyne [3 + 2] cycloaddition. J Am Chem Soc 125:4686–4687
70. Ovaa H, van Swieten PF, Kessler BM, Leeuwenburgh MA, Fiebiger E, van den Nieuwendijk AM, Galardy PJ, van der Marel GA, Ploegh HL, Overkleeft HS (2003) Chemistry in living cells: detection of active proteasomes by a two-step labeling strategy. Angew Chem Int Ed 42:3626–3629
71. Vocadlo DJ, Bertozzi CR (2004) A strategy for functional proteomic analysis of glycosidase activity from cell lysates. Angew Chem Int Ed Engl 43:5338–5342
72. Wright AT, Cravatt BF (2007) Chemical proteomic probes for profiling cytochrome p450 activities and drug interactions in vivo. Chem Biol 14:1043–1051

73. Cravatt BF, Wright AT, Kozarich JW (2008) Activity-based protein profiling: from enzyme chemistry to proteomic chemistry. Annu Rev Biochem 77:383–414
74. Saghatelian A, Jessani N, Joseph A, Humphrey M, Cravatt BF (2004) Activity-based probes for the proteomic profiling of metalloproteases. Proc Natl Acad Sci USA 101:10000–10005
75. Salisbury CM, Cravatt BF (2007) Activity-based probes for proteomic profiling of histone deacetylase complexes. Proc Natl Acad Sci USA 104:1171–1176
76. Salic A, Mitchison TJ (2008) A chemical method for fast and sensitive detection of DNA synthesis in vivo. Proc Natl Acad Sci USA 105:2415–2420
77. Kriaucionis S, Heintz N (2009) The nuclear DNA base 5-hydroxymethylcytosine is present in Purkinje neurons and the brain. Science 324:929–930
78. Tahiliani M, Koh KP, Shen Y, Pastor WA, Bandukwala H, Brudno Y, Agarwal S, Iyer LM, Liu DR, Aravind L, Rao A (2009) Conversion of 5-methylcytosine to 5-hydroxymethylcytosine in mammalian DNA by MLL partner TET1. Science 324:930–935
79. Song CX, Szulwach KE, Fu Y, Dai Q, Yi C, Li X, Li Y, Chen CH, Zhang W, Jian X, Wang J, Zhang L, Looney TJ, Zhang B, Godley LA, Hicks LM, Lahn BT, Jin P, He C (2011) Selective chemical labeling reveals the genome-wide distribution of 5-hydroxymethylcytosine. Nat Biotechnol 29:68–72

Top Heterocycl Chem (2012) 28: 185–232
DOI: 10.1007/7081_2011_68

Published online: 8 February 2012

Recent Developments in the Synthesis 1,4,5-Trisubstituted Triazoles

S. Mignani, Y. Zhou, T. Lecourt, and L. Micouin

Abstract Although the synthesis of 1,4- or 1,5-disubstituted 1,2,3-triazoles has been popularized by the recent developments of metal-catalyzed 3+2 cycloaddition of organic azides and terminal alkynes, the preparation of 1,4,5-trisubstituted triazoles is less popular. The focus of this chapter is the synthesis of these heterocycles, using condensation reactions and rearrangements, selective preparation of trisubstituted triazoles from disubstituted precursors, and formal [3+2] cycloaddition reactions involving internal alkynes.

Keywords Alkynes, Azides, C-H activation, Condensation, Cycloadditions, Diazo compounds, 1,4,5-Trisubstituted triazoles,

Contents

S. Mignani, T. Lecourt, and L. Micouin (✉)
UMR 8638, Université Paris Descartes, CNRS, Faculté de Pharmacie, 4 avenue de l'Observatoire, 75006 Paris, France
e-mail: laurent.micouin@parisdescartes.fr

Y. Zhou
State Key Laboratory of Fine Chemicals, School of pharmaceutical science and technology, Dalian University of Technology, Dalian 116024, People's Republic of China

1 Introduction

Triazole is an aromatic five-membered heterocyclic system consisting of two carbon atoms and three nitrogen atoms including one pair of isomeric chemical compounds: 1,2,3-triazole and 1,2,4-triazole. The triazole system is generally chemically stable because all ring atoms have sp^2-hybridation, and six electrons are delocalized in π–molecular orbitals.

It is well known that 1,2,3-triazole derivatives have found wide use in various domains such as pharmaceuticals, agrochemicals, dyes, photographic materials, and corrosion inhibitors [1, 2]. There are numerous examples of biologically active 1,2,3-triazoles reported in the literature. Certain compounds have been reported as antimicrobials [3, 4], antibacterials [5], fungicides [6, 7], anti-inflammatory compounds [6], analgesics [6], anti-HIV agents [8, 9], obesity regulators [10], anti-coccidiostat derivatives [11], antiallergic derivatives [12], potassium channel activators [13], antineoplastic agents [14], antianxiety agents [15], and anticancer compounds, inhibiting the tumor proliferation [16, 17] or invasion metastasis [18].

For a representative example, 1,2,3-triazole building block has been involved in pharmaceutical drugs such as the commercially available antibacterial Tazobactam **1**.

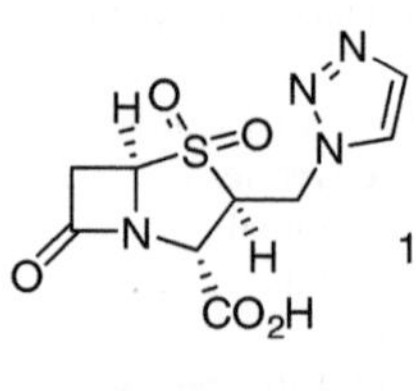

1

Recently, within the anticancer domain, a very interesting example has been described by Wang and coworkers [19]. The 1,2,3-triazole derivative NSC746457 is selectively inhibiting HDAC2, a histone deacetylase, and showed promising anticancer activities. Docking studies suggest that π-staking interactions between the triazole and Phe115 and Phe210 of HDAC2 and chelation of the hydroxamic acid to Zn^{2+} linked to Asp181, His183, and Asp269 of the active site could account for strong binding of NSC746457 to the active site of HDAC2 (Fig. 1) [20].

Most of bioactive 1,2,3 triazoles reported in the literature have either 1,4, or 1,5 disubstitution because they can be easily prepared using a thermal Huisgen

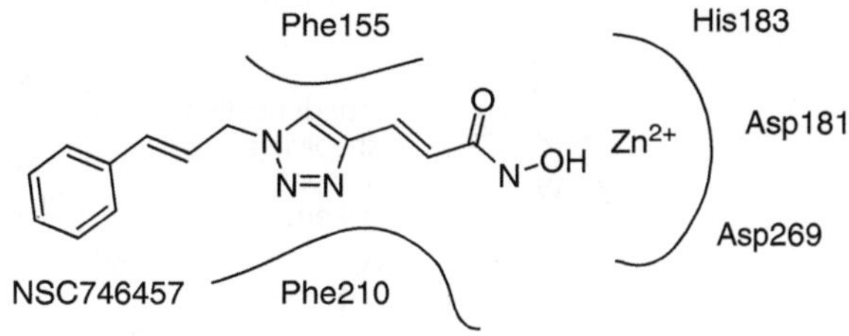

Fig. 1 Schematic representation of NSC746457 in the HDAC2 active site [20]

cycloaddition between azides and alkynes [21] or its popular metal-catalyzed variant, referred by Sharpless as "the cream of the crop" of click chemistry [22].

However, as the triazole can itself be important for interaction of the bioactive compound with its target, an additional substituent can improve the overall affinity and/or selectivity of the binding process. Thus, Goeminne et al. recently showed that the substitution pattern of trisubstituted 1,2,3-triazolyl alkyl ribitols was critical for displaying good inhibition of nucleoside hydrolase [23]. Interestingly, the 4-phenyl isomer **3** proved to be much more active than its regioisomer **2**.

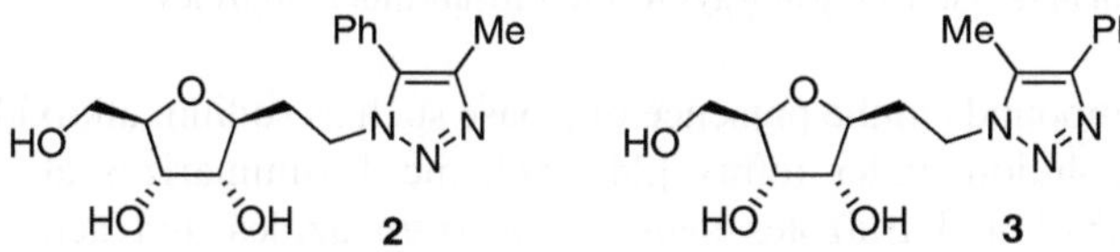

2

Ki = 32 μM

3

Ki = 2.3 μM

This example outlines the need for the development of efficient synthetic methods to prepare 1,4,5-trisubstituted triazoles in a selective manner.

In the present review, preparation of 1,2,3-triazoles with substituents at positions 1, 4, and 5 has been divided into three parts: condensation reactions and rearrangements, selective preparation of trisubstituted triazoles from disubstituted precursors, and formal [3+2] cycloaddition reactions involving alkynes (Fig. 2). Fused polycyclic 1,2,3-triazoles are out of the scope of this review. For simplicity, all the triazoles described in this review will be 1,2,3-triazoles and will be called "triazoles."

2 Synthesis of 1,4,5-Trisubstituted Triazoles by Condensations and Rearrangements

2.1 *Condensation Reactions with Active Methylene Compounds*

2.1.1 General Principle

One of the useful synthetic pathway for the preparation of 1,4,5-trisubstituted triazoles involves condensation of azides (aliphatic, aromatic, etc.) with active

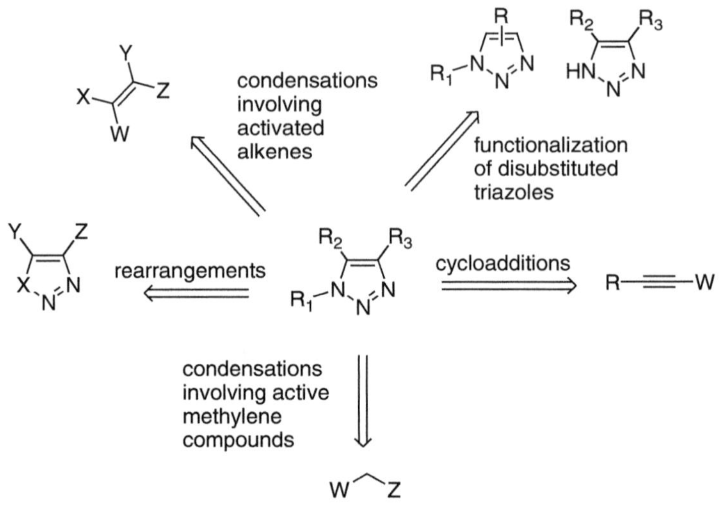

Fig. 2 Representative synthetic pathways to 1,4,5-trisubstituted triazoles

methylene compounds in the presence of a base such as sodium alkoxide or sodium in alcoholic solution under reflux [24]. Scheme 1 summarizes at a glance the different trisubstituted triazoles synthesized from azides and active methylene compounds.

This reaction is generally regiospecific. It is one of the most convenient methods to prepare triazoles with a C-heteroatom bound, generally at position 5, and/or an electron-withdrawing group (CO_2R, CO, CN) at position 4.

2.1.2 Synthesis of 4-Carboxy-5-Alkyl(Aryl) Triazoles

Cyclocondensation of aryl azides **4** with ethyl acetoacetate **5** in the presence of sodium ethoxide has been reported to lead to the corresponding 1-aryl-4-carboxy-5-methyl triazoles **6** with moderate to good yields (Scheme 2) [25].

A similar approach has been used to prepare triazoles **11** and **12**, with good antifungal activity against *Aspergillus flavus, Aspergillus fumigatus, Candida albicans, Penicillium marneffei, and Trichophyton mentagrophytes*, from 4-azido-8-(trifluoromethyl) quinoline **7** [7].

Condensation of **5** and **8** with **7** took place with sodium methoxide in methanol at 0°C to give **9** and **10** 61% and 75% yield, respectively (Scheme 3). A CF_3 group on the quinoline has been added because several triazoles with a fluorine atom on the aromatic groups already displayed very interesting biological activities, such as DNA cleavage [26] or potassium channel-activating properties [27].

Following a similar approach, triazole **13** with an aryl substituent at position 5 can be prepared from 4-azido-quinoline and 3-oxo-3-phenyl propionic acid. This compound strongly inhibited specific binding of diazepam to its target [15].

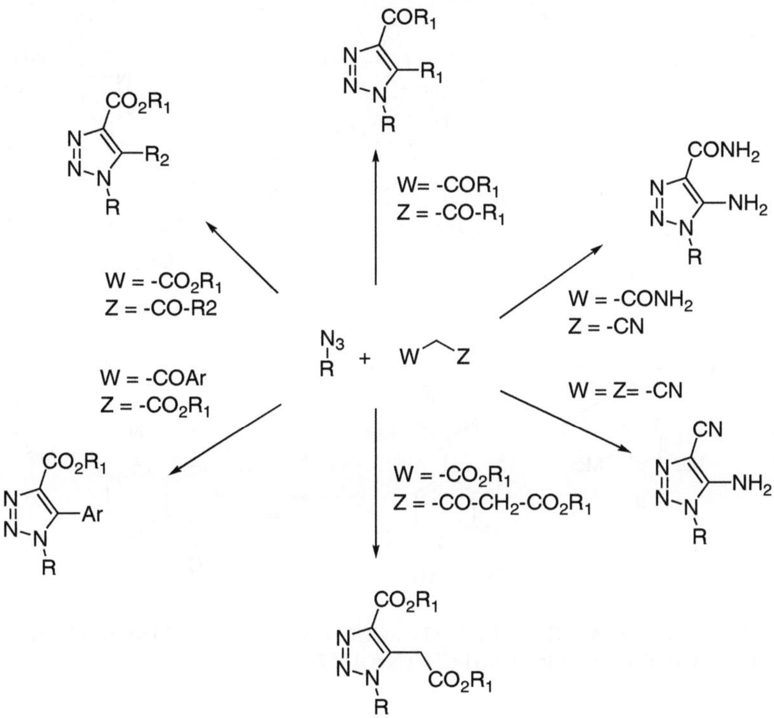

Scheme 1 Main 1,4,5-trisubstituted triazoles prepared from the corresponding active methylene compounds and azides

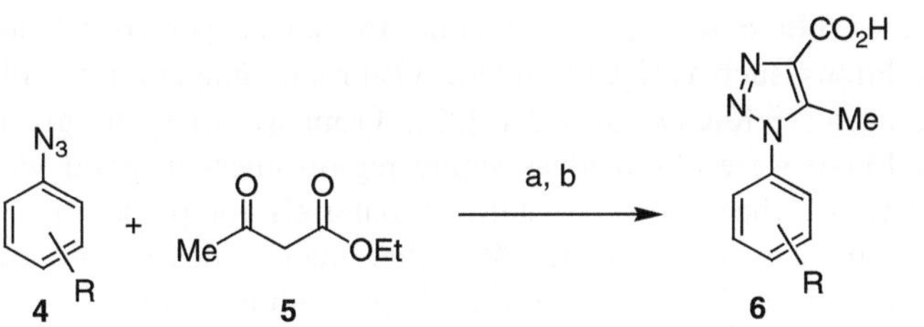

R = H, 4-Cl, 4-Me, 4-OMe, 4-Br, 3-Cl, 2-Cl, 2-Br, 2-OMe, 3-Me, 2-Me

Scheme 2 **a** EtONa, EtOH **b** NaOH, H_2O, 43–80% [25]

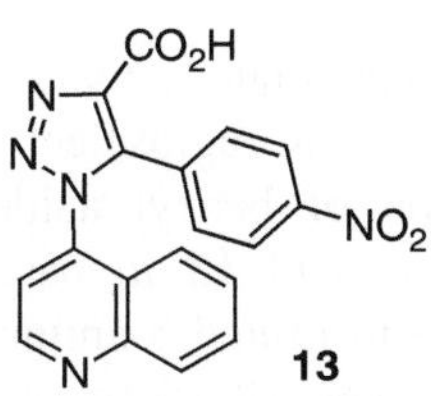

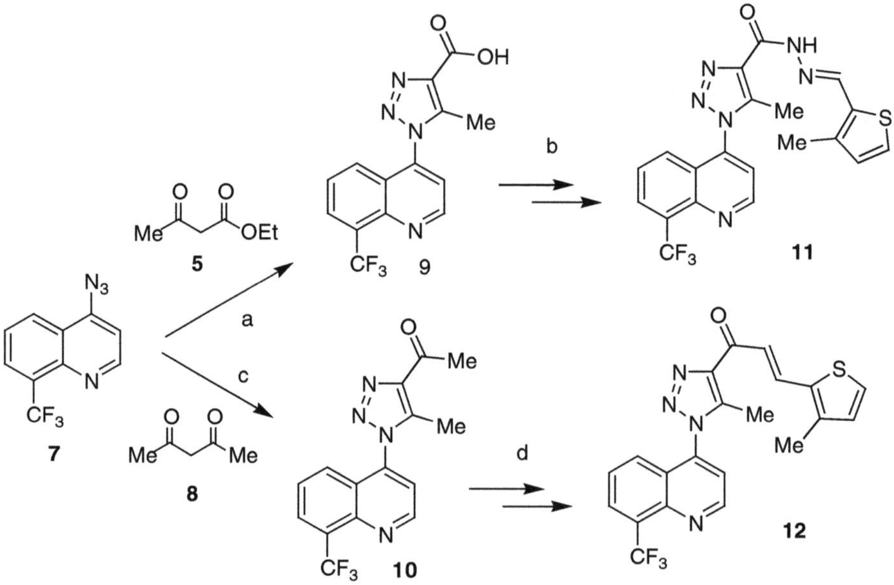

Scheme 3 **a** NaOMe, MeOH, 0°C; **b** (1) $SOCl_2$ (2) N_2H_4 (3) 3-Me-thienyl-CHO/MeOH; **c** NaOMe, MeOH, 0°C; **d** 3-Me-thienyl-CHO/NaOH [7]

Cyclocondensation of heterocyclic azides with either ethyl acetoacetate, acetylacetone, and derivatives has also been reported [28–32].

Reaction of aryl or heteroaryl azides with active methylene compounds usually requires a strong alkaline medium under reflux. In contrary, cyclocondensation of glycosyl azides **14a–e** with 3-oxoglutarate **15** can be performed under smooth reaction conditions, such as K_2CO_3 in DMSO at room temperature, with an optimal molar ratio azide:**15**:K_2CO_3 of 1:2:1 [33]. From a variety of glycosyl azides, compounds **16a–e** were obtained as single regioisomers in good yield. Diesters **16a–e** have further been used as starting materials for preparation of 8-aza-3-deazaguanine nucleosides (Scheme 4). Other nucleosides should be accessible such as 8-aza-3-deazaadenines, xanthines, hypoxanthines, etc.

2.1.3 Synthesis of 4-Carboxy-5-Amino(Hydroxy)-Triazoles

Condensation of azides and appropriate activated methylene compounds is the method of choice to prepare in a regioselective manner 5-heterosusbtituted triazoles. For instance, reaction of benzyl azide **17** with ethyl cyanoacetate, cyanoacetamide, or cyanoacetic acid **18** in basic alcoholic solution delivered the corresponding l-benzyl-4-substituted-5-amino-1,2,3-triazoles or l-benzyl-4-substituted-5-hydroxy-1,2,3-triazoles **19** (Scheme 5) [34, 35]. Sometimes, care

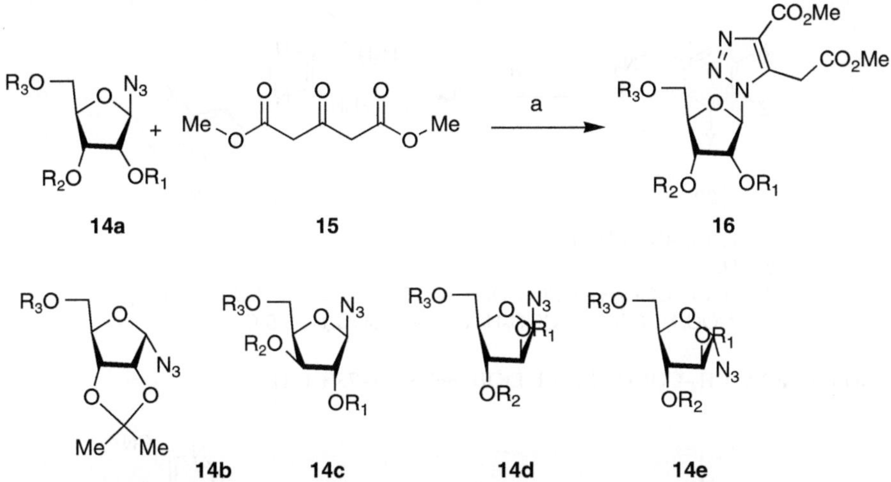

Scheme 4 **a** K_2CO_3, DMSO, rt, 80–95% [33]

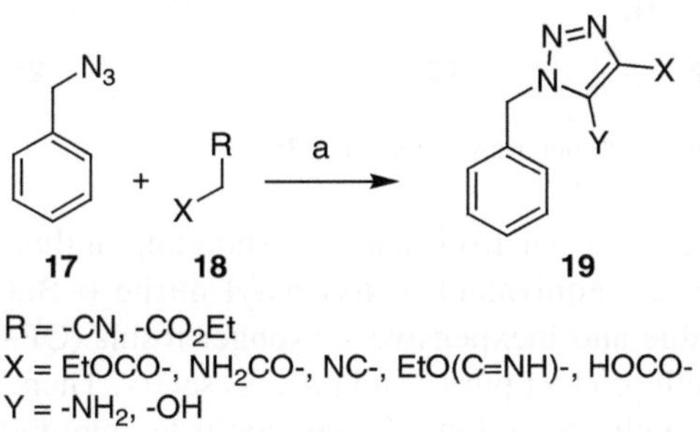

Scheme 5 **a** NaOH, EtOH, reflux, 3 h, 20–80% [34, 35]

must be taken because the final product can undergo Dimroth rearrangement (see below).

Substituted benzyl azides can also be used in such condensation reactions. Several 5-amino-4-carbamoyl-triazoles **21** were prepared as single isomers by condensation of the corresponding benzyl azides **20** with cyanoacetamide under basic conditions (NaOH/EtOH, reflux) in poor to reasonably good yields (Scheme 6) [11].

Not only benzylazides but also heteroaromatic azides can lead to interesting triazoles. For instance, Dimitrieva et al. reported the preparation of 5-amino-(pyrazolo[3,4-b]pyridin-3-yl)-1,2,3-triazole-4-carbonitrile **24** by condensation of an heterocyclic azide **22** with **23** (Scheme 7) [36].

Many other examples of preparation of 5-aminotriazoles following a similar approach have been reported in the literature (Fig. 3).

Recently, Smith et al. described a fully automated, multistep flow synthesis of a small collection of thirteen 5-amino-4-cyano-1,2,3-triazoles [41]. In situ generation

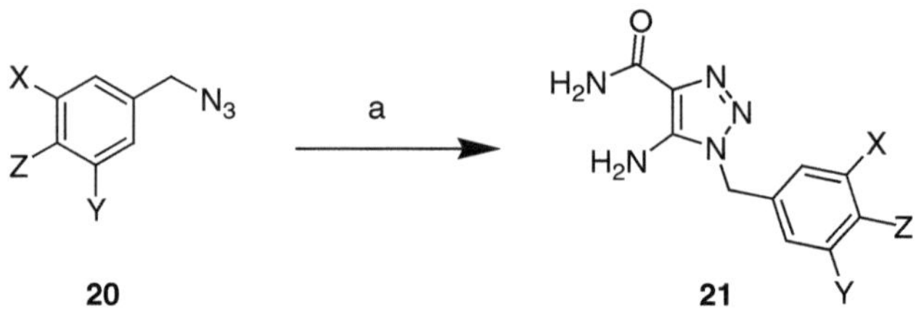

20 21

X = H, Cl, -COC_6H_5-4-Cl
Y = H, Cl
Z = H, Cl, -COC_6H_4-4-Cl, -COPh, -COC_6H_4-4-CN, -CO-C_6H_3-2,4-Cl_2, -COC_6H_3-3,5-Cl_2, -COC_6H_3-3,5-Cl_2, -COC_6H_3-3,5-Cl_2

Scheme 6 **a** NC-CH_2-$CONH_2$, NaOH, EtOH, reflux, 11–73% [11]

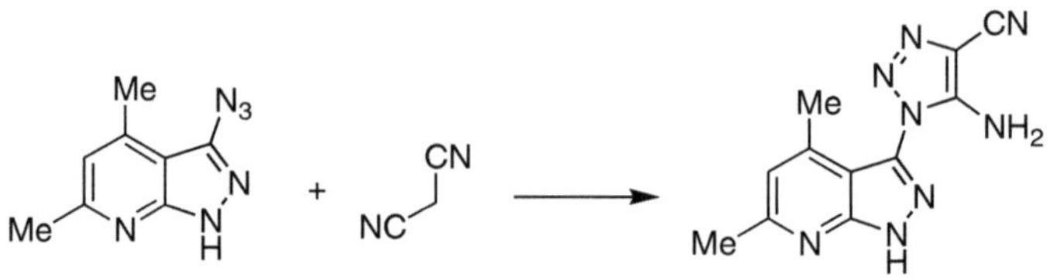

22 23 24

Scheme 7 Condensation with heterocyclic azides [36]

of aryl azides was carried out from the corresponding aniline with trimethylsilyl azide ($TMSN_3$) and 0.1 equivalent of *tert*-butyl nitrite (*t*-BuONO). Purifications using readily available and inexpensive scavenger resins (QP–SA and QP–DMA) greatly improved both product purity and process safety. Then, condensation of the azide intermediate with the anion of malononitrile immobilized on polymeric tetraalkylammonium (QP–TMA) and gave the desired 5-amino-4-cyanotriazoles in good isolated yields (53–85%). The whole process was carried out at 60°C in MeCN (Scheme 8).

2.2 *Condensation Reactions with Diazo Compounds*

Several syntheses of 1,4,5-trisubstituted triazoles from diazo compounds have been reported [1]. Thus, compound **26**, prepared in 79% yield from **25**, reacts with primary aliphatic or aromatic amines **27** in refluxing 1,2-dichloroethane with catalytic titanium (IV) chloride to give trisubstituted triazoles **28** in good yield (Scheme 9) [42].

Similarly, 5-aminoaryl-4-phenyl-triazoles **31** were prepared in moderate to good yields by condensation of various primary amines with 2-ethoxy-1-aryl-2-piperidino-1-ethenediazonium hexachloroantimonate **30** in diethylether (Scheme 10) [43].

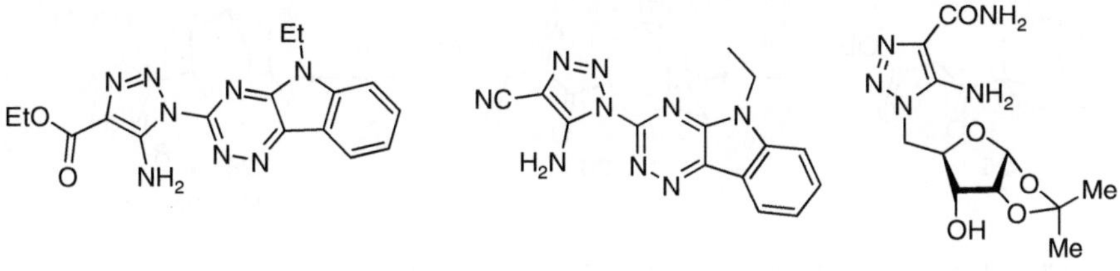

Temerik et al.[37] Temerik et al.[37] Ewing et al. [38]

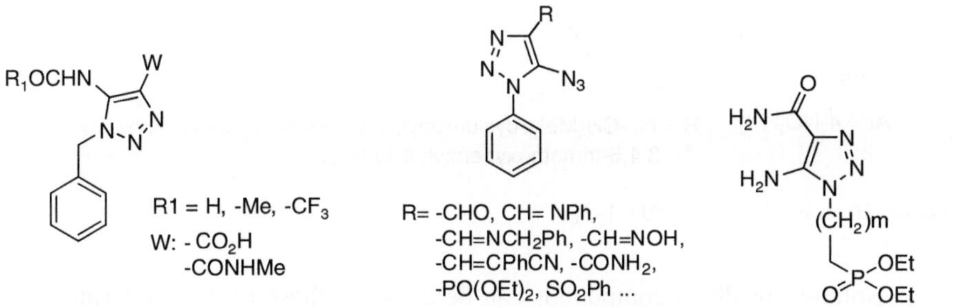

Kislyi et al.[40] L'abbé et al.[35] Beauchamp et al.[39]

Fig. 3 Representative examples of 5-amino-4-carboxamido(4-carboxyalkoxy)-1,2,3-triazole derivatives. Source L'abbé et al. [35], Temerik et al. [37], Ewing et al. [38], Beauchamp et al. [39], and Kislyi et al. [40]

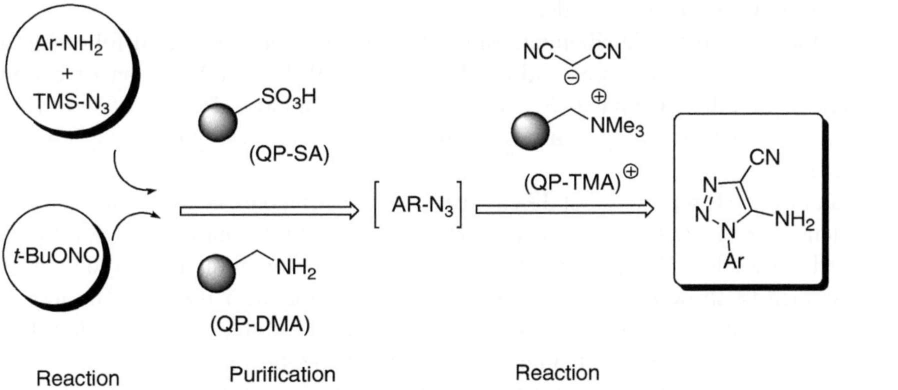

Ar: Ph, 4-Me-Ph, 4-OMe-Ph, 4-NO$_2$-Ph, 4-MeOCO-Ph, 2-I-Ph, 3-*Iso*-Pro-Ph, 2, 4-di-Cl-Ph, 3-CC-Ph, 3-OMe-Ph, 3,5-di-CF$_3$-Ph

Scheme 8 1,2,3-Triazole-automated multistep flow synthesis [41]

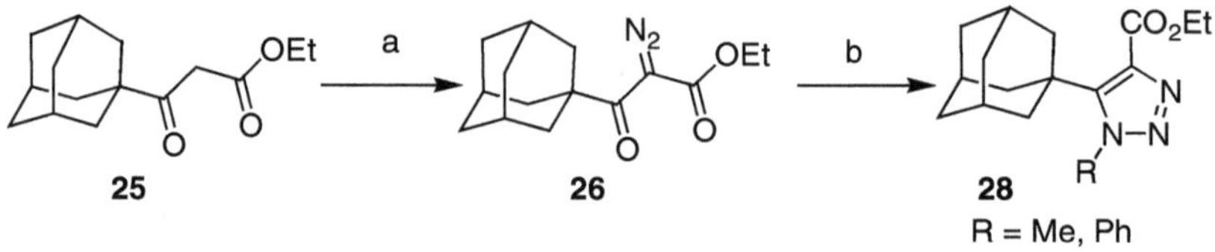

Scheme 9 **a** TsN_3, NEt_3, MeCN, rt; **b** R-NH_2 **27**, $TiCl_4$, $ClCH_2CH_2Cl$, reflux, 51–74% [42]

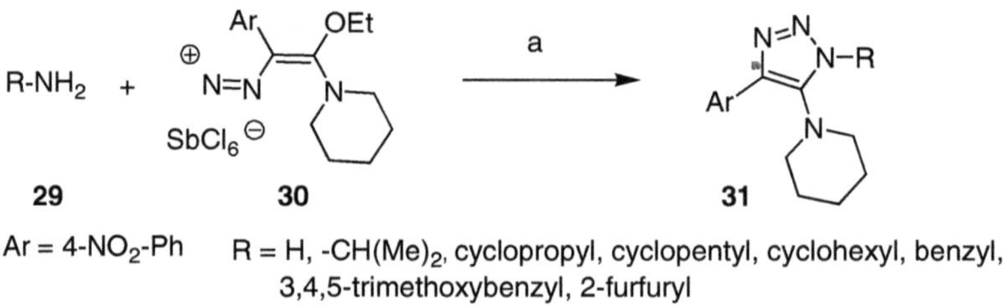

Scheme 10 **a** Et_2O 0°C, 44–80% [43]

As isolation of diazo compounds can be difficult, these reactive intermediates can be generated in situ by diazo transfer. The general principle for diazotation of β-amino-α,β-unsaturated ketones or esters is described in Scheme 11.

A versatile diazotation of enamines **32** with mesyl azide **33** (MsN_3) afforded the unique 4-carboxy-5-methyl-1,2,3-triazoles **34** in fairly good yields. The reaction works very well with alkyl amines but moderately with aromatic ones [44]. β-amino-α,β-unsaturated ketones are more reactive than β-amino-α,β-unsaturated esters, and mesyl azide is a better diazo transfer agent than tosyl azide. The reaction is carried out in acetonitrile by mixing enamines **32** with NaH, followed by the addition of MsN_3 in acetonitrile.

Syntheses of triazolic ribonucleosides **37** have also been reported following a similar approach. After preparation of enamines **36** by condensation of acetyl acetone **8** or ethyl acetonate **5** with 5-amino ribofuranose **35**, diazotation with MsN_3 delivered triazoles **37a** and **37b** in 74% and 70% yield, respectively (Scheme 12) [45].

Bakulev et al. reported that **38** exclusively gives triazoles **40** after formation of transient diazo compounds **39** with tosyl azide [46]. As the authors did not obtain isomeric triazoles **41**, resulting from cyclocondensation with the amide, these results clearly show that reaction of the acylhydrazine with the diazo moiety is highly favored (Scheme 13). Following the same strategy, 1-arylsulfonyl-5-hydroxy-triazoles **42** were obtained in good yields (62–68%).

A similar approach has been described using a diazo transfer reaction to trigger the formation of several 5-hydroxy triazoles in high yields (Scheme 14) [47].

Other representative examples of the synthesis of 1,2,3-triazole derivatives using inter- or intramolecular reaction of the corresponding diazo compounds are depicted in the Table 1.

Scheme 11 **a** NaH, MeCN [44]

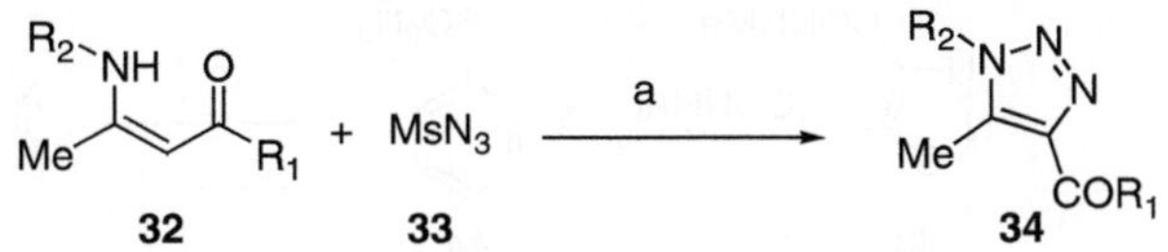

R_2 = Bn, Ph, *n*-Bu, 4-Me-Ph, *n*-decyl R_1 = Me, -OEt

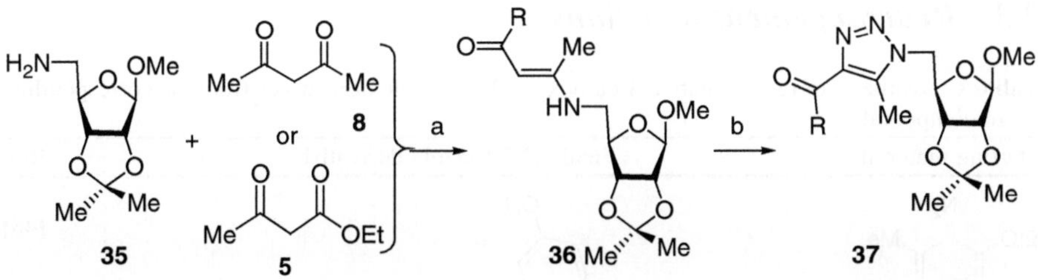

R = Me, -OEt

37a: R = Me
37b: R = -OEt

Scheme 12 **a** Montmorillonite K-10 clay, CH_2Cl_2, rt; **b** NaH, CH_3CN, rt, 70–74% [45]

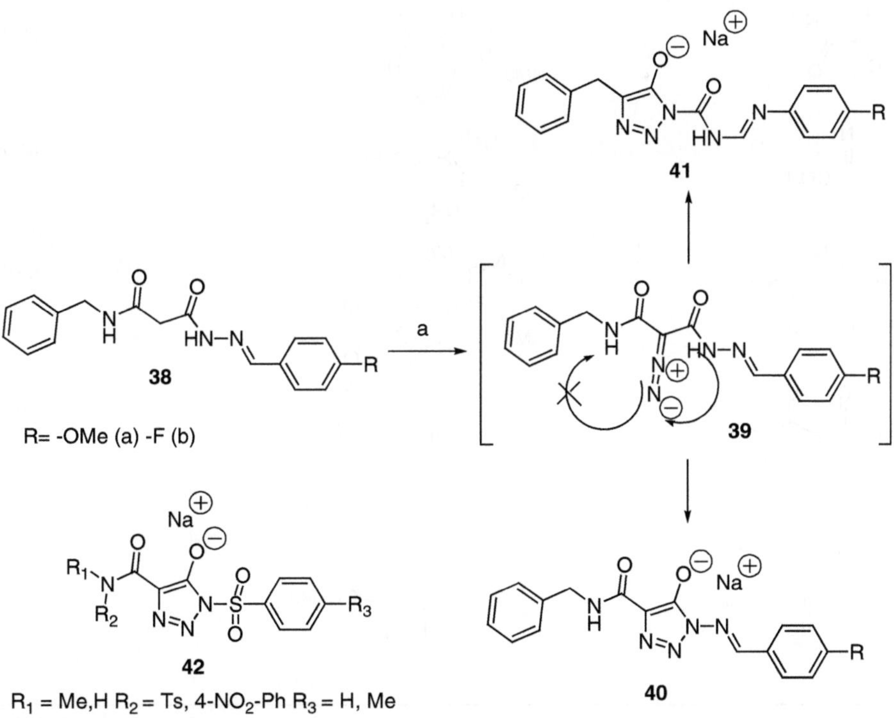

R= -OMe (a) -F (b)

R_1 = Me,H R_2 = Ts, 4-NO_2-Ph R_3 = H, Me

Scheme 13 Regioselective synthesis of 1-arylsulfonyl-5-hydroxy-triazoles [46]

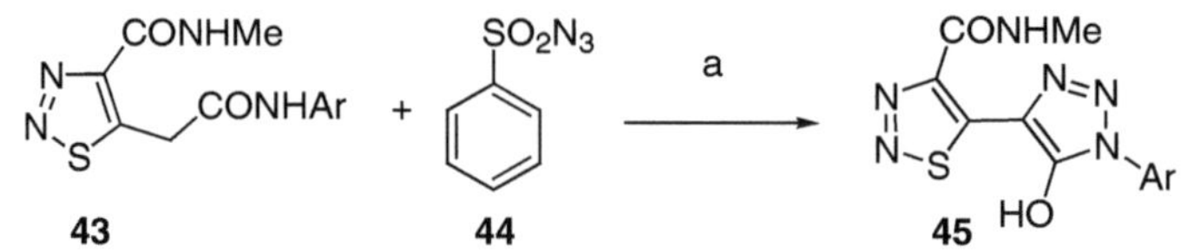

Ar: Ph, 4-Me-Ph, 2-MeO-Ph, 4-MeO-Ph, 3-Cl-Ph, 4-Cl-Ph, 2,6-Cl_2-Ph, 4-Br-Ph

Scheme 14 **a** EtONa, EtOH, rt, 58–79% [47]

2.3 Rearrangement Reactions

Table 1 Synthesis of representative 3-carboxy-1,2,3-triazole derivatives from the corresponding diazo compounds

Starting material	Final 1,2,3-triazole derivative	Ref.
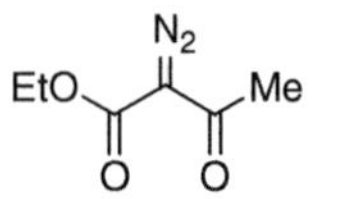	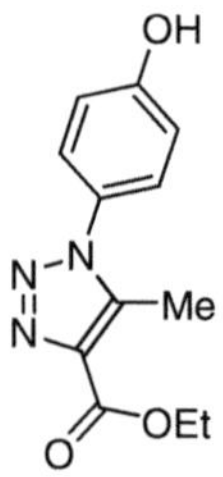	[48]
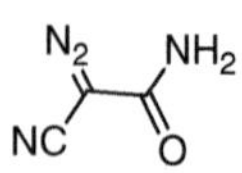	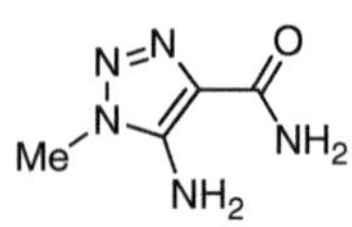	[49]
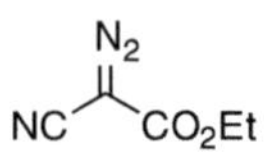	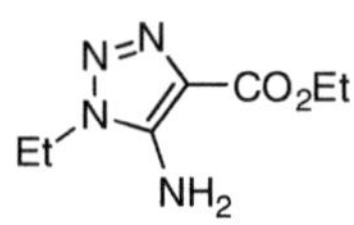	[49]
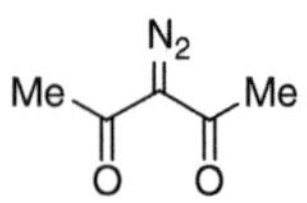	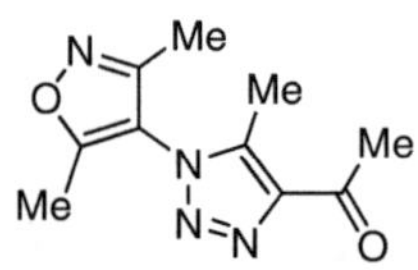	[50]
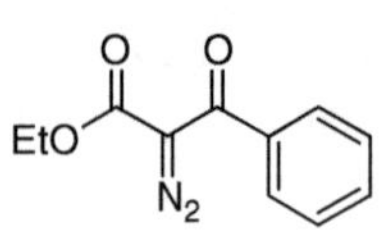	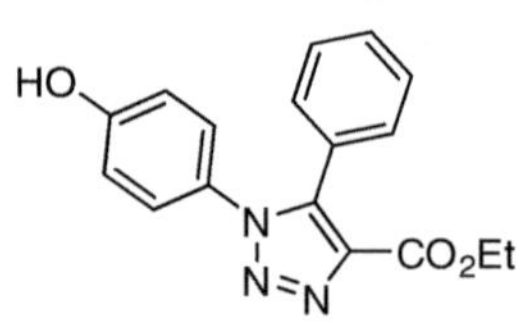	[48]
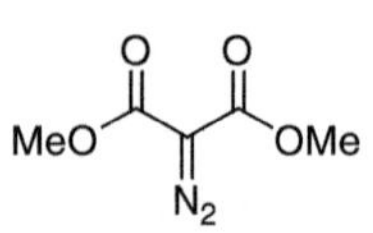	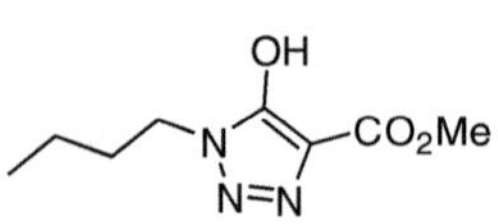	[51]

(continued)

Table 1 (continued)

Starting material	Final 1,2,3-triazole derivative	Ref.
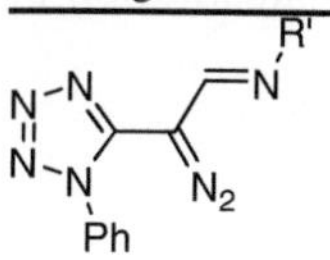	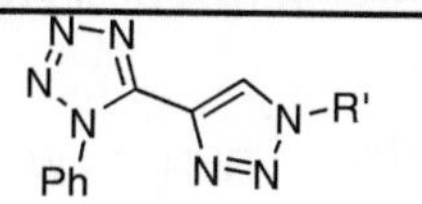 R' = Ph, 4-OMe-C_6H_4-,4-Cl-C_6H_4-, -Et,- CMe_3,-CH_2Ph	[52]

Within the huge field of heterocyclic rearrangements [53], the Dimroth rearrangement gives 5-mercapto-1,2,3-triazoles from 5-amino-1,2,3-thiadiazoles and allows interconversion of 5-amino-1,2,3-triazoles into a mixture of isomers under basic condition [54]. Thus, this reaction, discovered in 1909, involves a linear intermediate that can cyclize into two isomeric compounds (Scheme 15) [55]. This ring-chain tautomerization is reversible, but the equilibrium can be pushed toward an end if one of the two isomers is stabilized.

Interestingly, this reaction occurs not only under chemical medium but also under biological medium. Thus, in vitro Dimroth rearrangement of 1-(2 carboxyethyl) adenine to N6-(2-carboxyethyl) adenine in single-stranded calf thymus DNA

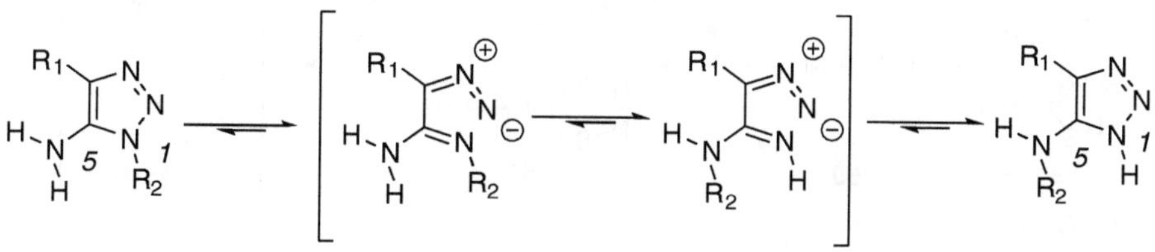

Scheme 15 Schematic representation of Dimroth rearrangement [55]

has been described [56].

2.3.1 Dimroth Rearrangement

Synthesis of 5-Mercapto-1,2,3-Triazole Derivatives from 5-Amino-1,2,3-Thiadiazoles

5-Amino-1,2,3-thiadiazoles **46** bearing an electron-withdrawing group at position 4 can be easily rearranged into the corresponding 5-mercapto-1,2,3-triazoles **48**. This transformation, involving ring-chain tautomerization with an α-diazo thiocarbonyl linear intermediate **47**, is favored because the S–N bond in **46** is weak and is pushed toward the right by formation of a thiolate salt (Scheme 16).

An expeditious synthesis of 5-arylthio-1,2,3-triazole derivatives **53** has been reported by V. A. Bakulev et al.[57, 58]. According to Scheme 17, substitution of

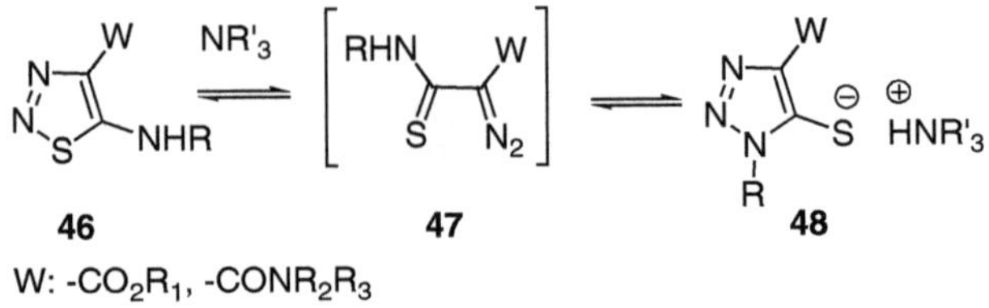

Scheme 16 Dimroth rearrangement of 5-amino-1,2,3-thiadiazoles

5-bromo-1,2,3-thiadiazoles **49** by various aromatic vicinal diamines **50** gave the corresponding 5-amino-1,2,3-thiadiazoles **51** in moderate yields. Following Dimroth rearrangement (EtOH, Et_3N, reflux), the resulting thiolate **52** was trapped by a second equivalent of **49** to give adducts **53** (Scheme 17).

In addition, 5-mercapto-1,2,3-triazole derivatives **55** bearing in 1-position different alkyl groups have been also prepared using the same synthetic pathways

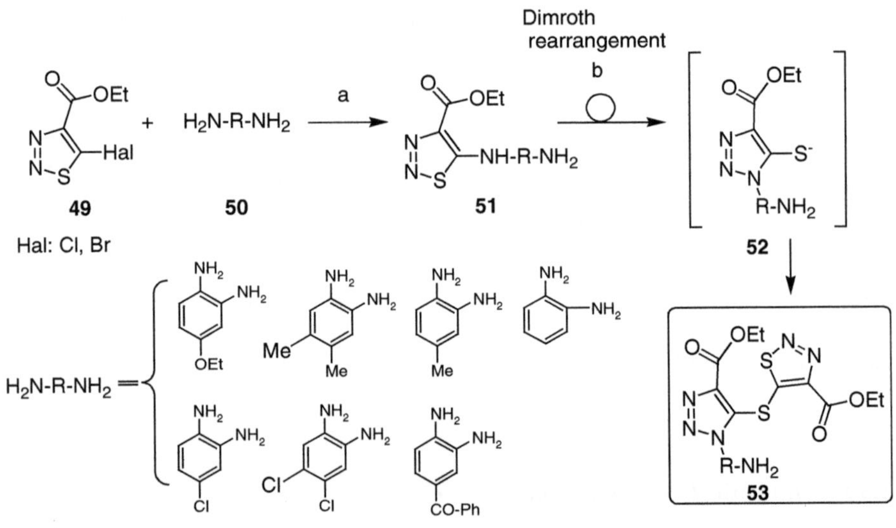

Scheme 17 **a** DMF, rt; **b** EtOH, reflux, Et_3N 86–93% [57, 58]

(Scheme 18) [59].

Further examples of 5-mercapto-triazoles **59** and bis-triazolo[1,3,6]thiadiazepines **60** have been reported by these authors [60]. Compounds **59** and **60**, resulting from one or two Dimroth rearrangements, respectively, were obtained in one step by condensation of 1,2-phenylenediamine with 5-chloro-1,2,3-thiadiazoles **56**. The product distribution is highly dependent upon the reaction time and the presence of Et_3N (Scheme 19). Without base, only **59** was obtained, whereas in the

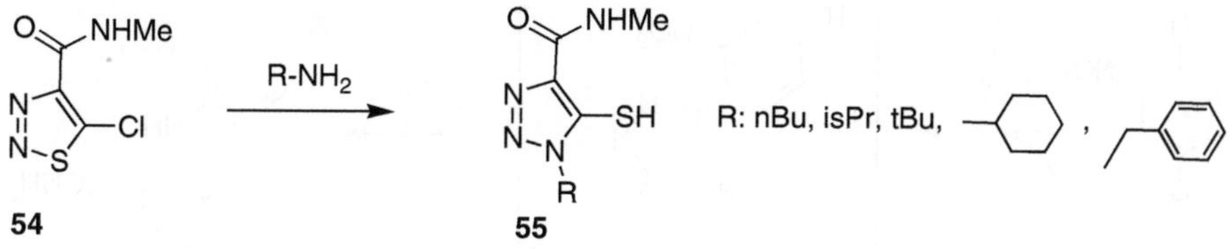

Scheme 18 5-Mercapto-1,2,3-triazole synthesis [59]

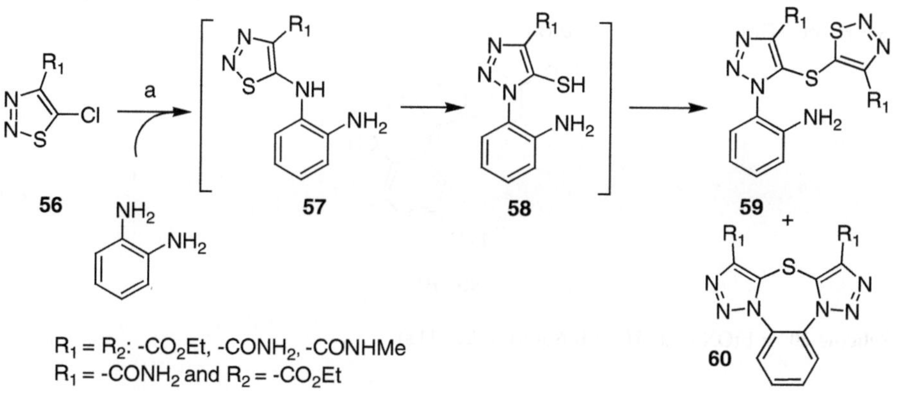

Scheme 19 **a** Et_3N, EtOH, reflux [60]

presence of 1 equivalent of phenylenediamine, 2 equivalents of thiadiazole **56**, and 3 equivalents of Et_3N, a 2:1 mixture of **59** and **60** was isolated after 3 h.

Synthesis of 4-Carboxamido-5-Amino-1,2,3-Triazole Derivatives from 5-Amino-1,2,3-Triazoles

5-Aminotriazoles can also be engaged in Dimroth rearrangement. A series of 5-(4′-substituted-2′-nitroanilino)-triazoles, with potential potassium channel-activating properties, has been prepared by Biagi et al. [13].

In anhydrous ethanol with sodium ethylate at room temperature, 1,3-dipolar cycloaddition between azides **61** and cyanoacetamide gave 4-carboxamido-5-amino-1,2,3-triazole intermediates **62**. In the same pot, addition of a small amount of 10% NaOH and heating to reflux delivered **63** after Dimroth rearrangement (Scheme 20). In this transformation, the greater acidity of the exocyclic N–H bond substituted by an electron-withdrawing group is the driving force that shifts the equilibrium toward formation of **63**. Finally, triazole **63a** displayed good vasorelaxant activity on endothelium-denuded rat aortic rings with a potency comparable to that recorded for the reference compound NS 1619.

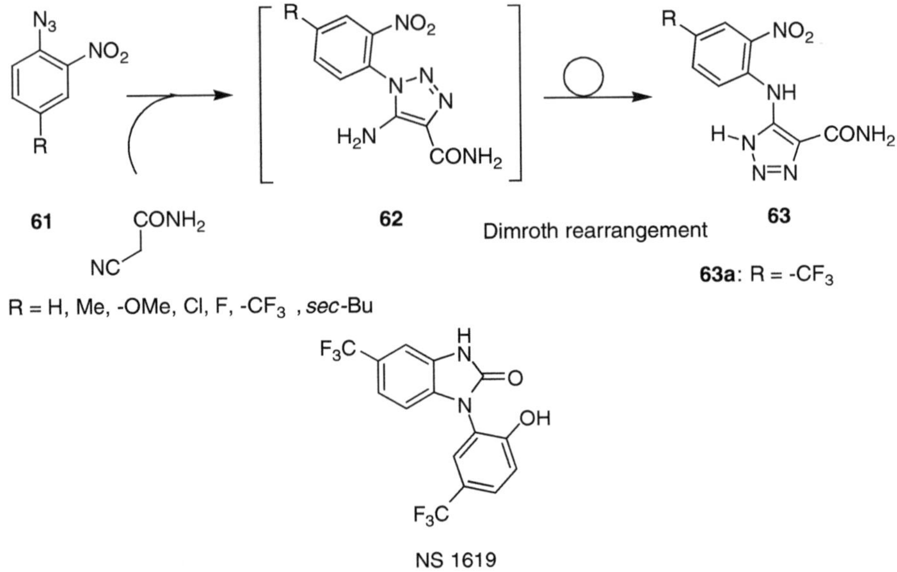

Scheme 20 **a** EtONa, EtOH, rt; **b** NaOH, reflux [13]

2.3.2 Other Rearrangements

Based on tandem Dimroth–Cornforth rearrangements, reaction of 1,2,3-thiadiazole-4-carboxaldehyde **64** with 1-amino-4-carbamoyl-1,2,3-triazole-5-olate **65** afforded the sodium *N*-(4-thiocarbonyl-1,2,3-triazol-1-yl)carbamoyl-1,2,3-triazol-5-olate **68** with moderate overall yield (Scheme 21) [61].

A similar tandem Cormforth rearrangement – diazo condensation gave compound **71**. Thus, reaction of diazomalonamohydrazide **69** with the 1,2,3-thiadiazole-4-carboxaldehyde **64** in ethanol at room temperature delivered diazo **70** in 72% yield. Addition of triethyl amine and heating at reflux finally give **71** in 40% overall yield (Scheme 22) [61].

Interestingly, 1-(imidazol-4-yl)-4-cyano/carboethoxy-5-amino/methyl-1,2,3-triazoles **73** were prepared in moderate yield by rearrangement of **72** (Scheme 23).

3 Synthesis of 1,4,5-Substituted Triazoles by Condensation with Alkenes

The cycloaddition of azides with alkenes is a favorable process, leading to dihydrotriazoles. However, in some cases, these dihydrotriazoles can evolve to the corresponding triazoles, by elimination, fragmentation, or retrocycloaddition reactions.

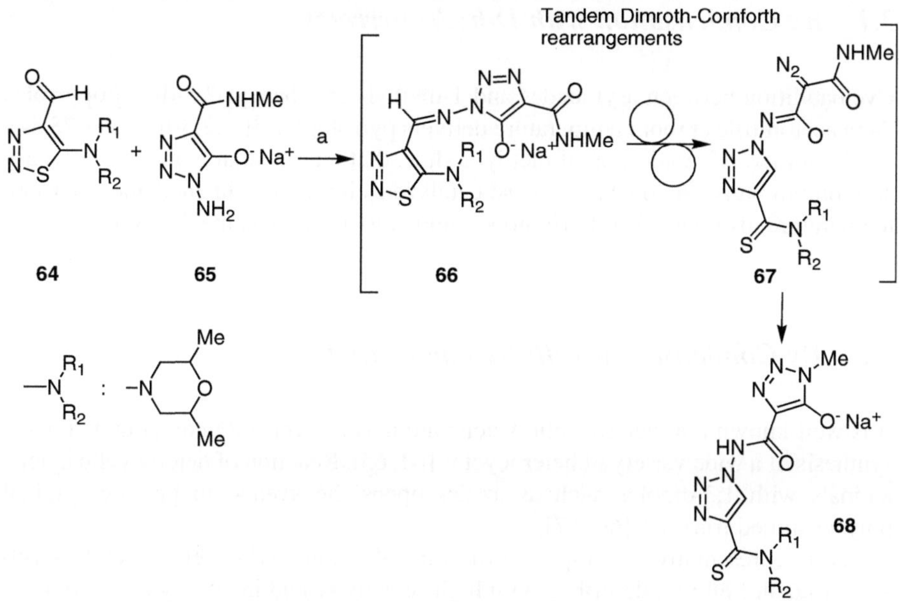

Scheme 21 **a** EtOH, reflux, 3 days, 45% [61]

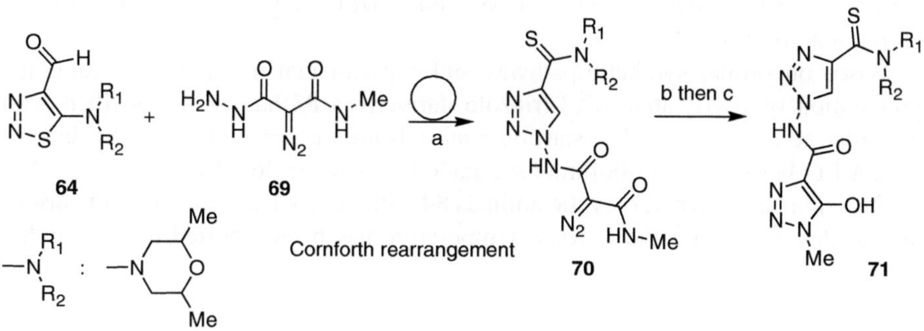

Scheme 22 **a** EtOH, rt; **b** EtOH, reflux; **c** EtOH, HCl, rt [61]

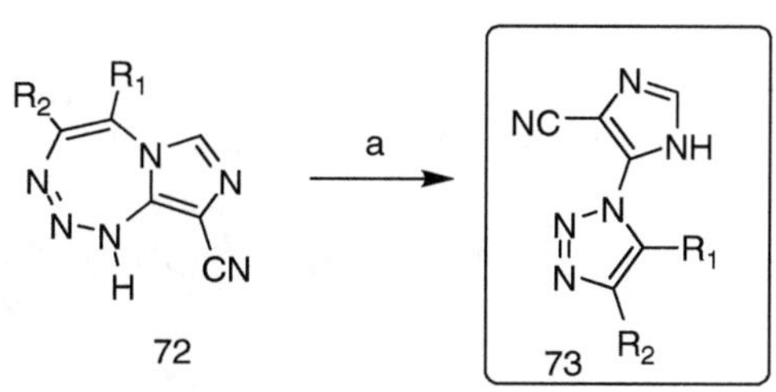

R_1 = -NH$_2$, -Me R_2 = -CN, -CO$_2$Et

Scheme 23 **a** Conc HCl, THF, rt [62]

3.1 By Condensation with Dihydropyrroles

Cycloaddition between aryl azides and 1-methyl- or 1-benzyl-2,5-dihydropyrroles **74** in acetonitrile at room temperature delivers pyrrolo[3,4-d]-1,2,3-triazoles **75** as a single isomer. After addition of methyl iodide, that intermediately gives quaternary ammonium salts **76** (65–95%), base-catalyzed elimination in methanol at room temperature afforded **77** in fairly good yields (40–60%, Scheme 24) [63].

3.2 By Condensation with Ketene Aminals

It is well known that heterocyclic ketene aminals are versatile compounds for the synthesis of a wide variety of heterocycles [64, 65]. Reaction of heterocyclic ketene aminals with 1,3-dipoles such as azides opens the avenue to prepare original polysubstitued triazoles [66, 67].

As a representative examples, synthesis of 1-glucosyl-4-heterocyclyl-5-aryl-1,2,3-triazoles **80** was described with high selectivity and in good yields [66, 68].

Condensation of heterocyclic ketene aminals **78** with 2,3,4,6-tetra-*O*-acetyl-β-D-glucopyranosyl azide **79** in anhydrous 1,2-dichloroethane at room temperature gave glucosylated 1,2,3-triazole derivatives **80** in very good yields (Scheme 25). Activity screening showed that glycosylated triazole **80** possessed antitumor and antiviral activities.

Based on similar synthetic pathway and experimental reaction conditions, the preparation of ribosylated 1,2,3-triazole derivatives **83** has been described with moderate to good yields. The starting materials are the ketene aminals **81** and the 2,3,5-tri-*O*-benzoyl-β-D-ribofuranosyl azide **82** (Scheme 26) [69].

Reaction between heterocyclic aminals **84** with benzoyl substituents and various aryl azides in 1,4-dioxane at room temperature has been reported to give highly

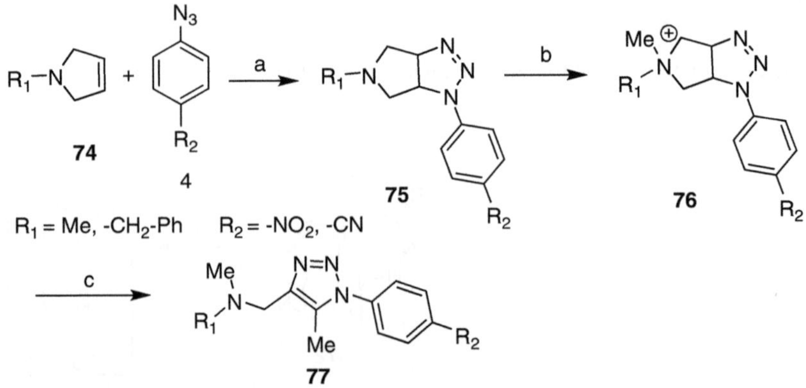

Scheme 24 **a** CH_3CN, rt, **b** MeI, CH_3CN, rt **c** NaOH, MeOH, rt [63]

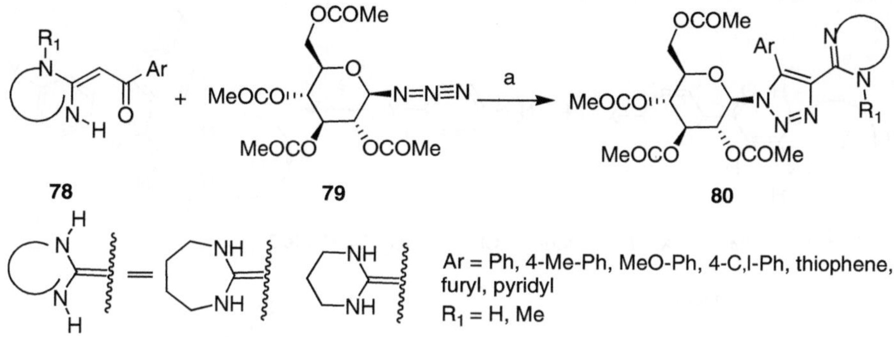

Scheme 25 **a** 1,2-Dichloromethane, rt, 91–95% [68]

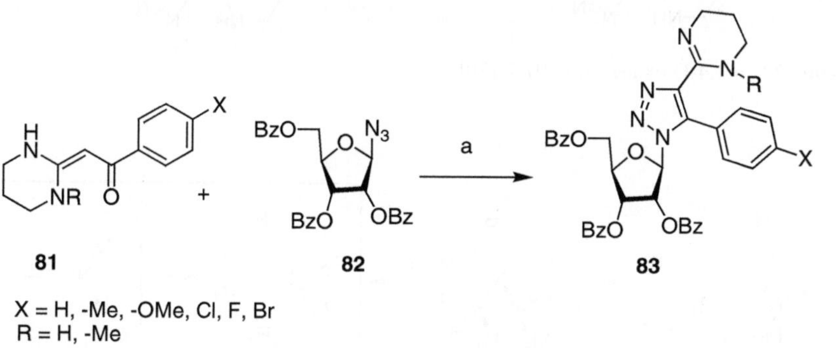

Scheme 26 **a** $CHCl_3$, rt, 50–75% [69]

substituted 1,2,3-triazoles **85** with poor to high yields (Scheme 27). This transformation was highly dependent upon the aromatic substituents X, Y of both the enediamines and azides.

While electron-withdrawing substituents on enediamines (Y = $-NO_2$ > $-Cl$ > H > $-OMe$) increased the rate of the reaction, electron-donating groups para to the azide (X = $-OMe$ > $-Me$ > H > $-Cl$) were deleterious to this transformation [70].

Very interestingly, Quast et al. described a convenient synthesis of 5-alkylamino-triazoles **89** in good to excellent yields (60–100%), using a base-mediated eliminative ring cleavage of the spirocycles **88** as the key reaction [71]. Thus, selective 1,3-dipolar cycloaddition between azides **87** and cyclic ketene *N*,*N*-acetals **86** afforded cycloadducts **88** [72]. Base-mediated eliminative ring cleavage of spirocycles **88**, that is facilitated by the inherent strain of the spirocycle and formation of an aromatic compound, gave 5-alkylamino-1,2,3-triazoles **89** (Scheme 28). This method is the only synthetic pathway, giving rise to these compounds, as a classical approach relying on a [3+2] cycloaddition between azide **87** and ynamines **90** is not possible, due to the instability of compound **90**.

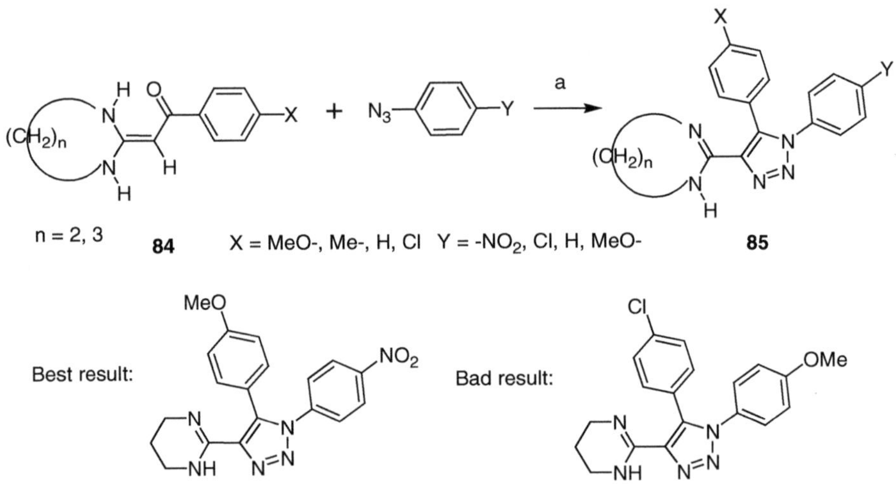

Scheme 27 **a** 1,4-Dioxane, rt, 8–97% [70]

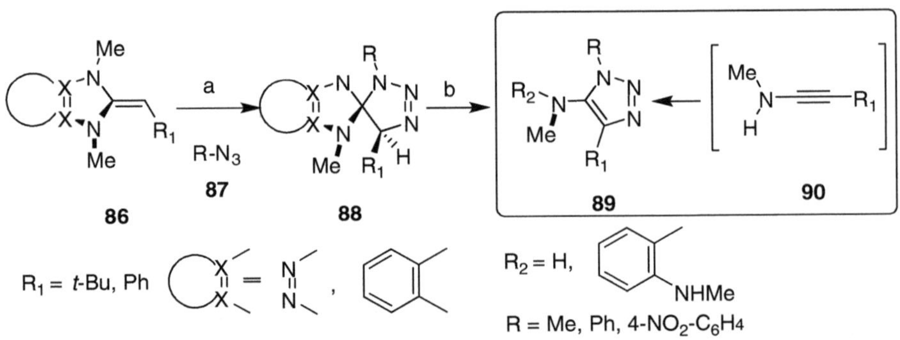

Scheme 28 **a** Ether or toluene, rt; **b** NaH, Et_2O, rt [72]

3.3 *By Condensation with Nitro Olefins*

The reaction of α-nitro-olefins with azides has only been reported scarcely by Carrié et al. Very interestingly, in contrary to Huisgen's cycloadditions of azides with internal alkynes that deliver mixture of regioisomers, condensation of methyl or phenyl azide **92** with nitro-olefins **91** gave triazoles **93** a single isomers after rearomatization by spontaneous loss of HNO_2 (Scheme 29) [73]. Triazoles **93** were obtained in good yield (> 80%) in benzene at 60°C for methyl azide and in refluxing toluene for phenyl azide.

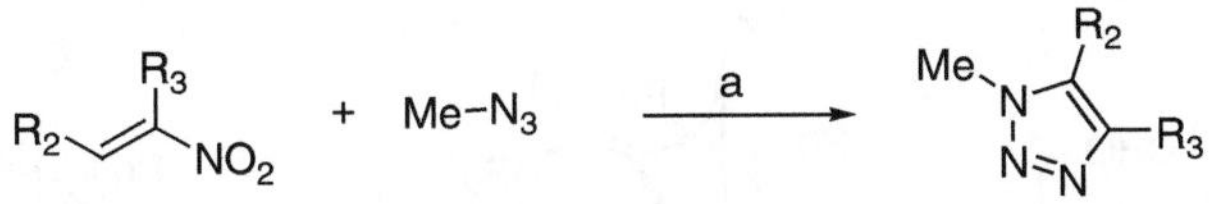

91 92 93

R_2 = Ph, Me, 4-NO_2-Ph,4-Me-Ph
R_3 = -CO_2Me, -CN, Ph, Me, Et, -NO_2

Scheme 29 **a** Refluxing benzene or toluene, > 80% yield [73]

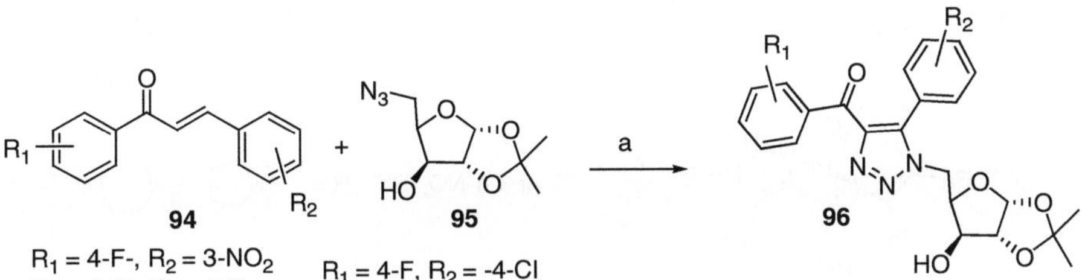

94 95 96

R_1 = 4-F-, R_2 = 3-NO_2
R_1 = 4-Cl, R_2 = 4-Br
R_1 = 4-F, R_2 = 4-Br
R_1 = H, R_2 = 4-Cl
R_1 = 4-F, R_2 = -4-Cl
R_1 = H, R_2 = 3-NO_2
R_1 = H, R_2 = H
R_1 = 4-Cl, R_2 = 4-Cl

Scheme 30 TBAHS (20 mol%), DMF, 100°C [74]

3.4 By Condensation with Chalcones

Recently, Huisgen regioselective [3+2] cycloaddition of various chalcones **94** with 1-azido sugars **95**, that gives 1-(5-deoxy-D-xylofuranos-5-yl)-4,5-disubstituted triazoles **96**, has been described by Tripathi et al. (Scheme 30) [74]. This reaction involves oxidation of a transient dihydrothiazole and proceeds well with catalytic tetrabutylammonium hydrogen sulfate (TBAHS, 20 mol%) in DMF at 100°C.

3.5 By Condensation with Strained Activated Alkenes

An interesting piece of work by Gelmi et al. showed that specific cycloaddition between 4-nitro-phenylazide and enamines **98a** and **98b** of 1,4,5,6-tetraphenyl-bicycloheptene-2-carboxaldehyde delivers spiro-bicyclo-heptenone-triazoles **99a** and **99b**. Enamines **98a** and **98b** were prepared in situ by addition of either morpholine or pyrrolidine to aldehydes **97a** and **97b**. Finally, rearrangement of spiro compounds **99a** and **99b** into triazole **100** occurred slowly in $CHCl_3$ at room temperature (Scheme 31) [75].

Rutjes et al. recently reported a tandem [3+2] cycloaddition-retro-Diels–Alder reaction that gives 1,4,5-substituted triazoles **104** and **105** (Scheme 32) [76]. Thus,

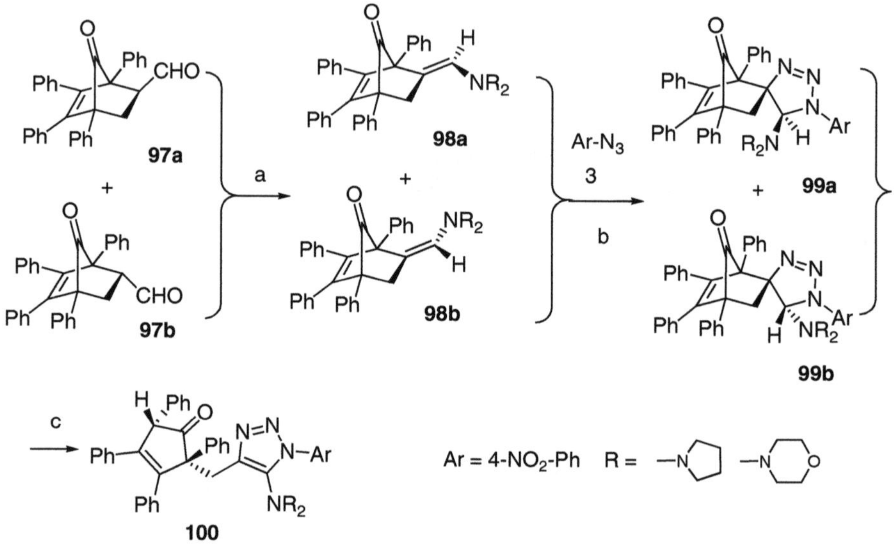

Scheme 31 **a** Morpholine or pyrrolidine, benzene, rt; **b** benzene, Ar-N_3, rt; **c** $CHCl_3$, rt [75]

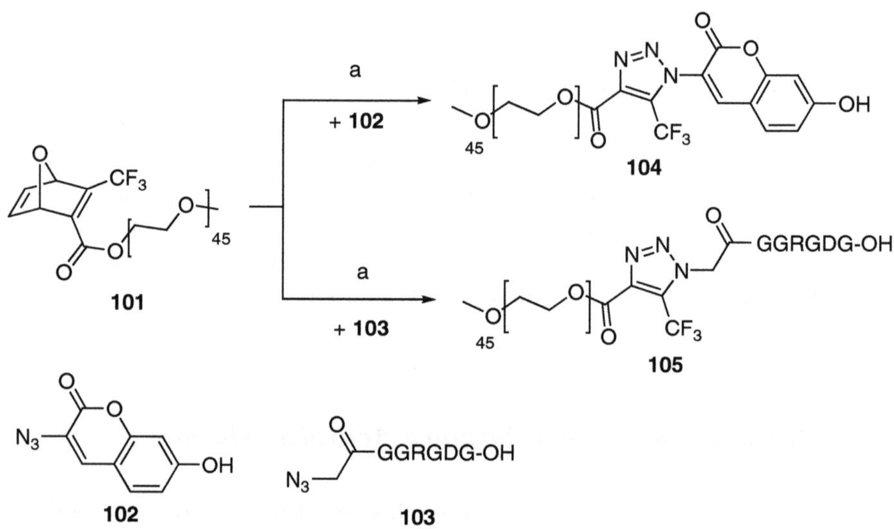

Scheme 32 **a** Water, 37°C, > 80% [76]

reaction between azido coumarin **102** or azide-functionalized hexapeptide GGRGDG **103** and PEGylated oxanorbornadiene **101** afforded 1,2,3-triazole **104** and **105**, respectively. This reaction represents a new bioconjugation method because it can be achieved in water at 37°C with high conversion (> 80%), as determined by NMR.

4 Synthesis of 1,4,5-Substituted Triazoles from Disubstituted Triazoles

As many reactions have been described to prepare disubstituted triazoles, one of the possible strategy to obtain 1,4,5-substituted triazoles is to functionalize these readily available precursors in a selective manner.

4.1 *Synthesis of 1,4,5-Trisubstituted Triazoles by N-Alkylation Reactions*

N-alkylation of 1,2,3-triazoles is readily achieved with various electrophiles under basic conditions [1]. In most of the cases, the reaction is unselective, with the N2 regioisomer being the main compound obtained [77–82]. The strong preference for N2 substitution has generally been explained by steric reasons and/or hard base character of the central nitrogen atom [83].

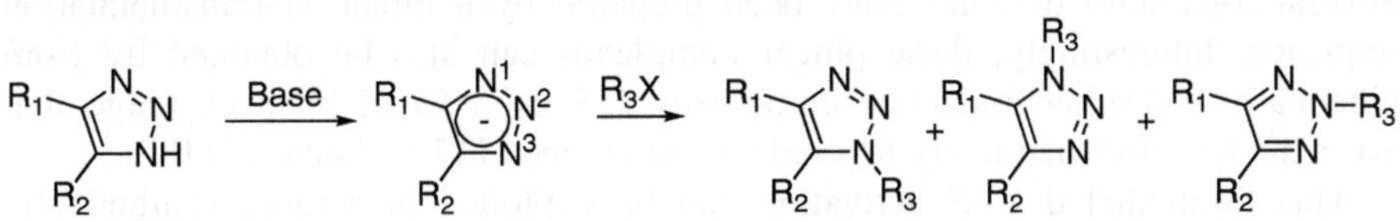

106 107 108 109

Furthermore, Au(I) catalyzed addition of triazoles to nonactivated alkynes has been reported to deliver mixture of N1 and N2 isomers, N2 being the major one [84]. This general lack of selectivity clearly outlines that the clean synthesis of 1,4,5-trisubstituted triazoles by *N*-alkylation is not possible.

4.2 *Synthesis of 1,4,5-Trisubstituted Triazoles from 1,4-Disubstituted Triazoles*

4.2.1 By Deprotonation

The C-5 position of 1,2,3 N-substituted triazoles is the most acidic one and can regioselectively be deprotonated with lithiated bases (Scheme 33) [83].

A similar strategy has been used for the preparation of chiral triazolium precursors [85], protease inhibitors [86] C5-iodinated [87] or phosphorylated [88] 1,4,5-trisubstituted triazoles (Scheme 34)

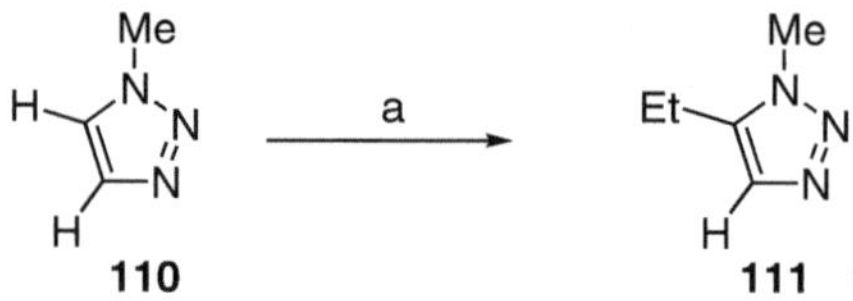

110 → 111 54.2%

Scheme 33 **a** BuLi, Hexane, THF, −40°C, 1 h then EtI, rt, 2 h [83]

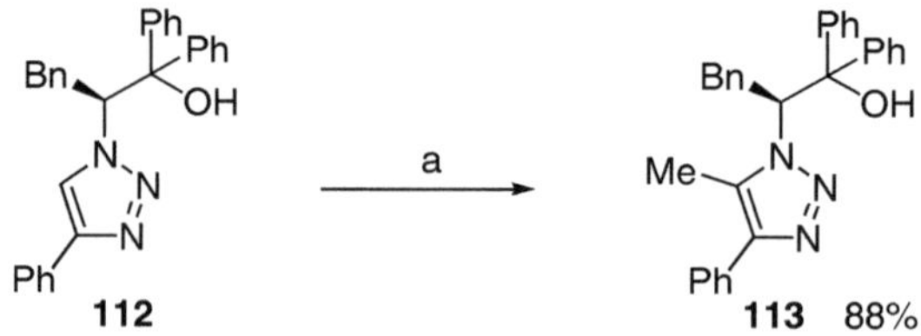

112 → 113 88%

Scheme 34 **a** BuLi, THF, −78 to 0°C, 5 min then MeI, 0°C, 1 h [85]

4.2.2 By Palladium-Catalyzed C–H Activation

Several 5-palladio triazoles have been prepared by a lithiation–transmetalation sequence. Interestingly, these pincer complexes can also be obtained by from kinetically controlled bidentate complexes **115** by a C_5–H bond cleavage that gives the thermodynamically favored tridentate ones **117** (Scheme 35) [89].

This Pd-mediated C–H activation can be exploited in a more synthetically oriented manner. Thus, a general access to 1,4,5-trisubstituted triazoles by a palladium-catalyzed direct arylation of 1,4-disubstituted triazoles **118** has been reported by the group of Gevorgyan [90]. This methodology allows for the introduction of both electron-deficient and rich aryl groups, with a good functional group tolerance. It has been latter reported that a similar arylation can be conducted under aerobic conditions, with a recyclable palladium catalyst [91] (Scheme 36)

The use of less expensive aryl chlorides in this transformation has been reported using microwave irradiation [92]. Best results were obtained with $P(c\text{-}C_6H_{11})_3$ as ligand in combination with carbonate bases. Notably, under these conditions, phenylchloride proved to be more reactive than bromide or iodide (Scheme 37).

Although a nonthermal microwave effect has been mentioned for this reaction, it has been latter reported that similar results could be obtained under milder (105–120°C) conventional heating [93]. The scope of this reaction is broad, leading a range of 1,4,5-trisubstituted triazoles. This protocol is applicable to intramolecular arylations. The reactivity order of leaving group was demonstrated to be Br > Cl > OTs. 1,2,3-triazoles bearing two N-alkyl substituents were less reactive under these reaction conditions, but a significant rate acceleration was obtained with addition of pivalic acid (30 mol%) in the reaction mixture (Scheme 38).

Although electron-rich aryl tosylates are electronically deactivated, the use of monophosphine biphenyl ligand X-Phos in combination with $Pd(OAc)_2$ enables a clean arylation of 1,4-disubstituted triazoles with these electrophiles (Scheme 39)[94].

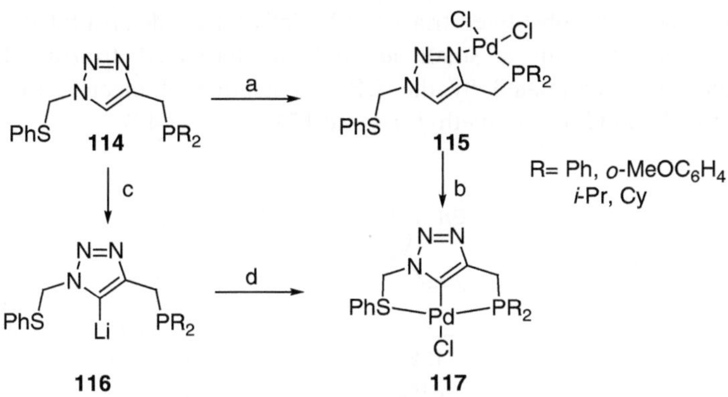

Scheme 35 **a** $(CH_3CN)_2PdCl_2$, THF; **b** Et_3N, Δ; **c** nBuLi, −78°C; **d** $(L)_2PdCl_2$ [89]

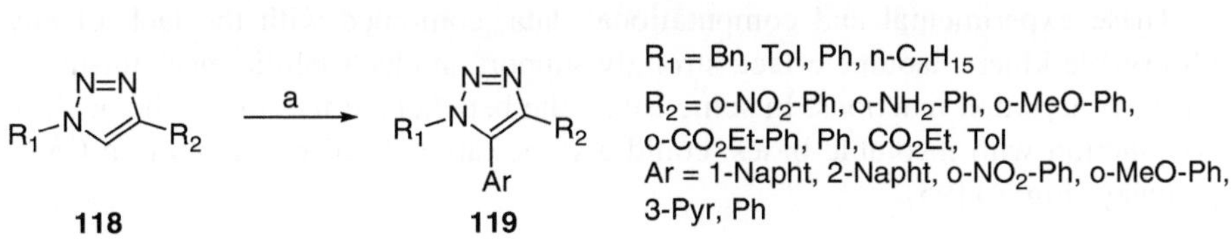

Scheme 36 **a** ArBr, $Pd(OAc)_2$ 5 mol%, Bu_4NOAc 2 equiv, NMP, 100°C, 61–99% [91]

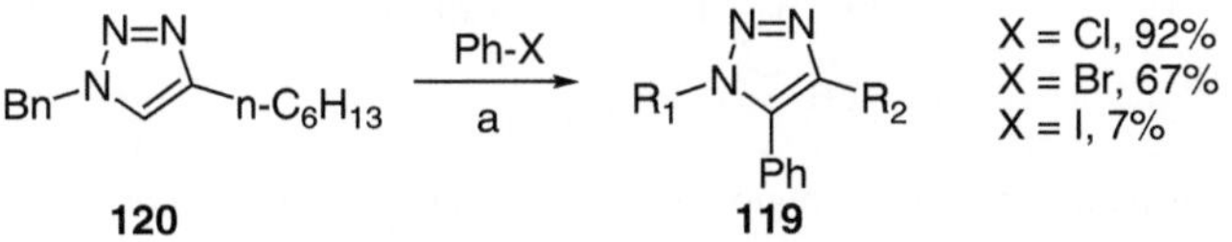

Scheme 37 **a** 0.025 mmol $Pd(OAc)_2$, 0.050 mmol $P(c\text{-}C_6H_{11})_3$, 1 mmol PhX, 1 mmol K_2CO_3, toluene (2 mL), DMF (0.4 mL), microwaves, 250°C, 15 min [92]

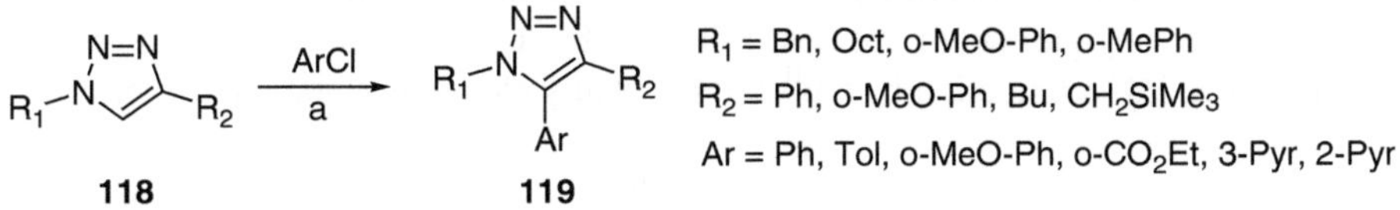

Scheme 38 **a** $Pd(OAc)_2$ 4 mol%, $P(c\text{-}C_6H_{11})_3$ 8 mol%, K_2CO_3, toluene, 105–120°C, 18–24 h, 37–96% [93]

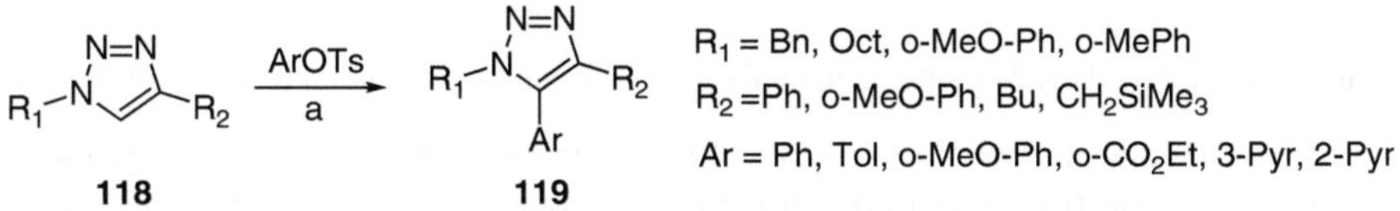

Scheme 39 **a** $Pd(OAc)_2$ 5 mol%, X-Phos 10 mol%, K_2CO_3, DMF/tBuOH, 100°C, 17–22 h, 62–99% [94]

Arylation of monosubstituted triazole **121** delivers 1,5-disubstituted compound as major product, with a small amount of diarylated triazole **123**. This regioselectivity is correlated by the DFT calculation of electrostatic potential charges at C-4 and C-5 of N-methyl triazole **124** (Scheme 40).

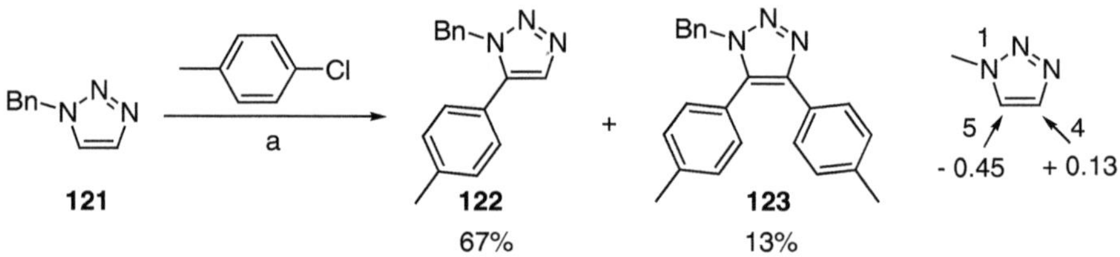

Scheme 40 **a** $Pd(OAc)_2$ 4 mol%, $P(c\text{-}C_6H_{11})_3$ 8 mol%, K_2CO_3, toluene, 120°C, 24 h

These experimental and computational data, combined with the lack of any observable kinetic isotope effect, strongly support an electrophilic mechanism for all these arylation reactions. In some cases, the beneficial effect of pivalic acid, in conjunction with insoluble bases, could also be rationalized by invoking a CMD pathway (Fig. 4) [95].

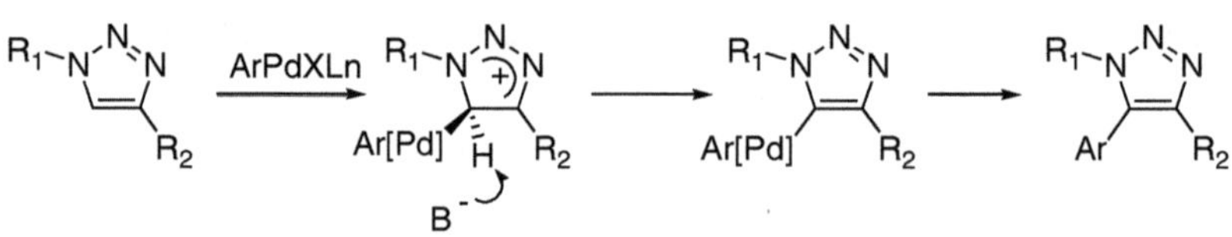

Fig. 4 Mechanistic rational for the regioselective Pd-catalyzed arylation of triazoles

Functionalization of the C-5 position of triazoles is not limited to arylation. A direct oxidative alkenylation of 1,4 triazoles has been reported (Scheme 41) [96]. Two mechanisms have been proposed. The first one involves an electrophilic attack of Pd^{II} on the triazole, followed by coupling with the alkene partner. The second one starts with a vinylpalladium species, followed by a vinylpalladation step. In both mechanisms, Pd^0 reoxidation by Cu^{II} and O_2 finally closes the catalytic cycle.

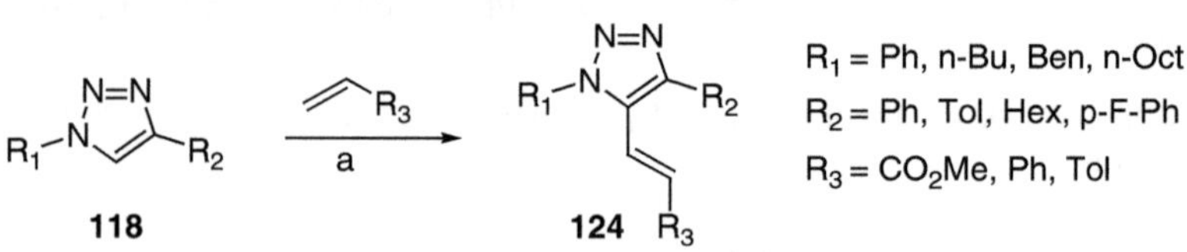

Scheme 41 **a** $Pd(OAc)_2$ 20 mol%, 2 equiv $Cu(OAc)_2$, 8 atm O_2, dioxane, 110°C, 42–90% [96]

Intramolecular arylation of triazoles by a palladium-catalyzed dehydrogenative pathway has also been reported. This closely related transformation enables efficient preparation of tricyclic triazoles without any prefunctionalization of arylating agents [97]. Under optimized conditions, this coupling can be conducted at ambient air pressure (Scheme 42).

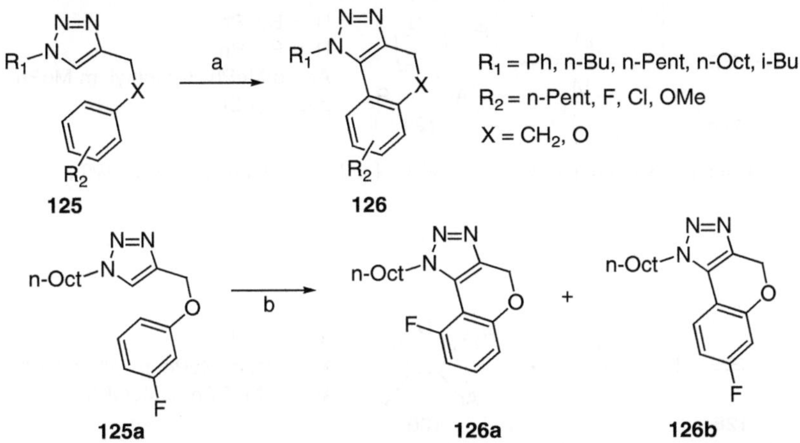

Scheme 42 **a** $Pd(OAc)_2$ 5 mol%, 1 equiv $Cu(OAc)_2$, air, toluene/PivOH 6/1, 140°C, 20 h 57–93%; **b** $Pd(OAc)_2$ 5 mol%, 1 equiv $Cu(OAc)_2$, air, toluene/PivOH 6/1, 140°C, 20 h 83% [97]

In case of possible regioisomers formation, it has been demonstrated that the less sterically hindered (less acidic) bond of fluorinated arene is preferentially heteroarylated.

Only one example of palladium-catalyzed benzylation of 1,2,3 triazoles has been reported by the group of late Fagnou (Scheme 43) [98]. The presence of pivalic acid as a cocatalyst suggests a CMD mechanism.

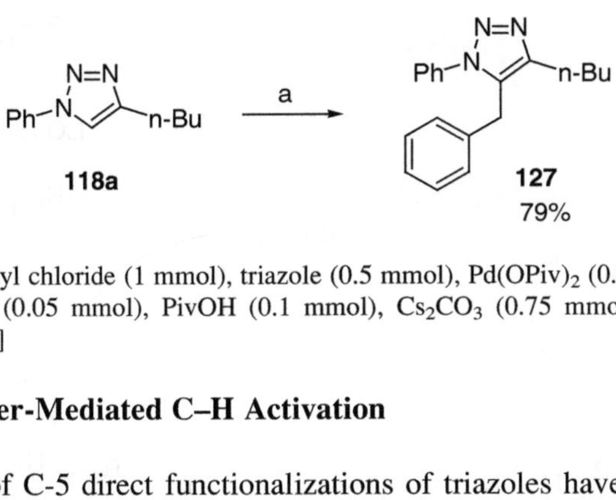

Scheme 43 **a** Benzyl chloride (1 mmol), triazole (0.5 mmol), $Pd(OPiv)_2$ (0.025 mmol), 2-Ph_2P-2'-(Me_2N)biphenyl (0.05 mmol), PivOH (0.1 mmol), Cs_2CO_3 (0.75 mmol), toluene (1 mL), 110°C, 16–20 h [98]

4.2.3 By Copper-Mediated C–H Activation

Although most of C-5 direct functionalizations of triazoles have been conducted under palladium catalysis, the use of copper in such a reaction might present an attractive alternative. Copper salts are indeed involved in the preparation of 1,4-disubstituted triazoles and could therefore be used in one-pot sequential synthesis of 1,4,5-trisubstituted heterocycles. A protocol based on the use of inexpensive CuI catalyst has been reported (Scheme 44) [99].

More interestingly, a sequential four component synthesis, with in situ generation of aryl azide is also possible. In this remarkable transformation, one C–C bond and three C–N bounds are formed in a highly regioselective fashion (Scheme 45). A related approach, using CuCl as a precatalyst, has also been reported [100].

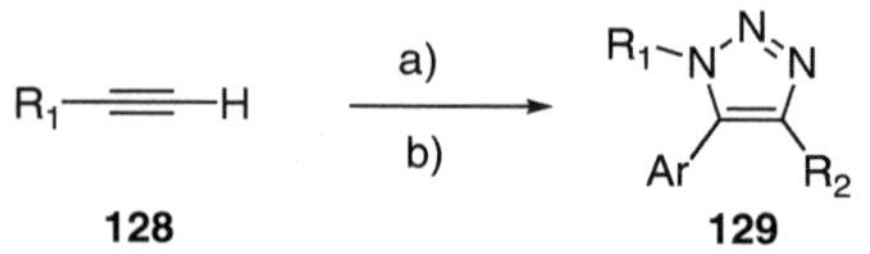

Scheme 44 **a** R_1N_3, CuI (10 mol%), DMF, 60°C, 4 h; **b** ArI 3 equiv, LiOt-Bu, DMF, 140°C, 20 h 63–84% [99]

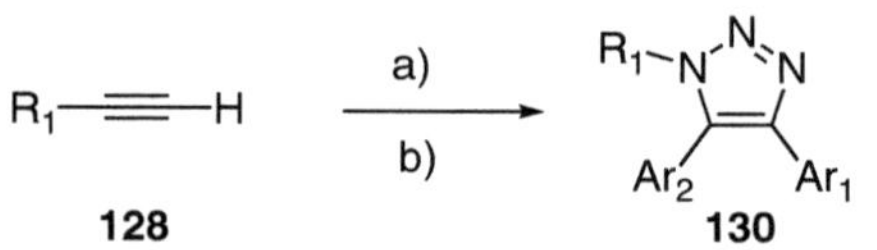

Scheme 45 **a** NaN_3, CuI (10 mol%), Ar_1I, DMEDA 15 mol%, DMF, 22°C, 2 h; **b** Ar_2I 3 equiv, LiOt-Bu, DMF, 140°C, 20 h 67–81% [100]

A mechanism involving an initial reversible deprotonation with LiOBu-*t*, followed by transmetalation into a 5-cuprated heterocycle has been proposed (Fig. 5) [101]

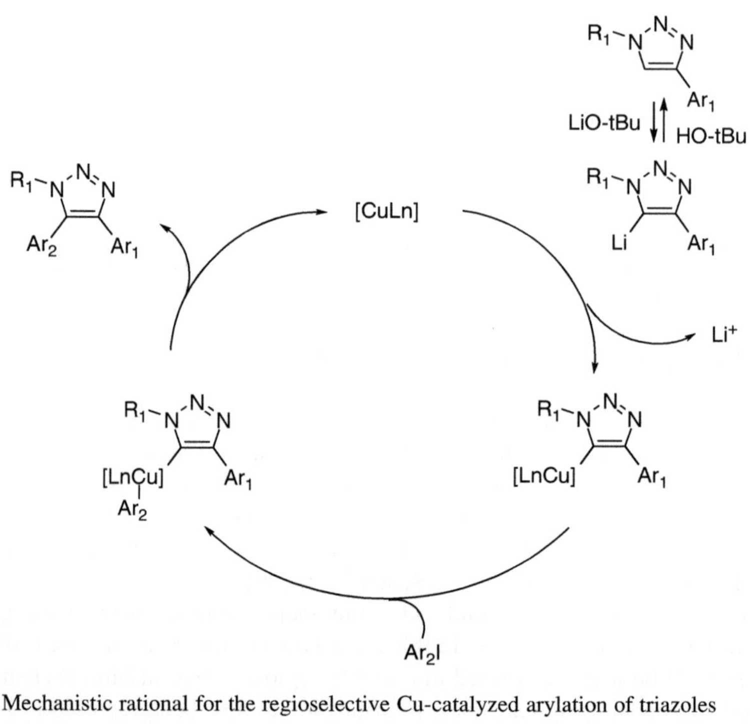

Fig. 5 Mechanistic rational for the regioselective Cu-catalyzed arylation of triazoles

4.3 Synthesis of 1,4,5-Trisubstituted Triazoles from 1,5-Disubstituted Triazoles

Access to 1,4,5-trisubstituted triazoles from 1,5-disubstituted triazoles is much less documented. One of the reason is probably that 1,4-triazoles can be easily prepared by copper-catalyzed 2+3 cycloaddition, whereas selective access to their 1,5 isomers by a ruthenium-catalyzed cycloaddition is still less popular. The lower acidity of C_4–H bond, and its lower nucleophilicity, is also a problem for such a strategy. Thus, compound **132** was obtained in only 36% yield from 1,5-disubstituted triazole **131** with high catalyst loading and prolonged heating (24 h), whereas it is obtained under much milder conditions from the 1,4-disubstituted precursor (Scheme 46) [90].

1,5-Disubstituted triazoles can nevertheless be deprotonated by strong bases. This route enables a modular synthesis of triazole-based monophosphine ligands (Scheme 47) [102, 103].

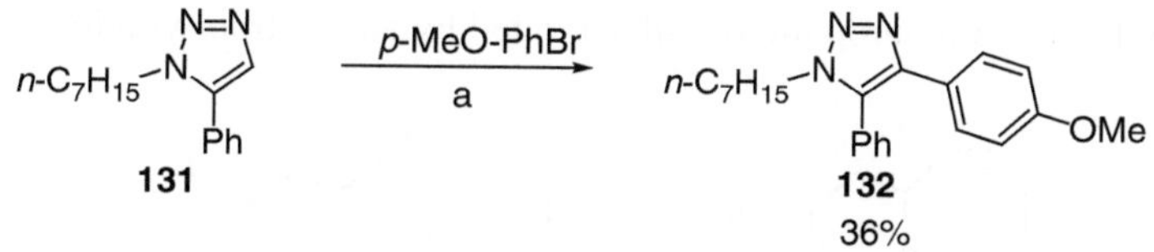

Scheme 46 a ArBr, $Pd(OAc)_2$ 10 mol %, Bu_4NOAc 2 equiv, NMP, 110°C [90]

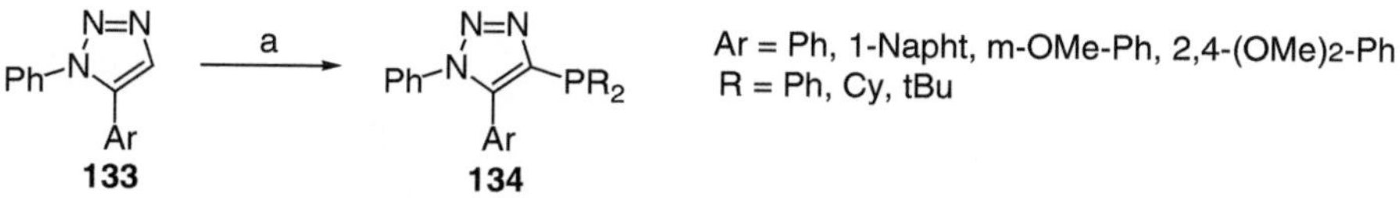

Scheme 47 a LDA, THF, 0°C, 1.5 h then $ClPR_2$, rt, 1 h. 58–93%

5 Synthesis of 1,4,5-Trisubstituted Triazoles by Condensation of Azides and Disubstituted Alkynes

5.1 Uncatalyzed Cycloadditions with Internal Alkynes

The first reaction between alkynes and azides has been reported at the end of the nineteenth century. Despite its high exothermicity, this cycloaddition suffers from a very high activation barrier. As a result, this reaction is generally slow, even at elevated temperature, and therefore delivers triazoles in only moderate yields. Furthermore, as the differences in HOMO–LUMO energies for both partners are in the same range, the uncatalyzed cycloaddition with internal alkynes generally delivers a mixture of regioisomers which might be in favor of triazoles where the electron-withdrawing groups are in 4-position and the electron-donating group are in 5-position (Scheme 48) [104].

One way to lower the activation barrier of this cycloaddition is to use strained alkynes. For instance, the reaction between cyclooctyne and phenylazide has been reported to be

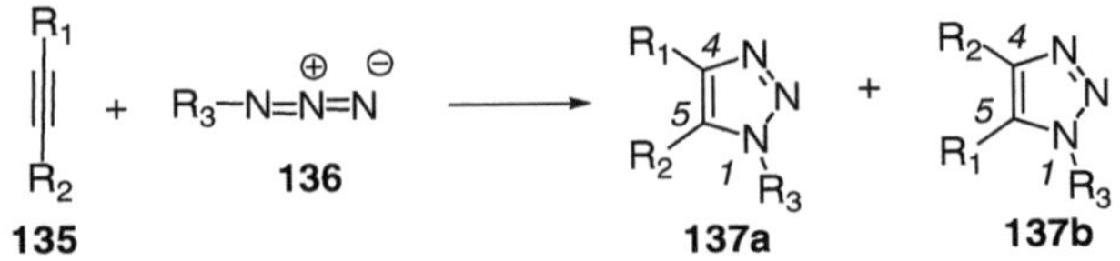

Scheme 48 Cycloaddition of internal alkynes with azides: regioselectivity issue

extremely violent. This high reactivity can be explained by the distortion of the bond angles of the sp-hybridized carbons, leading to a conformation closer to the one expected in the transition state [105]. This effect can even be improved by the introduction of strong electron-withdrawing groups nearby the triple bound, which lowers the LUMO energy level, leading to a greater reactivity (Fig. 6). This concept has been used in an extremely elegant approach by the group of Bertozzi, with the development of bioorthogonal copper-free click reactions as tools in chemical biology [106].

Azides can also react with undistorded alkynes under thermal activation, provided that the dipolarophile is still activated by an electron-withdrawing group.

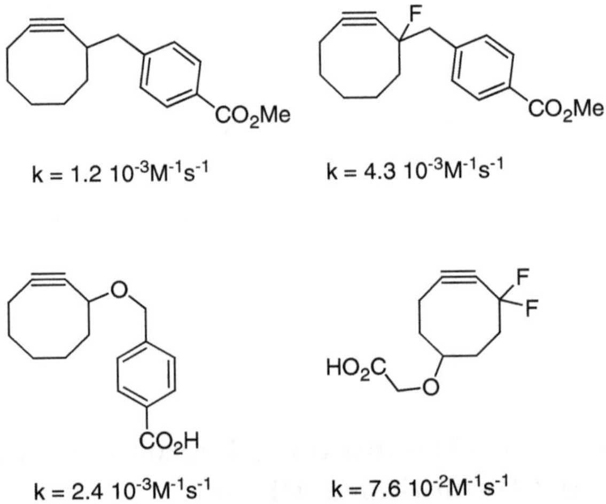

Fig. 6 Rate constants of different strained alkynes in 3+2 cycloadditions [106]

Cycloaddition of organic azides with symmetrical alkynes bearing electron-withdrawing group is obvious. Recently, Rode et al. reported for instance the synthesis of hybrid 1,2,3-triazolo-γ-lactams/lactones using Huisgen [3+2] cycloaddition between organic azides and an activated alkyne in water in quantitative yields (Scheme 49) [107].

Thermal cycloaddition with unsymmetrical internal alkynes is however in most of the cases not controlled [23, 108–111]. Steric effects on the regioselectivity of the azide-alkyne 1,3-dipolar cycloaddition reaction have been reported by Hlasta et al. [112]. The 1,2,3-triazoles derivatives **142** and **143** were synthetized via a 1,3-dipolar cycloaddition reaction of the azidomethyl-benzisothiazolinone **141** with various electron-deficient acetylenes **135a–k**. Interestingly, only one regioisomer **142a–e** was detected using 2-(trimethylsilyl)ethynyl derivatives (**135a–e**) in good yields (68–82%), whereas a mixture of stereoisomers **142f–k** and **143f–k** was obtained

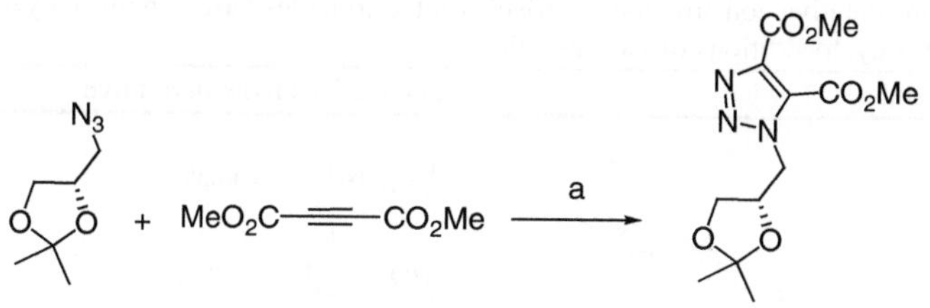

138 **139** **140**

Scheme 49 **a** H_2O, 70°C, quantitative [107]

with other acetylenes **135f–k** in good yields (> 80%). Steric effect of the trimethylsilyl group (TMS) and stabilization of a partial positive charge on the acetylene β-carbon by TMS in the transition state could explain the exclusive formation of **142b–e** (see frame). Regarding the **142f–k/143f–k** ratio, the major regioisomers **143f–k** have the electron-deficient group in 4-position (R_2), except for **135i** that delivers a 1:1 mixture of triazoles **142i** and **143i**, and for **135jk** that gives a ratio of ~2/1 (Scheme 50). Other examples of regioselective synthesis of 1,4,5-trisubstituted 1,2,3-triazoles starting from various 1-trimethylsilylacetylenes were presented by the same authors in moderate to good yields (33–82%, PhMe, reflux) [113].

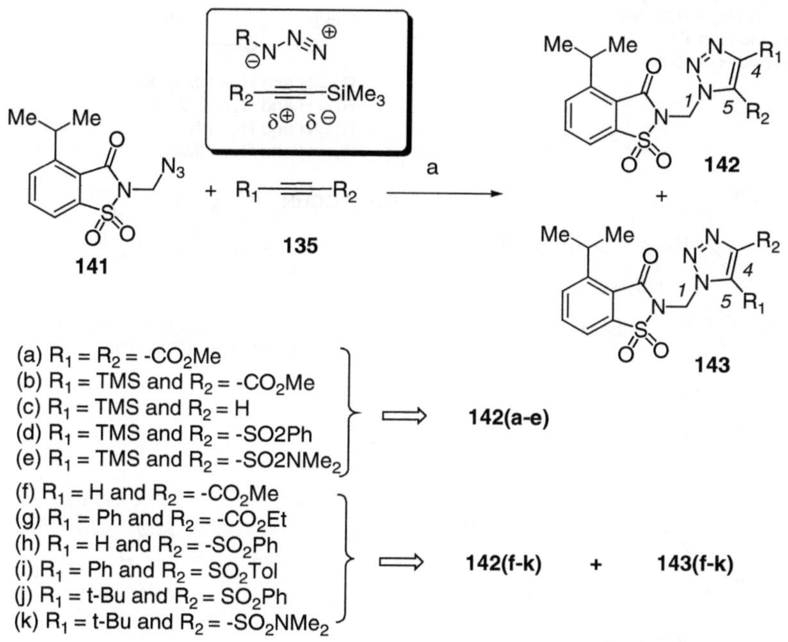

Scheme 50 **a** Refluxing benzene, 5 h to 6 days

Only one 1,2,3-triazole isomer was also obtained by Yamashita et al. [114] by the regioselective cycloaddition of (trimethylsilyl)acethylene with azido phospholane derivatives (see Table 2)

Table 2 Representative constructions of trisubstituted triazoles based on uncatalyzed directed or nondirected 1,3-cycloadditions of various azides

Starting material	Final 1,2,3-triazole derivative	Ref.
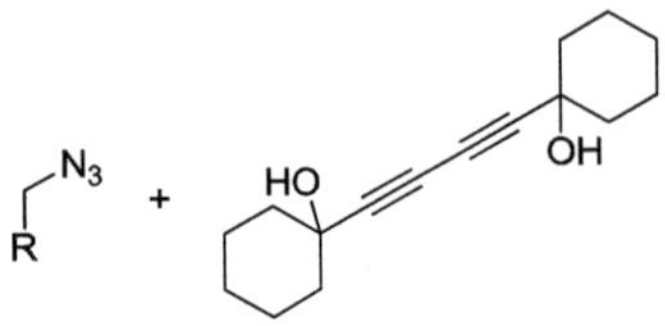 R = H, Bz, Ph, 4-Cl-Ph, 2,5-diMe-Ph, Me, -CH_2-$CONH_2$, -CH_2-CO_2Et, 3-NH_2-Ph	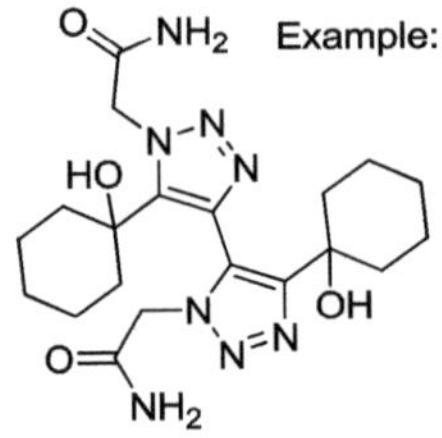 Mixture of regioisomers	[115]
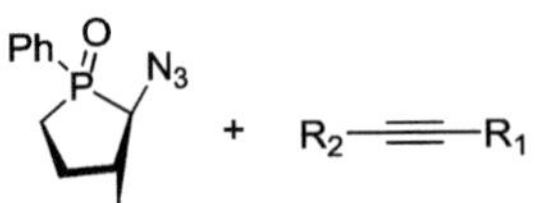 (DMF, reflux) R_1 = -CO_2Et and R_2 = -CO_2Et R_1 = -CO_2Me and R_2 = -CO_2Me R_1 = -CO_2H and R_2 = -CO_2H R_1 = H and R_2 = $SiMe_3$ R_1 = H and R_2 = CMe_2OH R_1 = H and R_2 = -CO_2Me R_1 = H and R_2 = Ph R_1 = H and R_2 = CMe_3	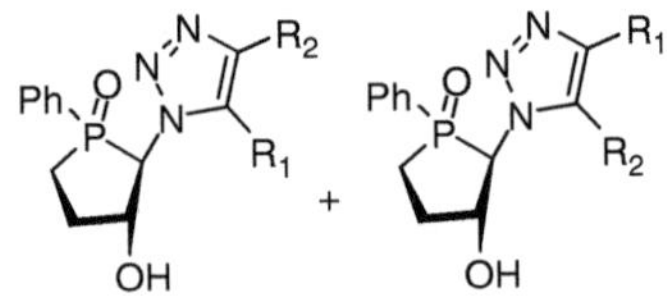 R_1 = -CO_2Et and R_2 = -CO_2Et R_1 = -CO_2Me and R_2 = -CO_2Me R_1 = -CO_2H and R_2 = -CO_2H R_1 = H and R_2 = $SiMe_3$ Minor Major R_1 = H and R_2 = CMe_2OH R_1 = H and R_2 = -CO_2Me R_1 = H and R_2 = Ph R_1 = H and R_2 = CMe_3	[114]
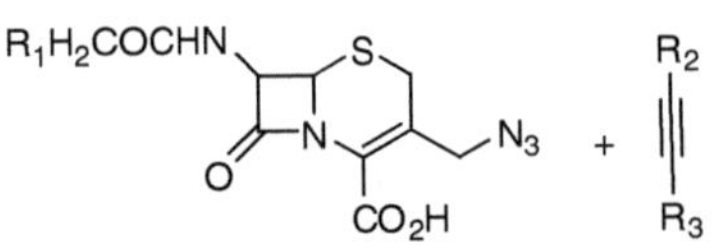 R_2 = R_3 = -CO_2Me (benzene, reflux) R_2 = R_3 = -CO_2H R_2 = R_3 = -CF_3 R_2 = -CO_2Me and R_3 = H R_2 = -CO_2H and R_3 = H R_2 = -CHO and R_3 = H R_2 = -COMe and R_3 = H	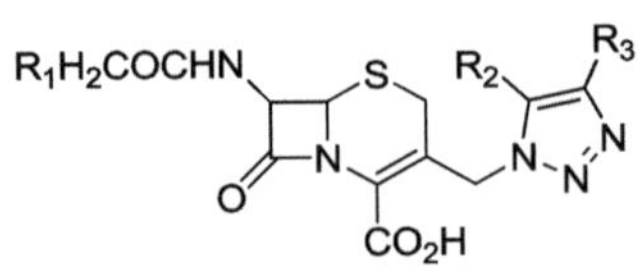 ~ 1/1 mixture of regioisomers for asymetrical acethylenesi	[4]
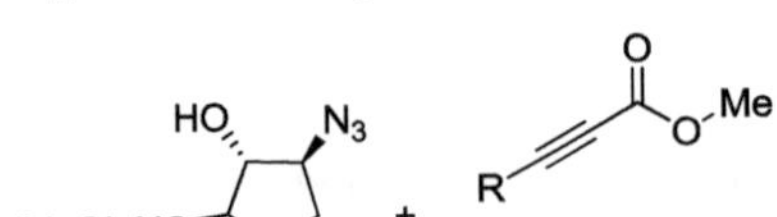 R = H, -CO_2Me (DME, reflux)	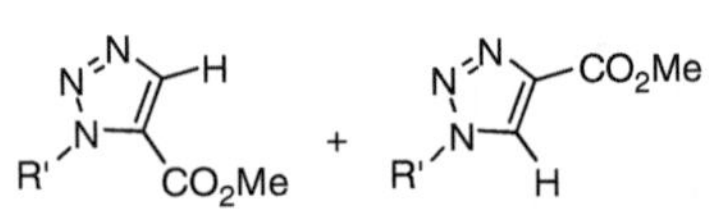 Minor Major 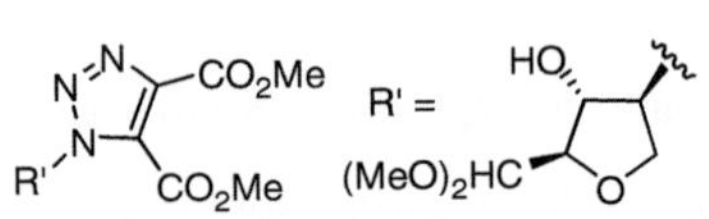	[116]

(continued)

Table 2 (continued)

Starting material	Final 1,2,3-triazole derivative	Ref.
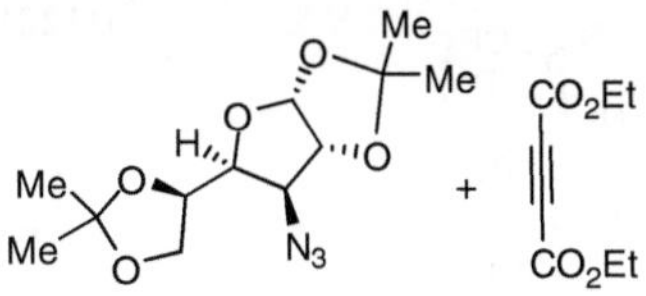 (Toluene,100°C)	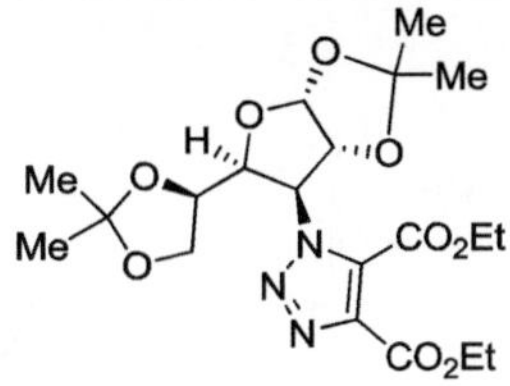	[117]
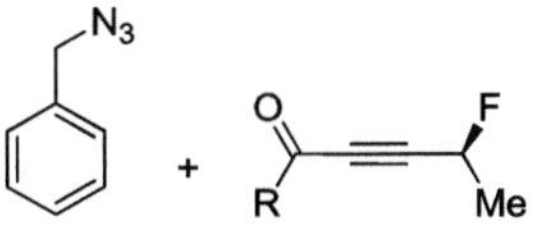 R = Ph, -CO_2Et (Toluene, 45°-60°C)	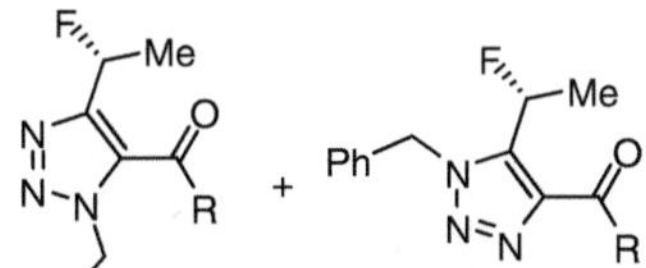 R = Ph,-CO_2Et Minor Major	[118]
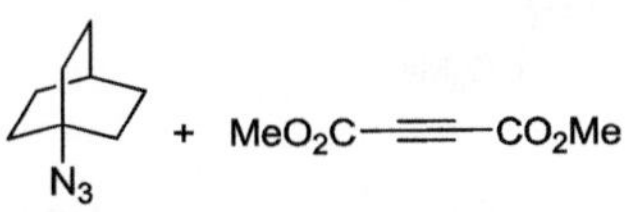 (benzene, 100°C, sealed tube)	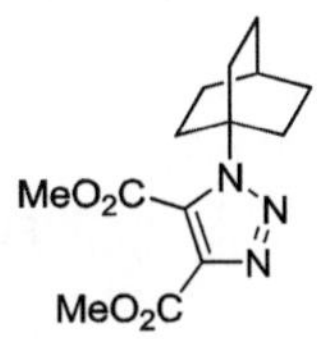	[119]
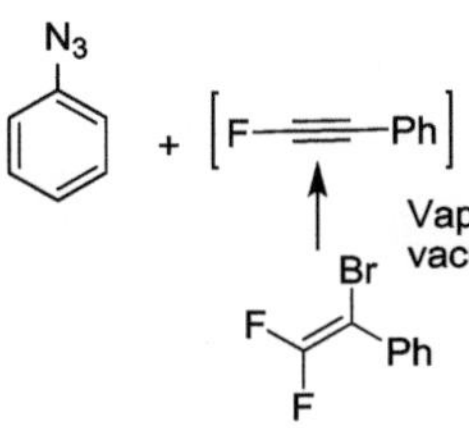 Vapor phase vacuum pyrolysis	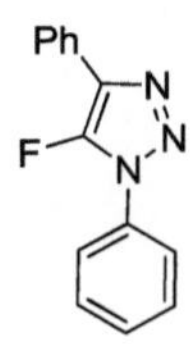	[120]
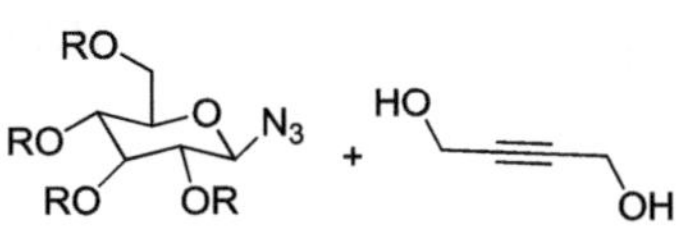 (Toluene-pyridine 9:1, reflux)	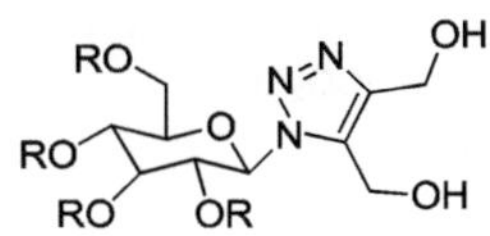	[14]
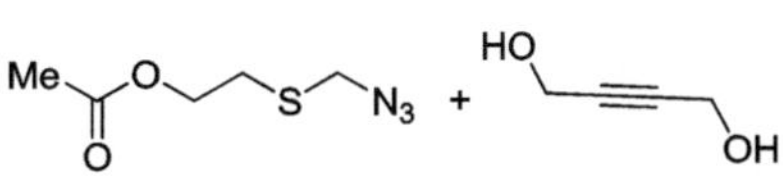 (Toluene-pyridine 9:1, reflux)	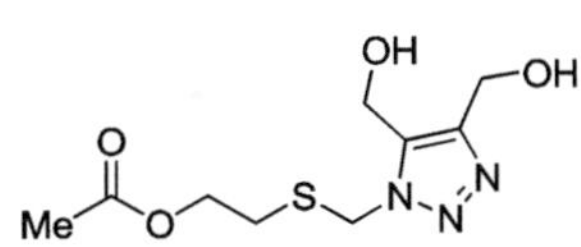	[121]

(continued)

Table 2 (continued)

Starting material	Final 1,2,3-triazole derivative	Ref.
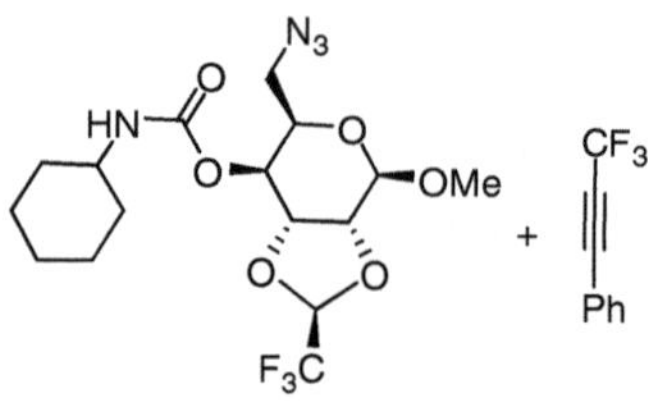	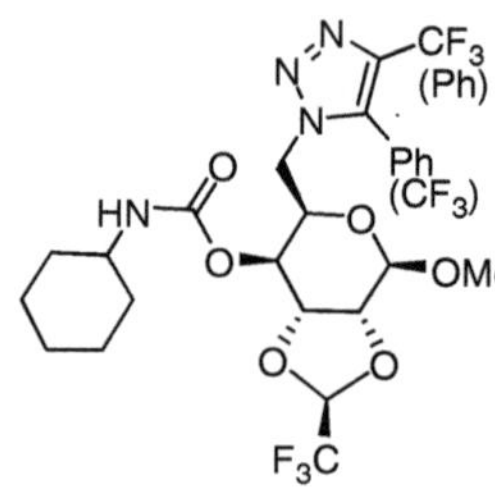	[122]
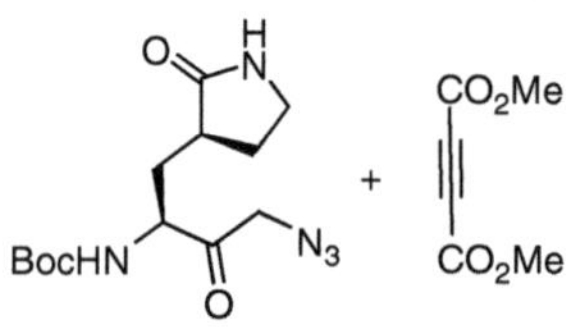 (Toluene, rt)	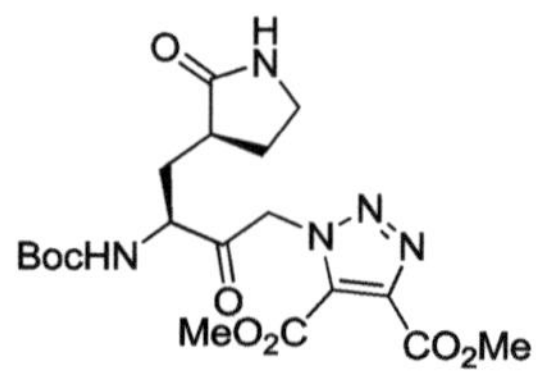	[123]
$Ar_1-C\equiv C-CF_3$ + Ar_2-N_3 (Tolene, reflux) AR_1 = 4Cl-Ph and Ar_2 = Ph Ar_1 = 2-Cl-Ph and Ar_2 = Ph Ar_1 = 4-NO2-Ph and Ar_2 = Ph Ar_1 = 4-Me-Ph and Ar_2 = Ph Ar_1 = 4-NH2-Ph and Ar_2 = Ph AR_1 = 4I-Ph and Ar2 = 2,6-Cl_2-Ph Ar_1 = 2-ClPh and Ar2 = 2,6-Cl_2-Ph Ar_1 = 4-NO2-Ph and 2,6-Cl_2-Ph Ar_1 = 4-Me-Ph and Ar2 = 2,6-Cl_2-Ph Ar_1 = 4-Cl-Ph and Ar2 = 4-NO_2-Ph	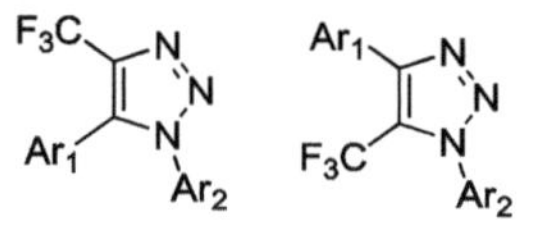 Major Minor	[124]
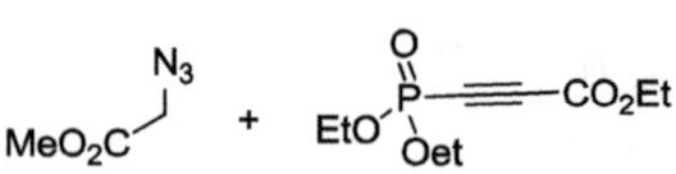 (Toluene, reflux)	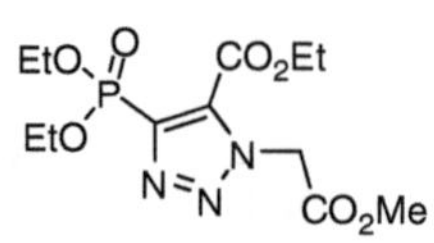 Minor + 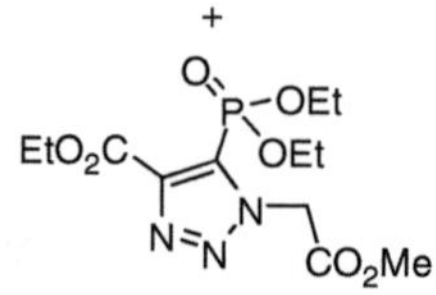 Major	[125]

(continued)

Table 2 (continued)

Starting material	Final 1,2,3-triazole derivative	Ref.
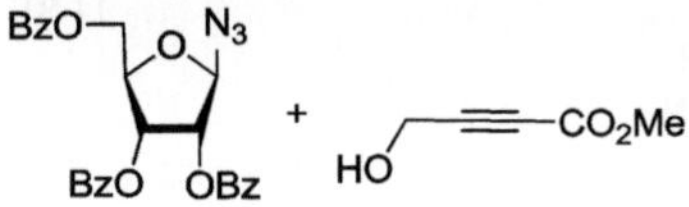 (Tolene, 50-55°C)	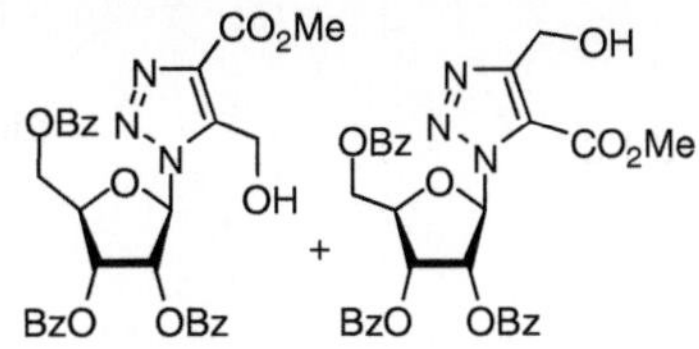 Major Minor	[126]
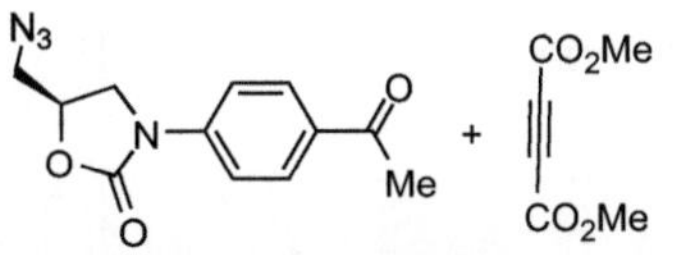 (Glyme, reflux)	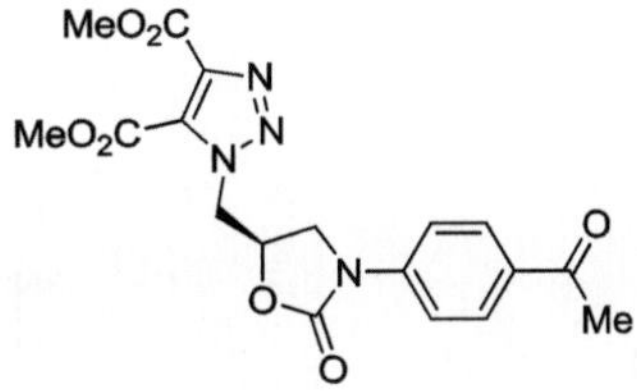	[5]
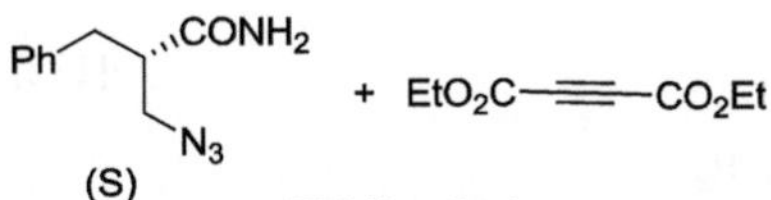 (EtOH, reflux)	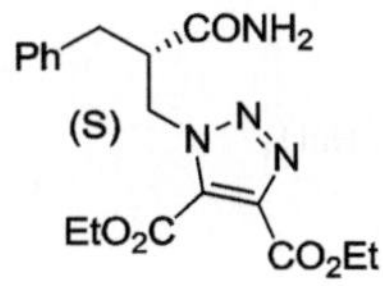	[127]
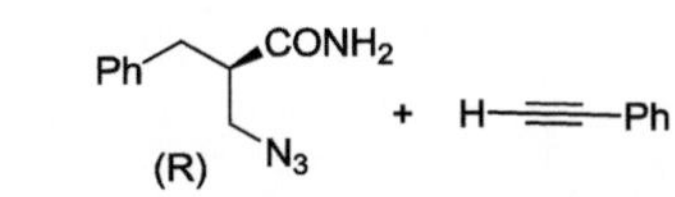 ($CuSO_4$, sodium ascorbate, t-BuOH/H2O, rt)	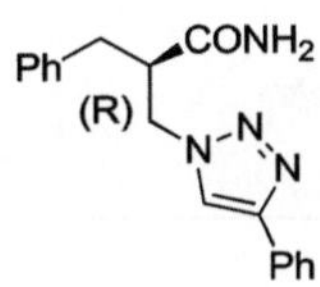	[127]
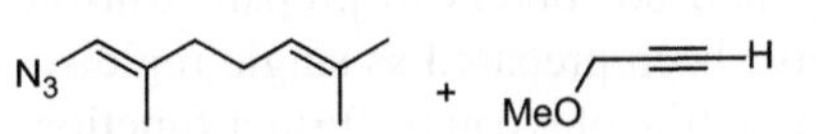 $CuSO_4$, sodium ascorbate, DMF,rt → (I) Cuwire, MeCN, rt → (I) + (II) (I)/(II)~ 1/1 Cu wire, MeCN, 70°C → (II)	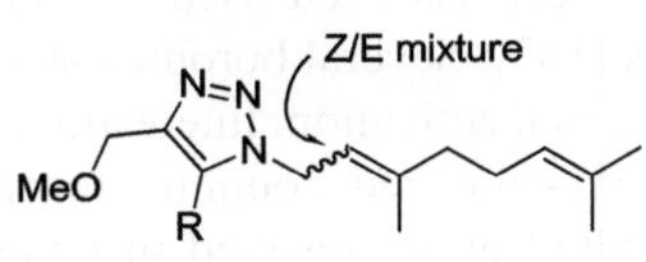 (I): R = H (I): R = -CC-CH_2-OMe	[128]
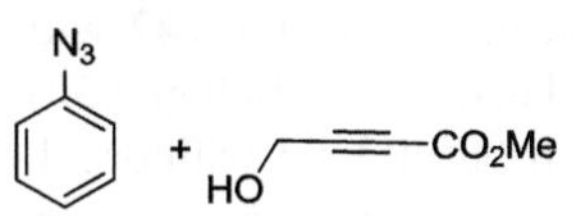 (Toluene, 55°C)	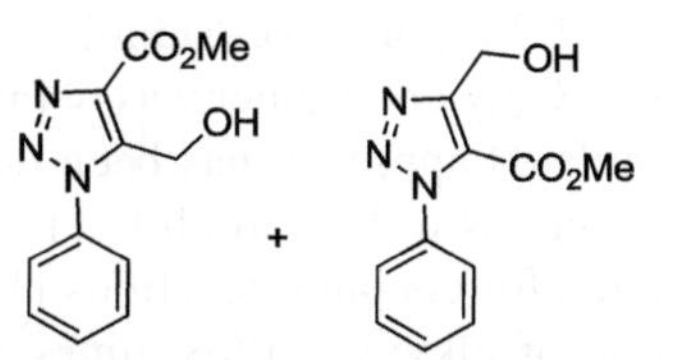 Major Minor	[129]

(continued)

Table 2 (continued)

Starting material	Final 1,2,3-triazole derivative	Ref.
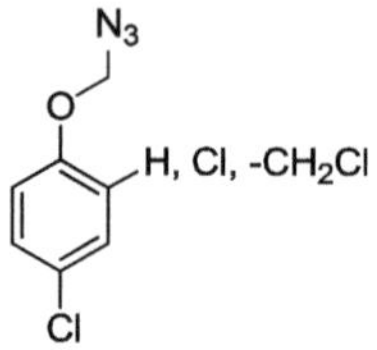	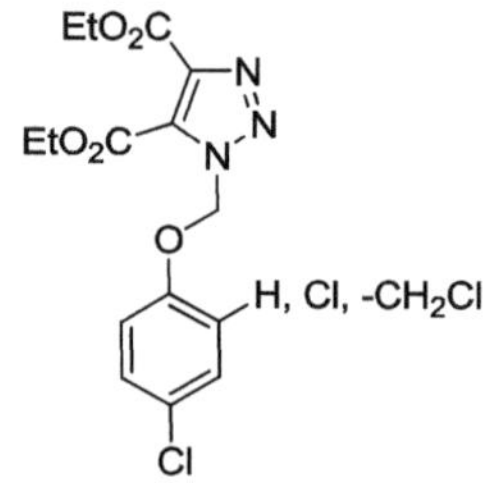	[130]
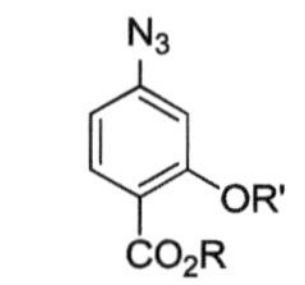 (Benzene, reflux)	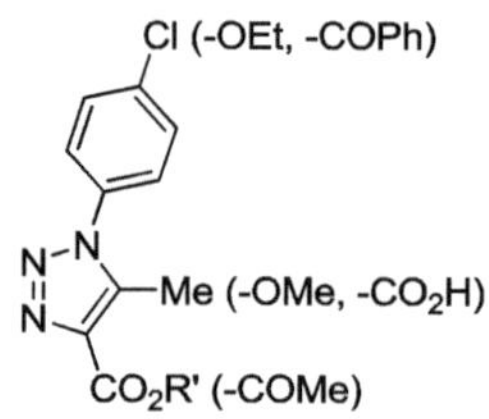	[131]
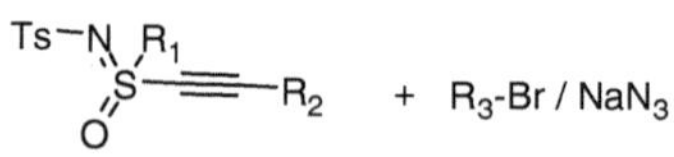 (H_2O-CH_2Cl_2, reflux) R_1 = 4-Me- Ph, Me, Ph R_2 = Bu, CH_2OMe R_3 = -NO_2-Bn, 4-MeO-Bn, $Ph(CH_2)_2$-, $Ph(CH_2)_3$-, Bn	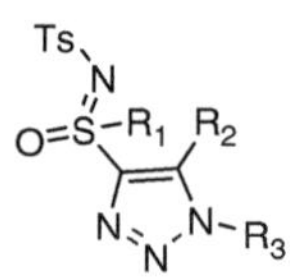	[132]

This silyl effect has been used by Harrity and coworkers to prepare triazole boronic esters [133]. Several boronic esters have been prepared as single regioisomers under thermal activation. Interestingly, these triazoles can be further functionalized using Pd-catalyzed coupling reactions. Replacement of the trimetylsilyl group by an alkyl group resulted in complete loss of the regioselectivity of the cycloaddition (Scheme 51).

Brimble et al. described an elegant synthesis of spiroacetal-triazoles **148** as privileged natural product-like scaffolds using "click chemistry" [134]. One of the key steps is the 1,3-dipolar cycloaddition of azido-spiroacetal derivative **147** with various alkynes. Only one isomer was obtained with a silylated akyne (Scheme 52).

A closely related approach has been described using tributyl(3,3,3-trifluoro-1-propynyl)stannane as a dipolarophile [135]. Once again, stanylated triazoles are useful precursors for coupling reactions [136] (Scheme 53)

The reaction of alkynyl sulfoximines with in situ-prepared organic azides in water–dichloromethane under reflux has been reported to afford 1,4,5-triazoles in a fully selective manner (Scheme 54) [132]. In addition, an instructive paper on the 1,3-dipolar cycloadditions of acetylenic sulfones in solution and on solid supports has been published by Back et al. [137].

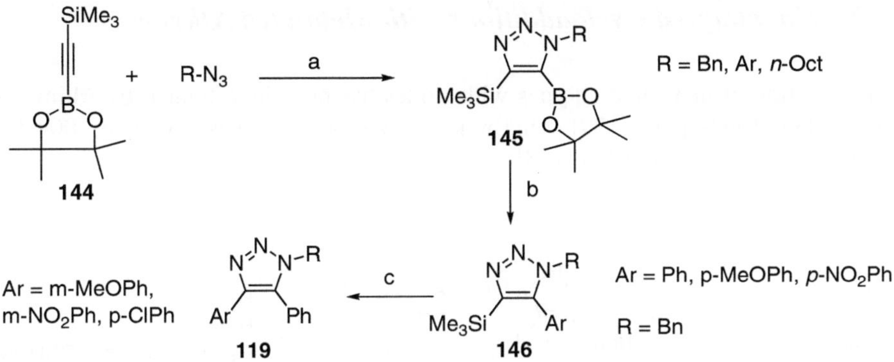

Scheme 51 **a** DCB, 110°C, 24 h 64–84% **b** 5 mol% $Pd_2(dba)_3$, 12 mol% t-$Bu_3P.HBF_4$, K_3PO_4, ArI, MeCN, 50°C, 16 h, 66–97% **c** NIS 5 equiv, then 5 mol% $Pd_2(dba)_3$, 12 mol% t-$Bu_3P.HBF_4$, K_3PO_4, ArBOPin, MeCN, 50°C, 59–75% [133]

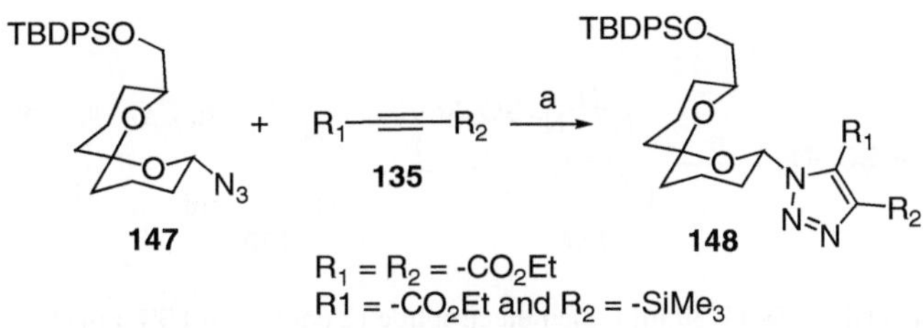

Scheme 52 **a** Toluene, sealed tube, reflux, 78–84% [134]

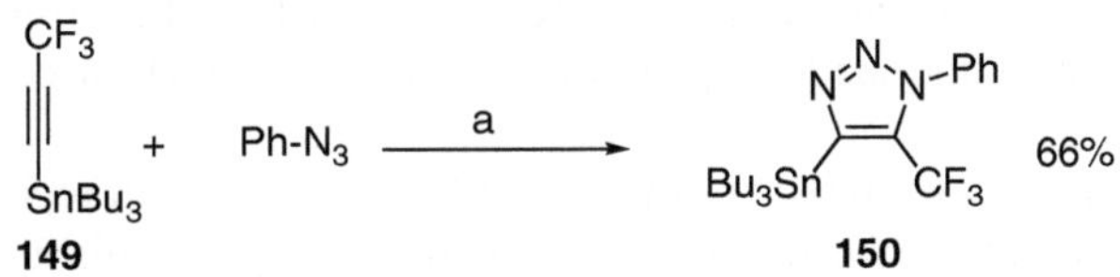

Scheme 53 **a** $CH(OMe)_3$, 80–85°C [135]

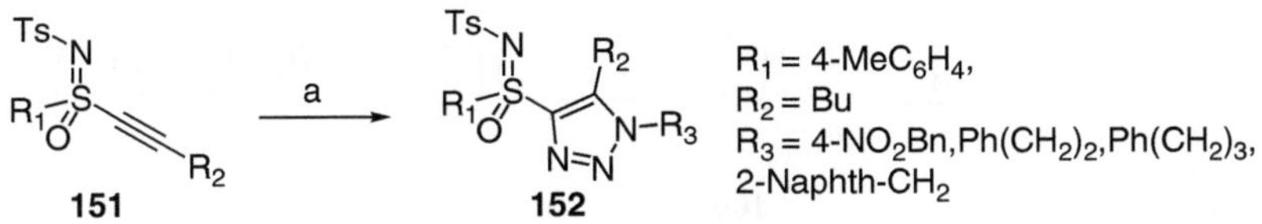

Scheme 54 **a** R_3Br, NaN_3 H_2O–CH_2Cl_2, reflux, 3 h 39–73%

5.2 *Uncatalyzed Cycloadditions with Metalated Alkynes*

The reaction of metalated alkynes with azides has been investigated by Akimova in the late 1960s [138, 139]. This reaction has been reinvestigated in 2004 by Krasinski et al. (Scheme 55) [118]

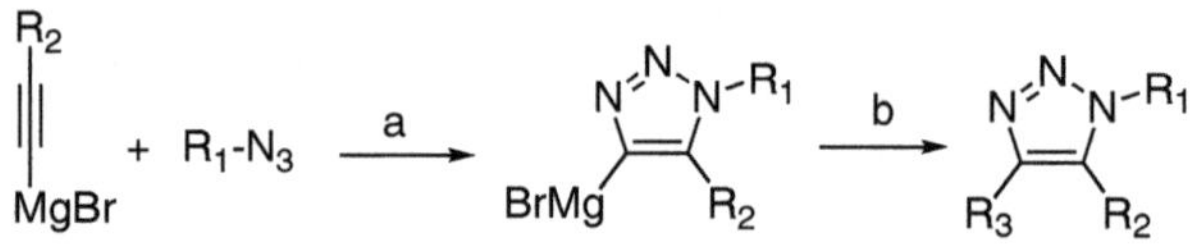

153 **154** **155**

R_1 = Ph
R_2 = Ph
R_3 = D, I, CO_2H, CONHPh

Scheme 55 **a** THF, 50°C, 1–24 h, then electrophile 77–95% [140]

A similar approach has been recently described using lithiated alkynes (Scheme 56) [141].

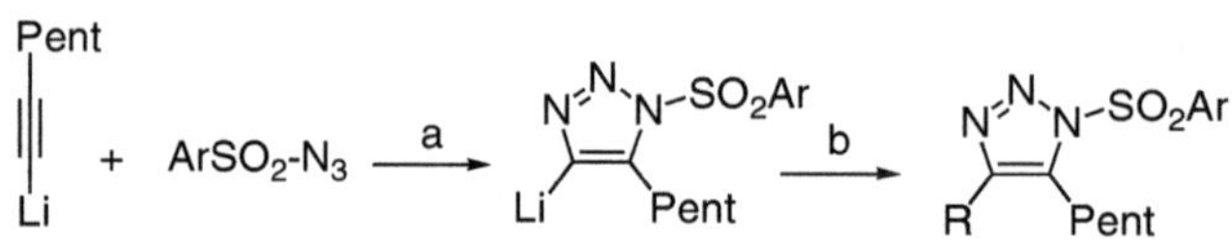

156 **157** **158**

R = PhCOPh, COnPr, CO_2Bn, TBS, Me

Scheme 56 **a** THF, –78°C, 30 min, then electrophile (2 equiv) 17–80% [141]

In both cases, the metallotriazole can react with several electrophiles, leading to 1,4,5-triazoles in a regioselective manner. The mechanism proposed for these transformations involves a nucleophilic addition of the metal acetylide, followed by ring closure to the metallotriazole. Preferential attack on the terminal rather than internal nitrogen of the azide explains the exclusive formation of the 4-metallo intermediate.

The reaction of alkynylzinc or zincates with azides has been reported to deliver metallotriazoles that can be further functionalized using a Negishi coupling reaction [142]. Although the mechanism has not been discussed, formation of a metalated heterocycle might be related to the one proposed by Krasinski (Scheme 57).

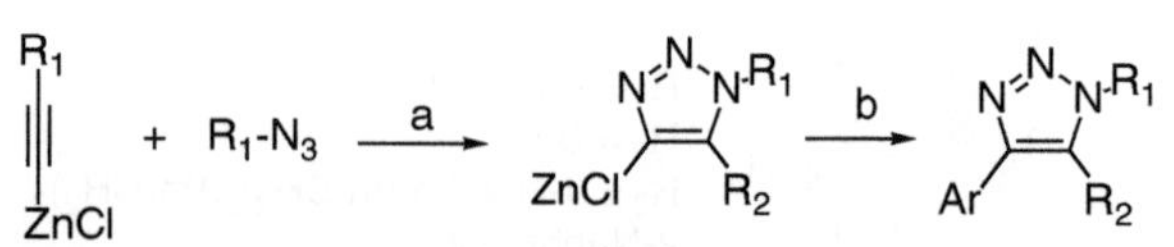

159 **160** **161**

R_1 = *p*-FPh, 2,4-F_2Ph
R_2 = Me, Ph
R_3 = Ar

Scheme 57 **a** THF, –78°C, 30 min, then electrophile (2 equiv) 17–80% [142]

5.3 Catalyzed Cycloadditions

The discovery of Cu(I) catalysis for the Huisgen 1,3-dipolar cycloaddition reaction has led to a tremendous number of synthetic developments [104, 143]. Even if the exact mechanism of the reaction is still debated, the process involves formation of a transient 5-cuprotriazole, which is protodemetalated to close the catalytic cycle. If stoichiometric or an excess of copper is used, one can conceive to trap this transient metalated intermediate by electrophiles (Fig. 7).

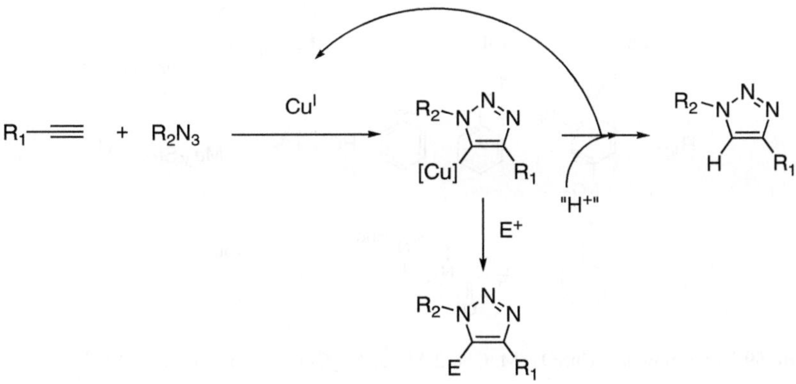

Fig. 7 Mechanistic rational for the regioselective Cu-catalyzed 3+2 cycloadditions with terminal alkynes

This strategy has been reported for the synthesis of several 1,4,5-trisubstituted triazoles [144] Interestingly, the electrophile is introduced at the beginning of the reaction. Regioselectivity of the reaction, opposite to the one observed using lithiated, magnesiated, or zincated alkynes, suggests that the mechanism is indeed a [3+2] copper-assisted cycloaddition that proceeds via a 5-cuprotriazole, and not a cyclometalation that would have deliver 4-substituted triazoles (Scheme 58)

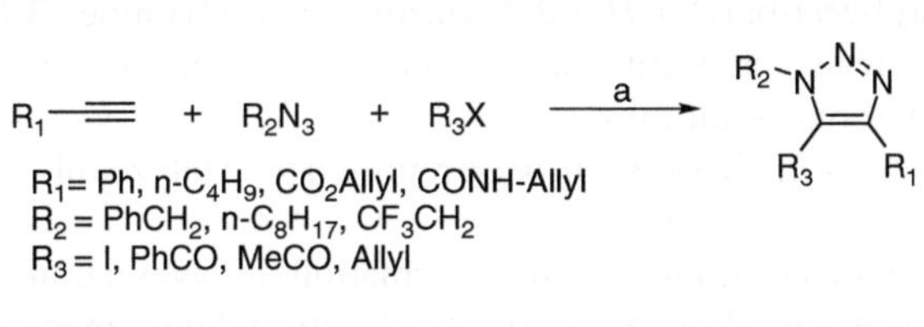

Scheme 58 **a** CuI 1 equiv, Et_3N 1 equiv, THF, R_3X 1–2.5 equiv, rt, 20 h 27–81% [144]

This stoichiometric method has been used for the synthesis of several 1,4,5-trisubstituted triazolyl nucleosides. In some cases, 5-halogenated compounds were obtained in mixture with the protonated ones [145]. A slight improvement of this methodology, by using a combination of CuI and N-bromosuccinimide that generates in situ a reactive iodonium electrophile, allowed clean preparation of 5-iodotriazoles [146].

One-pot regioselective synthesis of bistriazole derivatives **162** has been described by Burgess et al. [147]. Based on copper-assisted Huisgen reaction, condensation of acetylenes **128** with various azides **136** in presence of Cu powder/$CuSO_4$ and 2 M aq. Na_2CO_3 in acetonitrile under air atmosphere at room temperature afforded the bistriazoles **162** with moderate to good yields (Scheme 59).

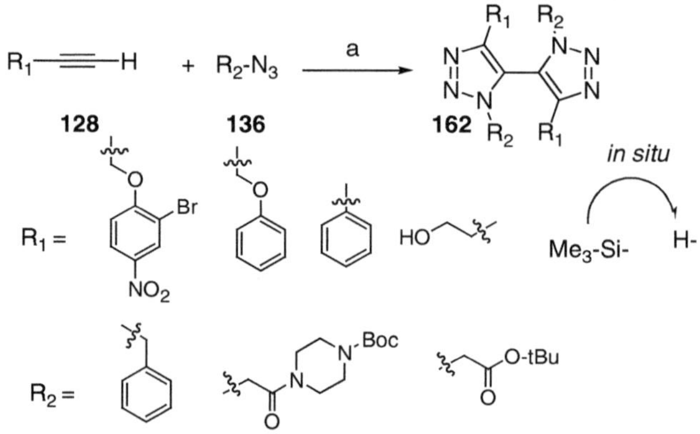

Scheme 59 **a** Cu powder, $CuSO_4$, MeCN, 2 M aq. Na_2CO_3, air, rt, 23–87% [147]

As generation of an alkynylcopper species was believed to be a requirement for copper-catalyzed huisgens-type cycloaddition, the use of copper salts to catalyze reaction of azides with internal alkynes was generally not considered as a convenient method for the preparation of trisubstituted triazoles. However, Hein, Fokin, and coworkers reported that copper(I) catalyzes cycloaddition of various azides **136** with 1-iodoalkynes **163** to give 5-iodo-1,2,3-triazoles **164** as the sole isomer in good yield [148]. Compounds **164** are versatile synthetic intermediates that are amenable to further chemical functionalizations. Remarkably, this cycloaddition is conducted at room temperature in presence of 5 mol% of CuI in THF with 5 mol% of tris((1-*tert*-butyl-1 *H*-1,2,3-triazolyl)methyl)amine (TTTA) as ligand (Scheme 60). Water-soluble salts have also been reported to catalyze this reaction in water under aerobic conditions.

The use of water-soluble salts has been reported to catalyze this reaction in water under aerobic conditions [149].

The proposed mechanism for this transformation involves a transient vinylidene-like transition state B (Fig. 8, left). A closely related mechanism has also been proposed for the copper-catalyzed cycloaddition of azides on mixed organoaluminum compounds.

Thus, it has been reported that aluminated triazoles can be prepared from the corresponding alkynides by a copper-catalyzed condensation. The aluminum bound can be further functionalized, leading to 1,4,5-trisubstituted triazoles [150]. No reaction occurs in the absence of copper, and the regioselectivity of the reaction clearly rules out a tandem addition-cyclometalation process (Scheme 61).

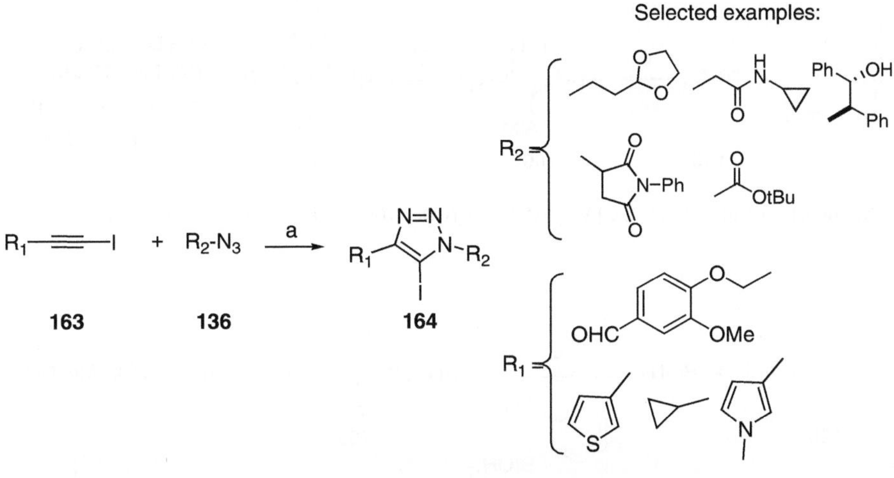

Scheme 60 **a** CuI (5 mol%), TTTA (5 mol%), THF, rt > 70% [148]

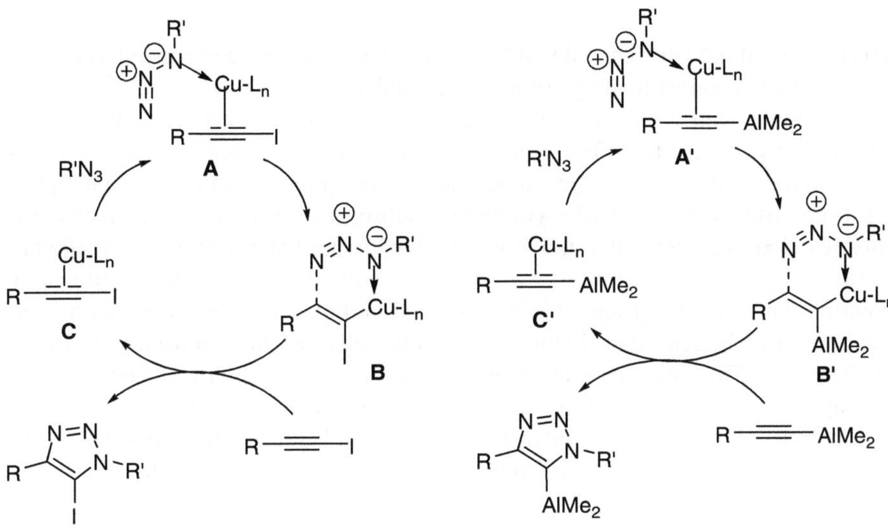

Fig. 8 Mechanistic rational for the regioselective Cu-catalyzed 3+2 cycloadditions with iodinated or aluminated alkynes

Although most of the catalytic activations of Huisgen reaction involve metal salts, a very interesting activation by boronic acids has been recently published [151]. This remarkable mild transformation enables the access to 4-carboxy-substituted triazoles without decarboxylation (Scheme 62).

The LUMO lowering of the dipolarophile by transient formation of a monoacylated hemiboronic ester not only enables a rate acceleration of the reaction with internal alkynes but also a good to excellent control of the regioselectivity of

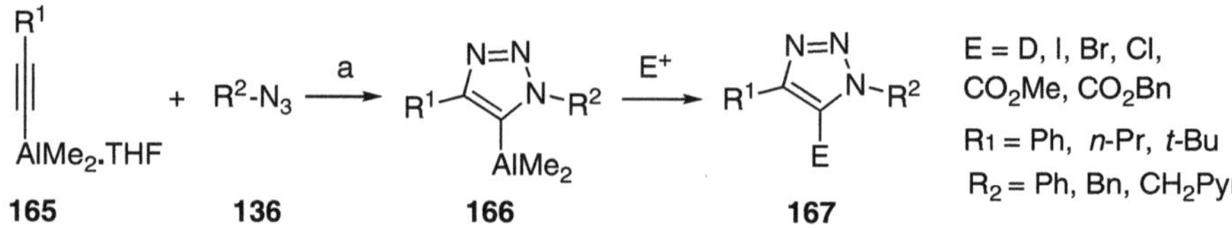

Scheme 61 **a** CuI (10 mol%), $[Me_2N(CH_2)_2]_2NMe$ (10%), toluene, rt, 48 h, 52–85% [150]

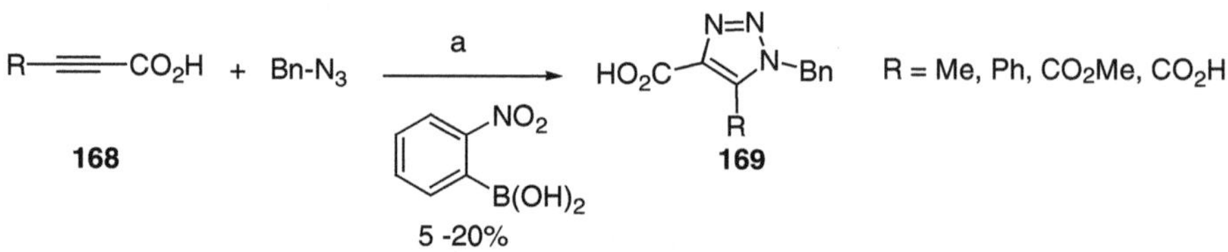

Scheme 62 **a** 25–40°C, 24–96 h, 68–73% [151]

this process. It has been shown that this regioselectivity can be enhanced, if needed, by increasing catalyst loading from 5 to 20 mol%.

Ruthenium salts are well known to catalyze several reactions involving alkynes. These catalysts have therefore been investigated in the Huisgen reaction. The first report of a Ru-catalyzed [3+2] cycloaddition with symmetrical internal alkynes [152] has been further extended to dissymmetrical substrates [153]. The regioselectivity of this reaction was generally not controlled with nonfunctional internal alkynes. However, perfect regiocontrol was achieved with dipolarophiles substituted by electron-withdrawing groups, giving rise to 4-carboxy triazoles, whereas alkynes bearing an hydrogen bond donors exclusively delivered the opposite regioisomer (Scheme 63). This directing effect was attributed to a strong hydrogen bond between the dipolarophile and the chloride ligand of the ruthenium catalyst [154].

The use of ruthenium ($CpRuCl(PPh_3)_2$) or palladium salts ($Pd(OAc)_2/PPh_3$, $PdCl_2$(dppf) and $Pd(PPh_3)_4$) to catalyze cycloadditions with internal alkynes has been recently compared.

Thus, ruthenium-catalyzed 1,3-dipolar cycloaddition of trifluorometylated propargylic alcohols **170a–d** with azides only gave 1,5-disubstituted triazoles **171a–d** in good yields (64–95%) [95], whereas palladium salts and uncatalyzed cycloadditions afforded a mixture of the two regioisomers **(171a + 171b)** (Scheme 64).

An experimental and theoretical investigation of electronic effects in ruthenium catalyzed cycloadditions with internal diarylalkynes has shown that 5-electron-donating group substituted triazoles should be favored [155]. This effect was further observed in the synthesis of 5-amino-triazoles from ynamides (Scheme 65) [156].

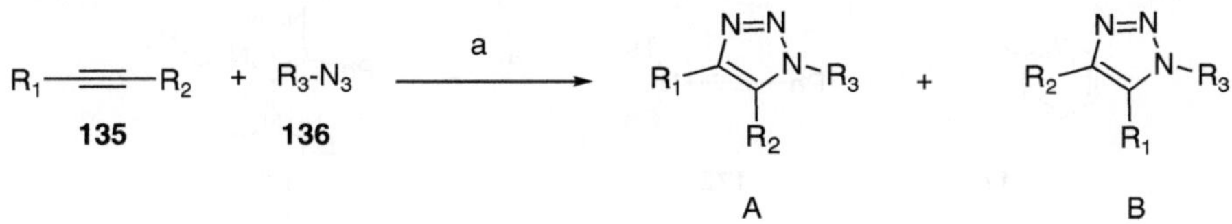

Substituents	Product
R_1 = Ph, R_2 = Me	A + B
R_1 = Ph, R_2 = n-Pr	
R_1 = Ph, R_2 = $CH(OEt)_2$	
R_1 = Ph, R_2 = CO_2Et	B
R_1 = Ph, R_2 = COMe	B
R_1 = n-Bu, R_2 = CO_2Me	B
R_1 = Et, R_2 = COMe	B
R_1 = Ph, R_2 = CH_2OH	A
R_1 = Me, R_2 = CH_2CH_2OH	A
R_1 = Me, R_2 = CH_2NEt_2	A
R_1 = Et, R_2 = CMe_2OH	A

Scheme 63 **a** 10 mol% Cp*RuCl$(PPh_3)_2$, PhH, 80°C 70–100% [154]

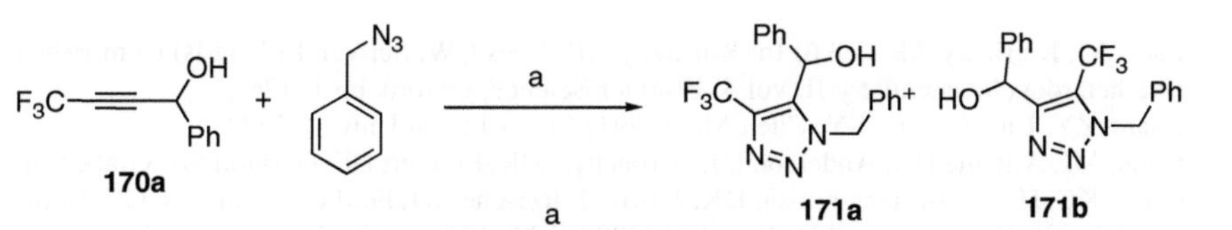

a	171a	171b
H_2O, 80°C	52%	48%
$Pd(OAc)_2/PPh_3$(5 mol%)	57%	81%
$PdCl_2$(dppf) (5 mol%)	54%	86%
$Pd(PPh_3)_4$ (5 mol%)	77	23%
$CpRuCl_2$ (10 mol%)	100%	0%

other examples:

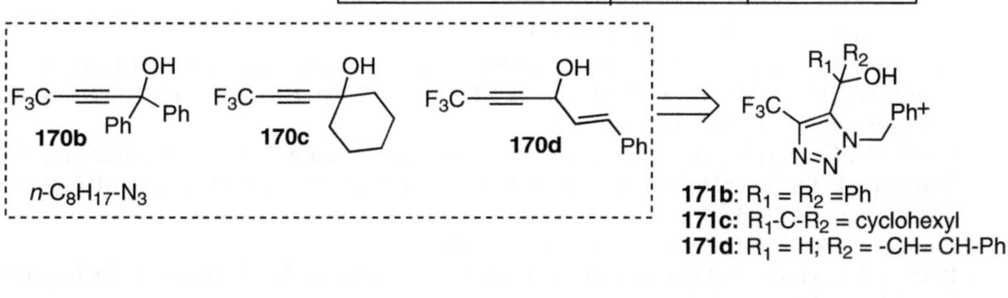

171b: R_1 = R_2 =Ph
171c: R_1-C-R_2 = cyclohexyl
171d: R_1 = H; R_2 = -CH= CH-Ph

Scheme 64 Catalyzed cycloaddition of azides with trifluoromethylated propargylalcohols : regioselectivity issue

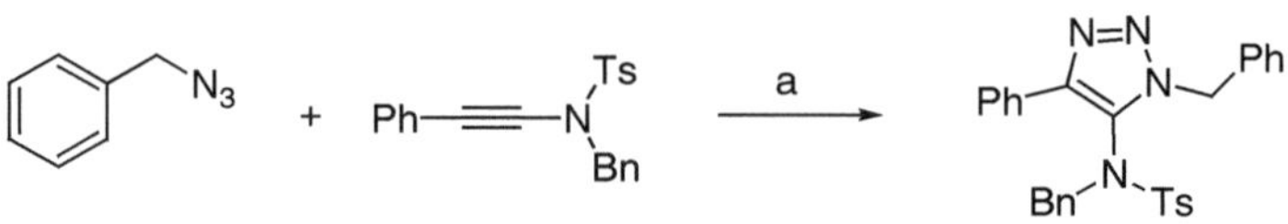

17 172 173
77%

Scheme 65 **a** 10 mol% Cp*RuCl(PPh_3)$_2$, toluene, 80°C [156]

6 Conclusion

1,4,5-Trisubstituted triazoles can be prepared using many synthetic pathways. In most of the cases, the nature of the functional groups introduced on the final heterocycle will dictate the synthetic route. Recent developments based on transition metal-catalyzed C–H activation or 3+2 cycloaddition are particularly promising for the general access to these structures. There is no doubt that these new approaches should enhance the chemical diversity of triazoles, leading to the discovery of compounds with important biological and/or physical properties.

References

1. Fan WQ, Katritzky AR (1996) In: Katritzky AR, Rees CW, Scriven EFV (eds) Comprehensive heterocyclic chemistry II, vol 4. Elsevier Science, Oxford, pp 1–126
2. Zhang ZY, Liu Y, Yang SY, Chen MQ (1991) Chem J Chin Univ 12:1344
3. Genin MJ, Allwine DA, Anderson DJ, Barbachyn MR, Emmert DE, Garmon SA, Graber DR, Grega KC, Hester JB, Hutchinson DK, Morris J, Reischer RJ, Ford CW, Zurenco GE, Hamel JC, Schaadt RD, Stapertand D, Yagi BH (2000) J Med Chem 43:953
4. Willner D, Jelenevsky A, Cheney LC (1972) J Med Chem 15:948
5. Gregory W, Britelli DR, Wang C-LJ, Wuonola MA, McRipley RJ, Eustice DC, Eberly VS, Bartholomew PT, Slee AM, Forbes M (1989) J Med Chem 32:1673
6. Dong HS, Wei K, Wang QL, Quan B (2001) Synth Commun 31:81
7. Holla BS, Mahalinga M, Karthikeyan MS, Poojary B, Akberali PM, Kumari NS (2005) Eur J Med Chem 40:1173
8. Alvarez R, Velazquez S, San F, Aquaro S, De C, Perno CF, Karlesson A, Balzarini J, Camarasa MJ (1994) J Med Chem 37:4185
9. De Clercq E (2002) Med Res Rev 22:531
10. Brockunie LL, Parmee ER, Ok HO, Candelore MR, Cascieri MA, Colwell LF, Deng L, Feeney WP, Forest MJ, Hom GJ, MacIntyre DE, Tota L, Wyvratt MJ, Fisher MH, Weber AE (2000) Bioorg Med Chem Lett 10:2111
11. Bochis RJ, Chabala JC, Harris E, Peterson LH, Barash L, Beattie T, Brown JE, Graham DW, Waksmunski FS, Tischler M, Joshua H, Smith J, Cplwell LF, Wyvratt MJ, Fisher MH (1991) J Med Chem 34:2843
12. Buckle DR, Rockell C (1984) J Med Chem 27:223
13. Biagi G, Calderonea V, Giorgia I, Livia O, Scartoni V, Baragatti B, Martinotti E (2000) Eur J Med Chem 35:715
14. Al-Masoudia NA, Al-Soudb YA (2002) Tetrahedron Lett 43:4021

15. Martini C, Marrucci W, Lucacchini A, Biagi G, Livi O (1988) J Pharm Sci 77:977
16. Norris P, Horton D, Levine BR (1996) Heterocycles 43:2643
17. Pearson WH, Bergmeier SC, Chytra JA (1990) Synthesis 2643
18. Kohn EC, Liotta LA. US Pat. 637145
19. Shen J, Woodward R, Kedenburg JP, Liu X, Chen MQ, Fang L, Sun D, Wang PG (2008) J Med Chem 51:7417
20. Hou J, Feng C, Li Z, Fang Q, Wang H, Gu G, Shi Y, Liu P, Xu F, Yin Z, Shen J, Wang P (2011) Eur J Med Chem 46:3190
21. Huisgen R (1961) Proc Chem Soc 357
22. Kolb HC, Finn MG, Sharpless KB (2001) Angew Chem Int Ed Engl 40:2004
23. Goeminne A, McNaughton M, Bal G, Surpateanu G, Van der Veken P, De Prol S, Verse W, Steyaert J, Apers S, Haemersa A, Augustynsa K (2007) Bioorg Med Chem Lett 17:2523
24. Dimroth O (1902) Ber 36:4041
25. Sun X-W, Xu P-F, Zhang Z-Y (1998) Magn Reson Chem 36:459
26. Manfredini S, Vicentini CB, Manfrini M, Bianchi N, Rutigliano C, Mischiati C, Gambari R (2000) Bioorg Med Chem 8:2343
27. Biagi G, Calderone V, Giorgi I, Livi O, Martinotti E, Martelli A, Nardi A (2004) Farmaco 59:397
28. Mikkhailychenko SN, Chesniuk AA, Koniushkin LD, Firgang SI, Zaplishny VN (2004) Chem Het Comp 40:1162
29. Olessen PH, Nielse FE, Pedersen EB, Becker J (1984) J Het Chem 21:1603
30. Mamedov VA, Valeeva VN, Antokhina LA, Doroshkina GM, Chernova AV, Nuretdinov IA (1993) Chem Het Comp 29:607
31. Savin L, Massarelli P, Corti P, Chiasserini L, Pellerano C, Bruni G (1994) Farmaco 49:363
32. Quan B, Zhuang S-X, Li C-C, Dong H-S (2005) Ind J Chem, Part B 44B:1717
33. Štimac A, Leban I, Kobe J (1999) Synlett 1069
34. Hoover JRE, Day AR (1956) J Am Chem Soc 78:5832
35. L'abbé G, Beenaerts L (1989) Tetrahedron 45:749
36. Dimitrieva IG, Kaigorodova EA, Dysdyuchenko LV, Strelkov VD (2008) Chem Het Comp 44:1267
37. Temeri HHA, Younes MI, Metwally SA (1993) Coll Czech Chem Commun 58:3017
38. Ewing DF, Goethals G, Mackenzie G, Martin P, Ronco G, Vanbaelinghem L, Villa P (1999) Carbohydrate Res 321:190
39. Beauchamp LM, Tuttle JV, Rodriguez ME, Sznaidman ML (1996) J Med Chem 39:949
40. Kislyi VP, Danilova EB, Semenov VV (2003) Russ Chem Bull, Int Ed 52:1770
41. Smith CJ, Nikbin N, Ley SV, Lange H, Baxendale IR (2011) Org Biomol Chem 9:1938
42. Masatomi O, Motohiro I, Toshiaki O, Shoji E (1993) Synthesis 793
43. Saalfrank RW, Ackermann E (1981) Chem Ber 3456
44. Romeiro GA, Pereira LOR, De Souza CBV, Ferreira VF, Cunha AC (1997) Tetrahedron Lett 29:5103
45. Cunha AC, Pereira LOR, De Souza ROP, De Souza CBV, Ferreira VF (2001) Nucleosides Nucleotides Nucleic Acids 20:1555
46. Rozin YA, Savel'eva EA, Morzherin YY, Dehaen W, Toppet S, Van Meervelt L, Bakulev VA (2002) J Chem Soc Perkin Trans 1 211
47. Bakulev VA, Tarasov EV, Morzherin YY, Luyte I, Toppet S, Dehaen W (1988) Tetrahedron 54:8501
48. Kadir D, Ahmet A (2002) Het Commun 8:385
49. Shafran YM, Bakulev VA, Mokrushin VS, Alekssev SG, Lebedev AT, Shabat D (1986) Chem Het Comp 22:926
50. Boulton AJ, Katritsky AR (1962) J Chem Soc 2083
51. Murray-Rust P, McManus J, Lennon SP, Poter AEA, Rechka JA (1984) J Chem Soc Perkin Trans 1 713
52. L'abbé G, Vandendriessche A, Toppet S (1988) Tetrahedron 44:3617

53. Babaev EV, Zefirov NS (1992) Bull Soc Chim Belg 101:67
54. L'abbé G (1984) J Het Chem 21:627
55. Dimroth O, Michaelis W (1927) Justus Liebig's Ann Chem 459:39
56. Segal A, Maté U, Solomon J (1979) J Chem Biol Interact 28:333
57. Volkova NN, Tarasov EV, Kodess MI, Van Meervelt L, Dehaen W, Bakulev VA (2003) Org Biomol Chem 1:4030
58. Volkova NN, Tarasov EV, Van Meervelt L, Toppet S, Dehaen W, Bakulev VA (2002) J Chem Soc Perkin Trans 1 1574
59. Morzherin YY, Tarasov EV, Bakulev VA (1994) Chem Het Comp 30:489
60. Volkova NN, Tarasov EV, Dehaen W, Bakulev VA (1999) Chem Commun 22:2273
61. Glukhareva TV, Morzherin YY, Savel'eva EA, Rozin YA, Tkachev AV, Bakulev VA (2001) Russ Chem Bull, Int Ed 50:268
62. Freitas AP, Proenca MF, Booth BL (1995) J Het Chem 32:457
63. Pocar D, Rossi LM, Trimarco P (1979) J Het Chem 17:267
64. Huang ZT, Wang M-X (1994) Heterocycles 37:1233
65. Wang M-X, Huang Z-T (2002) Prog Natural Sci 12:249
66. Yang Q, Li Z-J, Chen X-M, Huang Z-T (2002) Heteroatom Chem 13:242
67. Chakrasali RT, Ila H, Junjappa H (1988) Synthesis 851
68. Chen X-M, Li Z-J, Ren Z-X, Huang Z-T (1999) Carbohydrate Res 315:262
69. Ren Z-X, Chen X-M, Li Z-J, Huang Z-T (2003) Heteroatom Chem 14:487
70. Huang Z-T, Wang M-X (1992) J Org Chem 57:184
71. Quast H, Ach M, Hergenröther T, Regnat D (2006) Synthesis 1943
72. Quast H, Ach M, Balthasar J, Hergenrother T, Regnat D, Lehmann J, Banert K (2005) Helv Chim Acta 88:1589
73. Piet JC, Hetet GL, Cailleux P, Benhaoua H, Carrié R (1996) Bull Soc Chim Belg 105:33
74. Singh N, Pandey SK, Tripathi RP (2010) Carbohydr Res 345:1641
75. Almirante N, Gelmi ML, Marelli P, Pocar D (1986) Tetrahedron 42:57
76. Van Berkel SS, Dirks A, Debets MF, Van Delft FL, Cornelissen JJLM, Nolte RJM, Rutjes FPJT (2007) Chembiochem 8:1504
77. Adamson GA, Rees CW (1996) J Chem Soc Perkin Trans 1 1535
78. Mayot E, Lemière P, Gérardin Charbonnier C (2008) Eur J Org Chem 2232
79. Tullis JS, VanRes JC, Natchus MG, Clark MP, De B, Hsieh C, Janusz MJ (2003) Bioorg Med Chem 13:1665
80. Bochis RJ, Chabala JC, Harris E, Peterson LH, Barash L, Beattie T, Brown JE, Graham DW, Waksmunski FS, Tischler M, Joshua H, Smith J, Colwell LF, Wyvratt MJ Jr, Fishr MH (1991) J Med Chem 34:2843
81. Yan W, Wang Q, Chen Y, Petersen JL, Shi X (2010) Org Lett 12:3308
82. Iddon B, Nicholas M (1996) J Chem Soc Perkin Trans 1 1342
83. Ohta S, Kawasaki I, Uemura T, Yamashita M, Yoshioka T, Yamaguchi S (1997) Chem Pharm Bull 45:1140
84. Duan H, Yan W, Sengupta S, Shi X (2009) Bioorg Med Chem Lett 19:3899
85. Ohmatsu K, Kiyokawa M, Ooi T (2011) J Am Chem Soc 133:1307–1309
86. Witthing M, Tripp JC, Lin Y-C, Lindstrom W, Olson AJ, Elder JH, Sharpless BK, Fokin VV (2006) J Med Chem 49:7697
87. Diner P, Adersson T, Grotli M, Kjellen J, Elbing K, Hohmann S (2009) New J Chem 33:1010
88. Shen C, Zhang P-F, Chen X-Z (2010) Helv Chem Acta 93:2433
89. Schuster EM, Botoshansky M, Gandelman M (2009) Organometallics 28:7001
90. Chuprakov S, Chernyak N, Dudnik AS, Gevorgyan V (2007) Org Lett 9:2333
91. Ackermann L, Vicente R (2009) Org Lett 11:4922
92. Iwasaki M, Yorimitsu H, Oshima K (2007) Chem Asian J 2:1430
93. Ackermann L, Vicente R, Born R (2008) Adv Synth Catal 350:741
94. Ackermann L, Althammer A, Fenner S (2009) Angew Chem Int Ed Engl 48:201
95. Liégault B, Lapointe D, Caron L, Vlassova A, Fagnou K (2009) J Org Chem 74:1826

96. Jiang H, Feng Z, Wang A, Liu X, Chen Z (2010) Eur J Org Chem 1227
97. Ackermann L, Jeyachandran R, Potukuchi HK, Novak P, Büttner L (2010) Org Lett 12:2056
98. Lapointe D, Fagnou K (2009) Org Lett 11:4160
99. Ackermann L, Potukuchi HK, Landsberg D, Vicente R (2008) Org Lett 10:3081
100. Fukuzawa S, Shimizu E, Ogata K (2009) Heterocycles 78:645
101. Ackermann L, Potukuchi HK (2010) Org Biomol Chem 8:4503
102. Dai Q, Gao W, Liu D, Kapes LM, Zhang X (2006) J Org Chem 71:3928
103. Liu D, Gao W, Zhang X (2005) Org Lett 7:4907
104. Hien JE, Fokin VV (2010) Chem Soc Rev 36:1302
105. Schoenebeck F, Ess DH, Jones GO, Houk KN (2009) J Am Chem Soc 131:8121
106. Jewett JC, Bertozzi CR (2010) Chem Soc Rev 39:1272
107. Kumar I, Rana S, Cho JW, Rode CV (2010) Tetrahedron Asymmetry 21:352
108. Hager C, Miethchen R, Reinke H (2000) J Prakt Chem 342:414
109. Gao D, Zhai H, Parvez M, Back TG (2008) J Org Chem 73:8057
110. Back TG, Bethell RJ, Parvez M, Taylor JA, Wehrli D (1999) J Org Chem 64:7426
111. Grée D, Grée R (2007) Tetrahedron Lett 48:5435
112. Hlasta DJ, Ackerman JH (1994) J Org Chem 59:6184
113. Coats SJ, Link JS, Gauthier D, Hlasta DJ (2005) Org Lett 7:1469
114. Yamashita M, Reddy PM, Kato Y, Reddy VK, Suzuki K, Oshikawa T (2001) Carbohydr Res 336:257
115. Dornow A, Rombusch D (1958) Chem Ber 91:1841
116. Taleka RR, Wightman RH (1997) Tetrahedron 53:3631
117. Marco-Contelles J, Jiminez CA (1999) Tetrahedron 55:10511
118. Grée D, Grée R (2007) Tetrahedon Lett 48:5435
119. Tadashi S, Shoji E, Takashi O, Wakata T (1983) J Org Chem 48:4067
120. Okano T, Ito K, Ueda T, Muramatsu H (1986) J Fluor Chem 32:377
121. Al-Masoudi NA, Al-Soud YA, Abdul-Zahra A (2004) Heteroatom Chem 15:380
122. Hager C, Miethchen R, Reinke H (2000) J Prat Chem 342:414
123. Jain RP, Vederas JC (2004) Bioorg Med Chem Lett 14:3655
124. Meazza G, Zanardi G (1991) J Fluor Chem 55:199
125. Hall RG, Trippe S (1982) Tetrahedron Lett 23:2603
126. Earl RA, Townsend LB (1980) Can J Chem 58:2550
127. Ma D-Y, Wang D-X, Zheng Q-Y, Wang M-X (2006) Tetrahedron Asymmetry 17:2366
128. Praud-Tabaries A, Dombrowsky L, Bottzek O, Briand J-F, Blaches Y (2009) Tetrahedron Lett 50:1645
129. L'abbé G, Dehaen W (1988) Tetrahedron 44:461
130. Loubinoux B, Colin JL, Tabbache SJ (1984) Het Chem 21:1669
131. Settimo AD, Livi O, Biagi G, Lucacchini A, Caselli S (1983) Farmaco 38:725
132. Füger B, Sklute G, Marek I, Bolm GY, Bolm C (2008) Synlett 116
133. Huang J, Macdonald SJF, Harrity JPA (2009) Chem Commun 436
134. Choi KW, Brimble MA (2008) Org Biomol Chem 6:3518
135. Hanamoto T, Hakoshima Y, Egashira M (2004) Tetrahedron Lett 45:7573
136. Ito S, Hirata Y, Nagatomi Y, Satoh A, Suzuki G, Kimura T, Satow A, Maehara S, Hikichi H, Hata M, Ohta H, Kawamoto H (2009) Bioorg Med Chem Lett 19:5310
137. Gao D, Zhai H, Parvez M, Back TG (2008) J Org Chem 20:8057
138. Akimova GS, Chistokletov VN, Petrov AA (1967) Zh Org Khim 3:2241
139. Akimova GS, Chistokletov VN, Petrov AA (1968) Zh Org Khim 4:389
140. Kansinski A, Fokin VV, Sharpless KB (2004) Org Lett 6:1237
141. Meza-Avina ME, Patel MK, Lee CB, Dietz TJ, Croatt MP (2011) Org Lett 13:2984
142. Akao A, Tsuritani T, Kii S, Sato K, Nonoyama N, Mase T, Yasuda N (2007) Synlett 31
143. Meldal M, Tornøe CW (2008) Chem Rev 108:2952
144. Wu Y-M, Deng J, Li Y, Chen Q-Y (2005) Synthesis 1314
145. Malnuit V, Duca M, Manout A, Bougrin K, Benhida R (2009) Synthesis 2123

146. Li L, Zhang G, Zhu A, Zhang L (2008) J Am Chem Soc 130:3630
147. Angell Y, Burgess K (2007) Angew Chem Int Ed Engl 46:3649
148. Hein JE, Tripp JC, Krasnova LB, Sharpless KB, Fokin VV (2009) Angew Chem Int Ed 48:8018
149. Garcia-Alvarez J, Diez J, Gimeno J (2010) Green Chem 12:2127
150. Zhou Y, Lecourt T, Micouin L (2010) Angew Chem Int Ed Engl 49:2607
151. Zheng H, McDonald R, Hall DG (2010) Chem Eur J 16:5454
152. Zhang L, Chen X, Xue P, Sun HHY, Williams ID, Sharpless KB, Fokin VV, Jia GJ (2005) J Am Chem Soc 127:15998
153. Majireck MM, Weinreb SM (2006) J Org Chem 71:8680
154. Boren BC, Narayan S, Rasmussen LK, Zhang L, Zhao H, Lin Z, Guochen J, Fokin VV (2008) J Am Chem Soc 130:8923
155. Hou D-R, Kuan T-C, Li Y-K, Lee R, Huang K-W (2010) Tetrahedron 66:9415
156. Oppiliard S, Mousseau G, Zhang L, Guochen J, Thuéry P, Rousseau B, Cintrat J-C (2007) Tetrahedron 63:8094

Index

MIX
Papier aus verantwortungsvollen Quellen
Paper from responsible sources
FSC® C105338

If you have any concerns about our products,
you can contact us on
ProductSafety@springernature.com

In case Publisher is established outside the EU,
the EU authorized representative is:
Springer Nature Customer Service Center GmbH
Europaplatz 3, 69115 Heidelberg, Germany

Printed by Libri Plureos GmbH
in Hamburg, Germany